New Trends in the Physics and Mechanics of Biological Systems

École de Physique des Houches

Session XCII, 6–31 July 2009

New Trends in the Physics and Mechanics of Biological Systems

Edited by

Martine Ben Amar, Alain Goriely, Martin Michael Müller,
Leticia F. Cugliandolo

OXFORD
UNIVERSITY PRESS

OXFORD
UNIVERSITY PRESS

Great Clarendon Street, Oxford, OX2 6DP,
United Kingdom

Oxford University Press is a department of the University of Oxford.
It furthers the University's objective of excellence in research, scholarship,
and education by publishing worldwide. Oxford is a registered trade mark of
Oxford University Press in the UK and in certain other countries

First published 2011

Published in the United States of America by Oxford University Press
198 Madison Avenue, New York, NY 10016, United States of America

British Library Cataloguing in Publication Data
Data available

ISBN 978-0-19-960583-5

École de Physique des Houches
Service inter-universitaire commun à l'Université Joseph Fourier de Grenoble et à l'Institut National Polytechnique de Grenoble

Subventionné par l'Université Joseph Fourier de Grenoble, le Centre National de la Recherche Scientifique, le Commissariat à l'Énergie Atomique

Directeur:
Leticia Cugliandolo, Université Pierre at Marie Curie – Paris VI, France

Directeurs scientifiques de la session XCII:
Martine Ben Amar, LPS-ENS, Paris, France
Alain Goriely, OCCAM, Mathematical Institute, Oxford
Martin Michael Müller, Equipe BioPhysStat, Metz, France

Previous sessions

I	1951	Quantum mechanics. Quantum field theory
II	1952	Quantum mechanics. Statistical mechanics. Nuclear physics
III	1953	Quantum mechanics. Solid state physics. Statistical mechanics. Elementary particle physics
IV	1954	Quantum mechanics. Collision theory. Nucleon-nucleon interaction. Quantum electrodynamics
V	1955	Quantum mechanics. Non equilibrium phenomena. Nuclear reactions. Interaction of a nucleus with atomic and molecular fields
VI	1956	Quantum perturbation theory. Low temperature physics. Quantum theory of solids. Ferromagnetism
VII	1957	Scattering theory. Recent developments in field theory. Nuclear and strong interactions. Experiments in high energy physics
VIII	1958	The many body problem
IX	1959	The theory of neutral and ionized gases
X	1960	Elementary particles and dispersion relations
XI	1961	Low temperature physics
XII	1962	Geophysics; the earths environment
XIII	1963	Relativity groups and topology
XIV	1964	Quantum optics and electronics
XV	1965	High energy physics
XVI	1966	High energy astrophysics
XVII	1967	Many body physics
XVIII	1968	Nuclear physics
XIX	1969	Physical problems in biological systems
XX	1970	Statistical mechanics and quantum field theory
XXI	1971	Particle physics
XXII	1972	Plasma physics
XXIII	1972	Black holes
XXIV	1973	Fluids dynamics
XXV	1973	Molecular fluids
XXVI	1974	Atomic and molecular physics and the interstellar matter
XXVII	1975	Frontiers in laser spectroscopy
XXVIII	1975	Methods in field theory
XXIX	1976	Weak and electromagnetic interactions at high energy
XXX	1977	Nuclear physics with heavy ions and mesons
XXXI	1978	Ill condensed matter
XXXII	1979	Membranes and intercellular communication
XXXIII	1979	Physical cosmology

XXXIV	1980	Laser plasma interaction
XXXV	1980	Physics of defects
XXXVI	1981	Chaotic behavior of deterministic systems
XXXVII	1981	Gauge theories in high energy physics
XXXVIII	1982	New trends in atomic physics
XXXIX	1982	Recent advances in field theory and statistical mechanics
XL	1983	Relativity, groups and topology
XLI	1983	Birth and infancy of stars
XLII	1984	Cellular and molecular aspects of developmental biology
XLIII	1984	Critical phenomena, random systems, gauge theories
XLIV	1985	Architecture of fundamental interactions at short distances
XLV	1985	Signal processing
XLVI	1986	Chance and matter
XLVII	1986	Astrophysical fluid dynamics
XLVIII	1988	Liquids at interfaces
XLIX	1988	Fields, strings and critical phenomena
L	1988	Oceanographic and geophysical tomography
LI	1989	Liquids, freezing and glass transition
LII	1989	Chaos and quantum physics
LIII	1990	Fundamental systems in quantum optics
LIV	1990	Supernovae
LV	1991	Particles in the nineties
LVI	1991	Strongly interacting fermions and high Tc superconductivity
LVII	1992	Gravitation and quantizations
LVIII	1992	Progress in picture processing
LIX	1993	Computational fluid dynamics
LX	1993	Cosmology and large scale structure
LXI	1994	Mesoscopic quantum physics
LXII	1994	Fluctuating geometries in statistical mechanics and quantum field theory
LXIII	1995	Quantum fluctuations
LXIV	1995	Quantum symmetries
LXV	1996	From cell to brain
LXVI	1996	Trends in nuclear physics, 100 years later
LXVII	1997	Modeling the earths climate and its variability
LXVIII	1997	Probing the Standard Model of particle interactions
LXIX	1998	Topological aspects of low dimensional systems
LXX	1998	Infrared space astronomy, today and tomorrow
LXXI	1999	The primordial universe
LXXII	1999	Coherent atomic matter waves
LXXIII	2000	Atomic clusters and nanoparticles
LXXIV	2000	New trends in turbulence
LXXV	2001	Physics of bio-molecules and cells
LXXVI	2001	Unity from duality: Gravity, gauge theory and strings

LXXVII	2002	Slow relaxations and nonequilibrium dynamics in condensed matter
LXXVIII	2002	Accretion discs, jets and high energy phenomena in astrophysics
LXXIX	2003	Quantum entanglement and information processing
LXXX	2003	Methods and models in neurophysics
LXXXI	2004	Nanophysics: Coherence and transport
LXXXII	2004	Multiple aspects of DNA and RNA
LXXXIII	2005	Mathematical statistical physics
LXXXIV	2005	Particle physics beyond the Standard Model
LXXXV	2006	Complex systems
LXXXVI	2006	Particle physics and cosmology: the fabric of spacetime
LXXXVII	2007	String theory and the real world: from particle physics to astrophysics
LXXXVIII	2007	Dynamos
LXXXIX	2008	Exact methods in low-dimensional statistical physics and quantum computing
XC	2008	Long-range interacting systems
XCI	2009	Ultracold gases and quantum information
XCII	2009	New trends in the physics and mechanics of biological systems
XCIII	2009	Modern perspectives in lattice QCD: quantum field theory and high performance computing

Publishers

- Session VIII: Dunod, Wiley, Methuen
- Sessions IX and X: Herman, Wiley
- Session XI: Gordon and Breach, Presses Universitaires
- Sessions XII–XXV: Gordon and Breach
- Sessions XXVI–LXVIII: North Holland
- Session LXIX–LXXVIII: EDP Sciences, Springer
- Session LXXIX–LXXXVIII: Elsevier
- Session LXXXIX– : Oxford University Press

Preface

Over the last few decades, the life sciences have experienced a quiet but radical revolution. This revolution has transformed biology from a mostly observational and qualitative discipline into a fully quantitative science at all scales, ranging from the DNA code to proteins, cells, tissues, and organs. In the process, biology has become one of the ultimate scientific frontiers and exploded into many diverging sub-disciplines, all requiring the integration of mathematical, computational, and physical sciences. As such, biophysics and biomechanics is now an ever-growing branch of traditional physics. In particular, with the ability to measure forces, velocities, and displacements at the cellular and tissue levels, there is a clear need for mathematical theories and models that integrate the mechanical, chemical, and biological environments. For instance, it has now been understood that biological growth and development have an important mechanical component, playing a role in genetic programming, morphogenesis, and the regulation of physiological processes such as heart and artery remodeling. However, a unified theory of the growth of elastic tissues that addresses the fundamental coupling between geometric quantities and physical and chemical fields is still lacking. Theoretical progress on this problem will be needed both to answer basic scientific questions and to tackle fundamental challenges such as tumor growth. Similarly, the motion of biological organisms such as bacteria, sperm, and insects through fluids requires an understanding of fluid–solid interactions between self-propelled organisms and fluids that can have a gel-like or active response. With their training in mechanics, mathematics, and modeling, physicists have a unique opportunity to play a central role in the development of such modern biological theories. However, the classical training in the physical sciences does not introduce students to such problems, and fails to give them the proper state-of-the-art tools to address these challenges.

In July 2009, many experts in mathematical modeling in the biological sciences gathered in Les Houches for a four-week summer school on the mechanics and physics of biological systems. The goal of the school was to present to students and researchers an integrated view of the new trends and challenges in the physical and mathematical aspects of biomechanics. Although the scope of such a topic is very wide, we focused on problems where solid and fluid mechanics plays a central role. The school covered both the general mathematical theory of mechanical biology in the context of continuum mechanics and the specific modeling of particular systems in the biology of cells, plants, and microbes, and in physiology. The school was organized around five different main topics, all connected by the common theme of continuum modeling for biological systems. These collected lecture notes reflect the same organization, namely biofluidics, biogels, biomechanics, biomembranes, and morphogenesis. These lecture notes are not meant as a journal review of the topic but rather as a gentle tutorial introduction to readers who want to understand the basic problems of modeling biological systems from a mechanical perspective.

It is with great pleasure that we acknowledge the financial support of various organizations: the Centre Nationale de la Recherche Scientifique through its Formation Permanente program; the newly established Fondation Pierre-Gilles de Gennes; the Réseau des Systèmes Complexes (Paris Fédération); the Université Pierre et Marie Curie; the École Doctorale systèmes complexes (Paris); and the European Union through its Marie Curie Intra-European Fellowship (EIF) grant 042069, CancerBio-Mechanics, under the 6th Framework Programme. It is also our pleasure to thank the staff of the École for their support.

Martine Ben Amar (Laboratoire de Physique Statistique, École Normale Supérieure)
Alain Goriely (Mathematical Institute, University of Oxford)
Martin Michael Müller (Equipe BioPhysStat, Université Paul-Verlaine, Metz)
Leticia Cugliandolo (Laboratoire de Physique Théorique et Hautes Energies, Université Pierre et Marie Curie, Paris)

Contents

List of participants

Organizers

BEN AMAR Martine
LPS-ENS 24 Rue Lhomond, 75005 Paris, France

GORIELY Alain
OCCAM, Mathematical Institute, 24–29 St Giles', Oxford OX1 3LB, UK

MÜLLER Martin Michael
Equipe BioPhysStat, 1 boulevard Arago, 57070 Metz, France

Lecturers

BUSH John W. M.
Massachusetts Institute of Technology, Department of Mathematics, 77 Massachusetts Avenue, Room 2–392, Cambridge, MA 02139-4307, USA

CIARLETTA Pasquale
Laboratoire de Physique Statistique, Ecole Normale Supérieure, 24 Rue Lhomond, 75005 Paris, France

GOLDSTEIN Raymond E.
University of Cambridge, Centre for Mathematical Sciences, Wilberforce Road, Cambridge CB3 0WA, UK

HOSOI Anette
Massachusetts Institute of Technology, Room 3-262, 77 Massachusetts Avenue, Cambridge, MA 02139–4307, USA

JÜLICHER Frank
Max Planck Institute for the Physics of Complex Systems, Nöthnitzestr. 38, 01187 Dresden, Germany

LABOUESSE Michel
Institut de Génerique et de Biologie Moléculaire et Cellulaire, 1 rue Laurent Fries, 67404 Illkirch Cedex, France

MACKINTOSH Fred
Faculty of Sciences, Vrije University, 1081 HV, Amsterdam, Netherlands

MADDOCKS John
IMB, Faculté des Sciences de Base, EPFL, Station 8, 1015 Lausanne, Switzerland

OGDEN Ray W.
University of Glasgow, Department of Mathematics, 15 University Gardens, Glasgow G12 8QW, UK

PREZIOSI Luigi
Politecnico di Torino, Department of Mathematics, Corso Duca degli Abruzzi 24, 10129 Torino, Italy

SACCOMANDI Giuseppe
University of Perugia, Dipartimento Ingegneria Industriale, via G. Duranti, 06125 Perugia, Italy

SHELLEY Michael J.
The Courant Institute, New York University, 251 Mercer Street, New York, NY 10012, USA

SMITH Richard
University of Bern, Hochschulstrasse, CH-3012 Bern, Switzerland

TABOR Michael J.
Program in Applied Mathematics, University of Arizona, Tucson, AZ, USA

Participants

ALIEE Maryam
Max Planck Institute for the Physics of Complex Systems, Nothnitzestr. 38, 01187 Dresden, Germany

ALVARADO José
Institute AMOLF, Science Park 104, 1098 XG Amsterdam, Netherlands

AMUASI Henri Emmanuel
Technische University of Eindhoven, Department of Applied Physics, Jeroen Boschlaan 142, 5613 GC Eindhoven, Netherlands

BLANCH MERCADER Carles
Avenida Catalunya No 5 Atico 1er 43700 El Vendrell (Tarragona), Spain

CHATELAIN Clément
LPS, 24 Rue Lhomond, F-75005 Paris, France

COMER Jeffrey
University of Illinois, 1110 W. Green Street, Urbana-Champaign, IL 61801, USA

DE PASCALIS Riccardo
University of Salento, Mathematics Department, Palazzo Fiorini, Via per Arnesano, 73100 Lecce, Italy

DERVAUX Julien
LPS, 24 Rue Lhomond, F-75005 Paris, France

DMITRIEFF Serge
ESPCI, Laboratoire Gulliver, 10 Rue Vauquelin, F-75005 Paris, France

DUNSTAN Jocelyn
University of Chile, Department of Physics, Avenue Blanco Encalada 2008, Santiago, Chile

EBERT Maron Kai
Johannes Gutenberg University Mainz, FB Physik 08, Komet 331, Germany

EVANS Arthur
University of San Diego, 3476 Armstrong Street, San Diego, CA 92111, USA

GADELHA Hermes
Centre for Mathematical Biology, Mathematical Institute, University of Oxford, 24–29 St Giles, Oxford 0X1 3LB, UK

GALY Olivier
Laboratoire Physico-Chimie Curie, UMR 168, Institut Curie, 11 rue Pierre et Marie Curie, F-75005 Paris, France

GOMEZ-ORLANDI Javier
University of Barcelona, Department of ECM, C/Marti I Franques 1-Office 606, E–08028 Barcelona, Spain

GREULICH Philip
University of Köln, Institut für theoretische Physik, Zülpicher str. 77, 50937 Köln, Germany

GREVESSE Thomas
University of Mons Hainaut, 20 place du Parc, 7000 Mons, Belgium

GUERIN Thomas
Laboratoire Physico-Chimie Curie, UMR 168, 11 rue Pierre et Marie Curie, F-75005 Paris, France

HABIBI Mehdi
Institute for Advanced Studies in Basic Sciences, Department of Physics, Gavazang Street, Zanjan 45195, Iran

HAMEDANI RADJA Nima
Lorentz Institute for Theoretical Physics, University of Leiden, Nielsbohrweg 2, Leiden, 2333 CA, Netherlands

HERSCHLAG Gregory
University of North Carolina, Department of Mathematics, UNC, Phillipes Hall, CB# 3250, Chapel Hill, NC 27599, USA

HOHENEGGER Christel
University of Utah, Mathematics Department, 155 S 1400 E Room 233, Salt Lake City, UT 84112-0090, USA

KOURILOVA Hana
Charles University in Prague, MFF, Department of Macromolecular Physics, V Holeso-vickach 2, Praha 8, 180 00, Czech Republic

LAVRENTOVICH Maxim
Harvard University, Physics Department, 17 Oxford Street, Cambridge, MA 02138, USA

LOISEAU Etienne
Laboratoire des Colloïdes, Verres et Nanomatériaux, University of Montpellier 2, Batiment 11, C,C 026, Place Eugène Bataillon, F-34095 Montpellier, France

MAGGIONI Francesca
Department of Mathematics, Statistics, Computer Sciences & Applications, University of Bergamo, Via dei Camiana no 2, 24127 Bergamo, Italy

MARKOVA Olga
IBDML, Parc Scientifique de Luminy, F-13009 Marseille, France

MIAO Bing
Max Planck Intitute for Polymer Research, Ackermannweg 10, 55128 Mainz, Germany

MOULTON Derek
OCCAM, Mathematical Institute, 24–29 St Giles', Oxford OX1 3LB, UK

NICHOLAS Matthew P.
Albert Einstein College of Medicine, Department of Anatomy & Structural Biology, Forchheimer 610, 1300 Morris Park Avenue, Bronx, NY 10461, USA

NOUBISSIE Samuel
LPTA, Université de Montpellier II, Case Courrier 070, Batiment 13, Place Eugène Bataillon, F-34095 Montpellier Cedex 5, France

PETITJEAN Laurence
Laboratoire Physico-Chimie Curie, UMR 168, Institut Curie, 11 Rue Pierre et Marie Curie, F-75005 Paris, France

PUROHIT Prashant
Mechanical Engineering & Applied Mechanics, University of Pennsylvania, 220 South 33rd Street, Philadelphia, PA 19104, USA

QUINT David
Department of Physics, Syracuse University, University Avenue, Syracuse, NY 13244, USA

RANFT Jonas
Laboratoire Physico-Chimie Curie, UMR 168, Institut Curie, 11 Rue Pierre et Marie Curie, F-75005 Paris, France

RIGOZZI Michelle Kyone
BSS, Cavendish Laboratory, Madingley Road, University of Cambridge, Cambridge, UK

RODRIGUES FERREIRA Elizabete
Université Libre de Bruxelles, Boulevard du Triomphe, Campus Plaine, CP 218/1, Dept. Mathématique, Service mécanique et math appliquées, 1050 Ixelles, Brussels, Belgium

SILVA Susana
Center of Computational Physics, University of Coimbra, Rua Larja, 3004-516 Coimbra, Portugal

SPAGNOLIE Saverio
Department of Mechanical & Aerospace Engineering, 9500 Gilman Drive, La Jolla, CA 92093, USA

SPORER Susan
Institute of Theoretical Physics 1, Uni Erlangen, Staudtstr. 7, 91058 Erlangen, Germany

STOOP Norbert
Institute of Building Materials, ETH Zurich, HIF E20, Schafmattstrasse 6, 8093 Zurich, Switzerland

SVOBODOVA Miroslava
University of West Bohemia, Department of Mechanics, FAV, Univerzitni 22, 306 14 Pilsen, Czech Republic

TKACIK Gasper
Department of Physics & Astronomy, University of Pennsylvania, 209 South 33rd Street, Philadelphia, PA 19104, USA

TRAVASSO Rui
Center of Physics, University of Coimbra, Department of Physics, Rua Larga, 3004-516 Coimbra, Portugal

VAZQUEZ MONTEJO Pablo Agustin
Nuclear Sciences Institute, National Autonomous University of Mexico, Apartado Postal 70-543, Office Number 67, Circuito Interior, 04510, D.F.; Mexico

VINCENT Olivier
Laboratoire de Spectrométrie Physique, Equipe DYFCOM, 140 Avenue de la Physique, F-38402 Saint-Martin d'Hères, France

VITALE Guido
Mathematics Department, Politecnic of Torino, CSO Duca degli Abruzzi, 24 Turin, Italy

YOLCU Cem
Department of Physics, 5000 Forbes Avenue, Carnegie Mellon University, Pittsburgh, PA 15213, USA

ZASTROW Lena Rebecca
Department of Studies of Structures, University Roma Tre, Via Corrado Segre 4/6, 00146 Roma, Italy

Others

Monsieur Le Président
Université Joseph Fourier, Bâtiment Administratif Sciences BP 53X, F-38041 Grenoble Cedex, France

BLAIZOT Jean-Paul
CEA Saclay, Institut de physique théorique, Direction des Sciences de la matière, F-91191 Gif-sur-Yvette, France

CARDIN Philippe
Laboratoire de Géophysique Interne et Tectonophysique, Université Joseph Fourier (Grenoble I), Oberservatoire de Grenoble, BP 53, F-380741 Grenoble Cedex 9, France

COMBES Françoise
LERMA, 61 Avenue de l'Oberservatoire, F-75014 Paris, France

CUGLIANDOLO Leticia
LPTHE, Tour 24-5ème étage, 4 place Jussieu, F-75232 Paris Cedex 05, France

DEWITT Cécile
Department of Physics, University of Texas, Austin, TX 78712, USA

FOURCADE Bertrand
Institut Albert Bonniot, Centre de Recherche INSERM/UJF U 823-Equipe DYSAD, Site Santé, La Tronche, BP 170, F-38042 Grenoble Cedex 9, France

FRAPPAT Luc
LAPTH, Chemin de Bellevue, F-74941 Annecy-le-Vieux, France

GARDETTE Murielle
Ecole de Physique des Houches, La Côte des Chavants, F-74310 Les Houches, France

GIAMARCHI Thierry
Département de Physique, Université de Genève, 24 Quai E. Ansermet, 1211 Genève, Switzerland

GUERY-ODELIN David
Université Paul Sabatier, Laboratoire de Collisions-Agrégats-Réactivité, 118 Route de Narbonne (Bât. 3R1b4), F31062 Toulouse Cedex 9, France

HIPPERT Françoise
LMGP-Laboratoire de Matériaux et Génie physique, INP Grenoble Minatec, BP 257, 3 parvis Louis Néel, F-38016 Grenoble, France

JOLICOEUR Thierry
LPMC/ENS, 24 Rue Lhomond, F-75231 Paris Cedex 5, France

JULIEN Marc-Henri
LNCMI-G CNRS 25 rue des Martyrs, BP166, F-38042 Grenoble Cedex 9, France

KARYOTAKIS Yannis
Laboratoire d'Annecy-le-Vieux de Physique des Particules, Chemin de Bellevue, BP 110, F-74941 Annecy-le-Vieux, France

LEBOEUF Patricio
LPTMS, Bâtiment 100, Rue G. Clémenceau, F-91405 Orsay Cedex, France

PEBAY-PEYROULA Eva
IBS, 41 Avenue des Martyrs, F-38042 Grenoble Cedex, France

PEYRARD Michel
Laboratoire de Physique, ENS, 46 Allée d'Italie, F-69364 Lyon Cedex 07, France

PINTON Jean-François
Laboratoire de Physique de l'ENS-Lyon, 46 Allée d'Italie, F-69364 Lyon Cedex 07, France

PROTASSOV Konstantin
Université Joseph Fourier, Bâtiment Administratif Sciences, BP 53X, F-38041 Grenoble Cedex, France

RENEVIER Hubert
Laboratoire des matériaux et du génie physique, UMR 5628, Grenoble INP-Minatec, 3 parvis Louis Néel, BP 257, F-38016 Grenoble Cedex I, France

ROUSSET Brigitte
Ecole de Physique des Houches, La Côte des Chavants, F-74310 Les Houches, France

VANNIMENUS Jean
LPS/ENS, 24 Rue Lhomond, F-75231 Paris Cedex 05, France

VIAL Claude
Cadist De Physique, SICD1, Bibliothèque Universitaire de Sciences, BP66, F-38402 Saint Martin d'Hères, France

ZINN-JUSTIN Jean
S.Ph.T, CE Saclay, F-91191 Gif-sur-Yvette, France

Part I

Biofluidics

1
Locomotion at low Reynolds numbers

Anette E. Hosoi

Department of Mechanical Engineering, MIT

1.1 Introduction

The following three lectures are intended as an introduction to locomotion at low Reynolds numbers rather than as a comprehensive review. For students who desire a more in-depth treatment, I have included references throughout the text. These notes are written in an informal tone to match the lectures, which included a mix of chalk talks and PowerPoint presentations. Many of the PowerPoint slides have been reproduced here, although some copyrighted images have been omitted (deleted images have been replaced by references directing the reader to the original sources).

1.2 Lecture 1: Swimming

Over the three days of these lectures, we discussed various forms of low-Reynolds-number locomotion, including:

Day 1: overview & swimming;
Day 2: crawling (snails);
Day 3: burrowing (clams).

However, before we consider these specific forms of locomotion, it is useful to consider what types of biological organisms, generally speaking, operate in a low-Reynolds-number regime.

1.2.1 Reynolds numbers in biology

The Reynolds number is a dimensionless group that characterizes the ratio of inertial to viscous forces. It is defined as

$$Re = \frac{\rho U L}{\mu} = \frac{U L}{\nu},$$

where ρ is the density of the medium the organism is moving through, μ is the dynamic viscosity of the medium, ν is the kinematic viscosity, U is a characteristic velocity of the organism, and L is a characteristic length scale. When we discuss swimming *biological* organisms, we are usually referring to creatures that are moving through water (or through a fluid with material properties very close to those of water). This means that the material properties μ and ρ are fixed[1] and the Reynolds number is roughly determined by the *size* of the organism.

In general, the characteristic size of the organism and the characteristic swimming velocity are related. As a rule of thumb, the characteristic locomotion velocity U in biological organisms is related to L by $U \sim L/\text{second}$; for example for people, $L \sim 1$ m and we move at $U \sim 1$ m/s; bugs are about $L \sim 1$ mm, and they move at about $U \sim 1$ mm/s; for microorganisms, $L \sim 100$ μm and $U \sim 100$ μm/s. Obviously, this is a very, very, very, very rough estimate, and one does not have to think very hard to come up with exceptions (as is always the case in biology!). However, it serves as a good starting point for estimating the Reynolds numbers for various biological organisms

[1]For water, $\nu \approx 10^{-2}\,\text{cm}^2/\text{s}$ and $\rho \approx 1\ \text{g/cm}^3$.

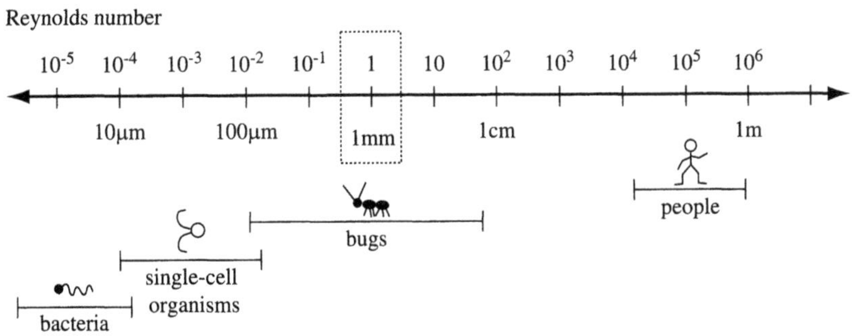

Fig. 1.1 Typical Reynolds numbers for various biological organisms. Reynolds numbers are estimated using the length scales indicated, the "rule of thumb" in the text, and material properties of water.

as illustrated in the sketch in Fig. 1.1. Note that even for organisms as small as ants, the Reynolds number is still on the order of 1 (which is not very low). In this lecture, we will focus on $Re \ll 1$, which is relevant to single-cell organisms and bacteria.

At this point it is worth mentioning a few biological details about single-celled organisms. A good place to start is the well-known image sketched by the fluid dynamist James Lighthill (made in the days before PowerPoint). I haven't included the image here, for fear of copyright infringement, but it is worth looking up and readily available in Lighthill (1976). Lighthill's figure represents a fluid dynamicist's view of biology as he has classified the organisms first and foremost not by any biological metric, but by how they swim. Organisms within the central circle all consist of a head (which is generally modeled as a sphere) propelled by N tails, where N is a small number (generally 1, 2, or 3). Organisms outside the circle use some other mechanism or don't locomote at all.

On top of this division, there is a second (more traditional) biological classification. The semicircle dividing the top of the image from the bottom marks the dividing line between eukaryotic cells and prokaryotic cells. Roughly speaking, prokaryotic cells are bacteria and eukaryotic cells are everything else (people, fungi, grasshoppers, elephants, algae, ...). This distinction between eukaryotic and prokaryotic swimmers is quite important, because the structure of the tails (which determines the mechanics of swimming) is different in the two groups. For eukaryotes, the tail is made of flagellin and has a characteristic 9+2 microtubule structure. This has two important consequences:

1. By sliding the microtubules relative to one another, the organism can exert a local bending moment anywhere along the length of the tail, which means that eukaryotic organisms can *select the shape of the tail as a function of time* (i.e. they can control the kinematics). This is in contrast to prokaryotic organisms, which have a single motor that connects the tail to the head and hence can only apply a localized torque at the base of the tail.

Table 1.1 The structure of tails

Eukaryotes	**Prokaryotes**
Flagellin	Tubulin
Distributed torque	Concentrated torque
9+2 microtubule structure	
Diameter of tail $\sim$ 300 nm	

2. Since all eukaryotic tails have the same 9+2 microtubule structure, they all have roughly the same cross-sectional diameter. The length of the tail may vary from organism to organism, but the diameter is approximately 250–400 nm for all species (this includes the cilia in our lungs, the tails of green algae, the tails of spermatozoa, etc.).

These differences between prokaryotic and eukaryotic tails are summarized in Tabel 1.1

1.2.2 The physics of low-*Re* swimming

At this point we need to introduce our next reference, Purcell (1977). This is one of the most well-known papers on swimming at low Reynolds numbers and I highly recommend it for anyone who is looking for a gentle introduction to the topic. For the purposes of this lecture, we will define swimming as follows:

> **Swimming:** To undergo cyclic deformations that result in a net translation and/or rotation (with no externally applied forces or torques).

The governing equations for low-Reynolds-number flow around an organism undergoing such deformations are the Stokes equations:

$$\nabla p = \mu \nabla^2 \mathbf{u}, \nabla \cdot \mathbf{u} = 0.$$

In addition, the forces and torques on the organism must balance (since we are considering an inertialess world, there is no acceleration, and hence $m\mathbf{a} = 0$):

$$\mathbf{F} = \mathbf{0}, \mathbf{M} = \mathbf{0}.$$

A few features of swimming at low Reynolds number are readily apparent from these equations:

1. Usually when we balance forces in locomotion studies, we balance some propulsive force against drag (e.g. vortex shedding or inertia). However, at low Reynolds numbers the only thing that can balance drag is ... drag! Hence propulsion comes from the *anisotropy* in the drag force. At low Reynolds numbers, the drag force on an object moving through a fluid is linearly proportional to the velocity: $\mathbf{F} \propto \mathbf{U}$, or

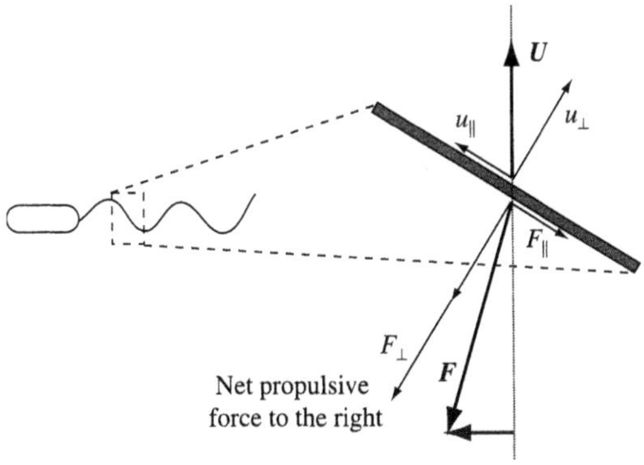

Fig. 1.2 Net propulsive force generated by a deforming tail.

$$F_{\perp} = C_{D\perp} u_{\perp}, \qquad F_{\|} = C_{D\|} u_{\|},$$

where the subscripts indicate directions parallel to and perpendicular to the surface of the object (as shown in Fig 1.2), and C_D is the drag coefficient. For a slender object, $C_{D\perp} \approx 2C_{D\|}$. This difference in the magnitude of the parallel and perpendicular drag coefficients can result in a net propulsive force as illustrated in the figure.

Challenge to students: Can you think of "objects" for which $C_{D\perp}/C_{D\|} > 2$? Or examples of slender objects for which the drag coefficient in the "thin" direction is greater than the drag coefficient in the "thick" direction? (For a sphere, $C_{D\perp}/C_{D\|} = 1$.)

2. Time does not appear explicitly in the Stokes equations (which are linear), and, consequently, they are time-reversible. This reversibility is beautifully demonstrated in G. I. Taylor's movie on low-Reynolds-number flows, which is now available online at

web.mit.edu/fluids/ww.shapiro/ncfmf.html.

(The demonstration of reversibility begins approximately 27 minutes into the film). Reversibility implies that a swimmer that undergoes a reciprocal deformation (i.e. a sequence of deformations that is symmetric when time is reversed) can never generate a net translation. This inability of reciprocal "swimmers" to swim at low Reynolds number is commonly known as the *Scallop Theorem* (since, as in Purcell's paper, scallops are often used as the canonical example of a swimmer that can only undergo reciprocal deformations). In "Life at low Reynolds number," Purcell (1977) suggests a number of ways to beat the Scallop Theorem that are *intrinsic* to the swimmer (e.g. by introducing chirality into the swimmer or by additional degrees of freedom).

Challenge to students (originally posed by Purcell): A three-link swimmer consists of three rigid pieces (two "arms" and a "body") connected by hinges; time-varying torques can be independently applied at each of the hinges.

Imagine that the swimmer is activated in the following sequence (starting with both arms up): (1) arm 1 moves down, (2) arm 2 moves down, (3) arm 1 moves up, (4) arm 2 moves up. Which direction does this three-link swimmer go? (See Purcell (1977) for an illustration of the swimmer; see Becker *et al.* (2003) for the solution.)

However, in addition to *intrinsic* solutions, there has been a recent surge in interest in developing strategies to beat the Scallop Theorem that are *extrinsic* to the swimmer. The key idea here is to make things *soft.* For example, one can:

(a) Put the swimmer near a "soft" interface (e.g. a free surface, membrane, or elastic solid). The flow around the swimmer induces deformations in the interface, and the resulting change in geometry introduces a source of nonlinearity that can break the symmetry (Trouilloud *et al.* 2008).
(b) Give the surrounding fluid a "soft microstructure" which leads to a viscoelastic constitutive relationship. (Other people will talk about this at this summer school, so I will not discuss swimming in viscoelastic fluids here.)

In addition to making things soft, one can also beat the Scallop Theorem by *swimming with friends.* See, for example, Lauga and Bartolo (2008).

3. In optimization calculations, we need to choose a metric to quantify what is being optimized. Common goals in locomotion studies include:

(a) maximize speed for a given power ("sprinter") or
(b) maximize efficiency for a given speed ("endurance").

Since time does not appear explicitly in the Stokes equations (which are linear), *speed and efficiency are equivalent measures of performance at low Reynolds numbers* The efficiency is generally defined as

$$\eta = \frac{\text{Rate of useful work}}{\text{Rate of viscous dissipation}} = \frac{\text{Force} \cdot \text{Velocity}}{\Phi} = \frac{\alpha V^2}{\Phi}.$$

The second criterion, maximizing efficiency for a given speed, corresponds to maximizing η. Likewise, maximizing the speed for a given power (fix Φ and maximize V) also corresponds to maximizing η. These two criteria, associated with speed and endurance, are equivalent at low Reynolds numbers because time can be scaled out of the equations and the performance of the stroke is determined purely by geometric considerations. Typical efficiencies for low-Reynolds-number swimmers are 1–2%; I would consider 10% to be a fantastically good efficiency for a low-Reynolds-number swimmer. In general, if authors cite numbers that are as high as (or higher than) 10%, they have used a different definition of η.

As an aside, Purcell worried about these low efficiencies (why would nature select organisms that are so inefficient?). To understand this, he calculated the power *per unit mass* for microorganisms, which he found to be on the order of 0.5 W/kg, which is lowish relative to larger organisms. Purcell likened this to "driving a Datsun in Saudi Arabia": it's not a great car, but it's cheap to fill up the tank.

1.2.3 Simple swimmers

In the next part of the lecture we will review a simple model for a specific low-Reynolds-number swimmer, the 2D swimming sheet as analyzed by Taylor (1951).

For his swimmer, Taylor considered an infinite, inextensible sheet undergoing small-amplitude sinusoidal deformations. The vertical displacement of the sheet is prescribed by a traveling wave $y_0 = b\sin(kx - \sigma t)$ (Fig.1.3) moving with a wave speed $c = \sigma/k$. Recall that the Stokes equations can be rewritten using the stream function as

$$\nabla^4\psi = 0, \tag{1.1}$$

where ψ is the stream function and is related to the velocity field $\mathbf{u} = (u, v)$ by $u = -\partial\psi/\partial y$, $v = \partial\psi/\partial x$. This equation must be solved subject to the following boundary conditions:

- no slip at the sheet: $\mathbf{u} = \mathbf{u}_{\text{sheet}}$ at $y_0 = b\sin(kx - \sigma t)$; and
- $\mathbf{u} = \mathbf{0}$ as $x \to \infty$ or $\mathbf{u} = (-V, 0)$ in a frame moving with the sheet, where V is the swimming velocity.

The general solution to eqn (1.1) is given by

$$\frac{1}{\sigma}\psi = \sum_{n\,\text{odd}}^{\alpha} (A_n y + B_n)e^{-ny}\sin nz + \sum_{n\,\text{even}}^{\alpha} (C_n y + D_n)e^{-ny}\cos nz - \frac{Vy}{\sigma},$$

where $z \equiv kx - \sigma t$. (Note that in addition to A_n, B_n, C_n, and D_n, V is also unknown and must be found by applying the boundary conditions.) Taylor then assumed $bk \ll 1$, expanded in powers of bk, and, half a page of algebra later, found $V = 0$ at lowest order (i.e. no swimming). The next-order correction is found many, many, many pages of algebra later (which I will not write out here) to be

$$\frac{V}{C} = \frac{2\pi^2 b^2}{\lambda^2}\left(1 - \frac{19}{4}\frac{\pi^2 b^2}{\lambda^2}\right).$$

Thus the sheet travels $\approx\lambda/84$ in one cycle (which is consistent with what is observed in live microorganisms).

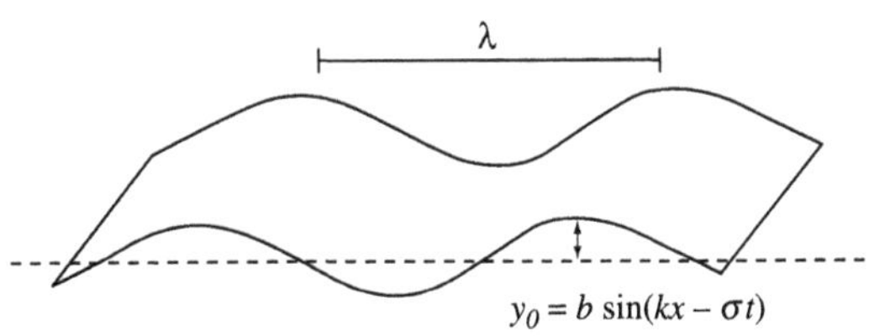

Fig. 1.3 2D swimming sheet.

1.2.4 Slightly less simple swimmers

In the final part of this lecture, I will discuss recent work from our group (primarily done by Daniel Tam) on slightly more realistic microswimmers. This part of the lecture was given in PowerPoint format and the slides have been included here as figures. We consider a swimmer that consists of a spherical head attached to a slender tail. The swimmer is allowed to select the kinematics of the tail (as is the case with eukaryotic flagella), and we seek the optimal curvature of the tail as a function of s, the distance along the flagellum, and of t, the time. The tail is modeled using slender-body theory and the head is modeled as a singularity distribution (for details, see the references in the slide shown in Fig. 1.4).

To test the model, we first consider an organism with no head. The optimal wave form is found numerically to be a sawtooth, consistent with the analytic optimal solution found by Lighthill for an infinitely long swimming snake (see the table in Fig. 1.4 for a quantitative comparison).

Next we consider an organism with a spherical head, and find that the optimal strokes have several characteristics that are consistent with what is observed in nature:

- The optimal strokes contain approximately one wavelength.
- The optimal wave shapes are *not* sinusoidal; rather, they consist of localized regions of high curvature connected by segments of low (approximately zero) curvature.
- The optimal curvature gradually decreases from head to tail.

Swimmer model

- Flagellum: Slender-body theory - find Stokeslet distribution (Keller and Rubinow 1976)
- Head: Exact singularity distribution (Chwang and Wu 1975)
- Head–flagellum interaction: Faxen's laws (Happel and Brenner 1973)
- Find optimal curvature along the tail

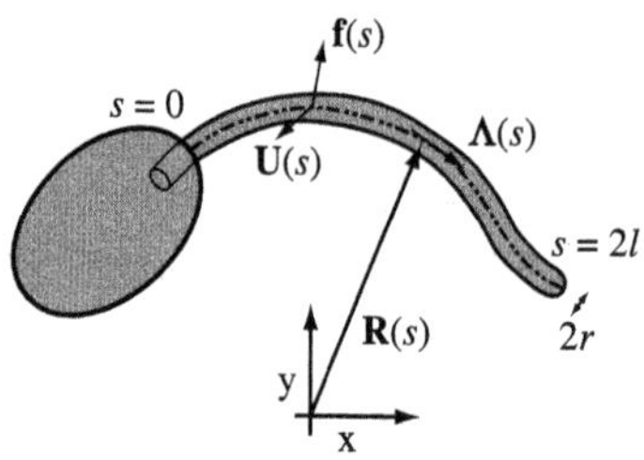

No head:

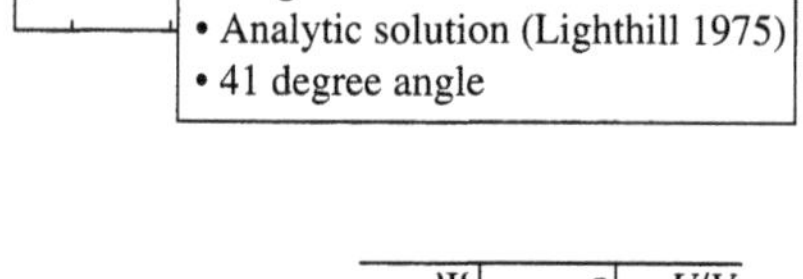

	Ψ	ε	U/V
Analytical solution	40°	0.0857	0.29
Computed solution	~41°	~0.08	~0.25

Fig. 1.4 Single-tail swimmer model.

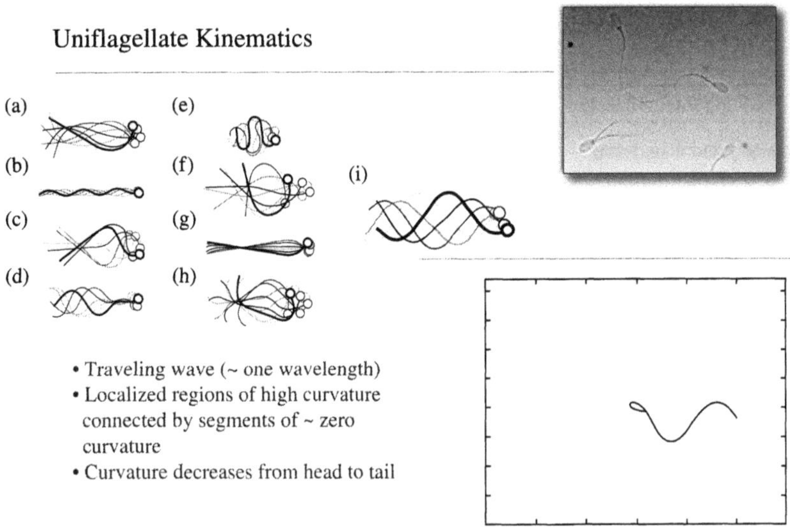

Fig. 1.5 Initial guesses for the optimal stroke pattern. All initial conditions converge to (i).

These optimal strokes, along with sample initial guesses used in the optimization routine, are shown in Fig 1.5. While this qualitative comparison with live organisms is promising, it is difficult to make quantitative comparisons with biological data, as finding high-resolution kinematic data is challenging. However, although detailed kinematic data is lacking, there is quite a bit of data available on *morphology*.

To determine whether optimal morphologies for swimming have evolved in nature, we consider a specific microorganism with a well-defined objective function, the spermatozoon. It can be argued that spermatozoa have one primary objective: to deliver a package of genetic material to the egg. Hence we consider the following question: For a given head size (containing the genetic cargo), what is the optimal length of the tail for propulsion? One can argue that such an optimal length should exist by considering two extreme tail lengths L. As $L \to 0$, the organism cannot move and the swimming efficiency goes to zero. As $L \to \infty$, all of the energy goes into moving the infinitely long tail rather than the precious genetic payload, and again the efficiency goes to zero. Since there are tail lengths for which the efficiency is not zero (and the optimal efficiencies appear to vary smoothly with L), there must be at least one optimal value between these two extremes. Figure 1.6 shows computed optimal efficiencies for various values of head-to-tail length ratio. Note that every point on this curve corresponds to an optimization calculation, and the optimal kinematics for short tails may not be the same as the optimal kinematics for long tails. Our computations indicate that the optimal value of $L/2R$ is approximately 12, which corresponds to the peak in the histogram, containing data from over 400 mammalian species! We can now postulate that the morphologies of spermatozoa have evolved into optimal geometries for low-Reynolds-number swimming.

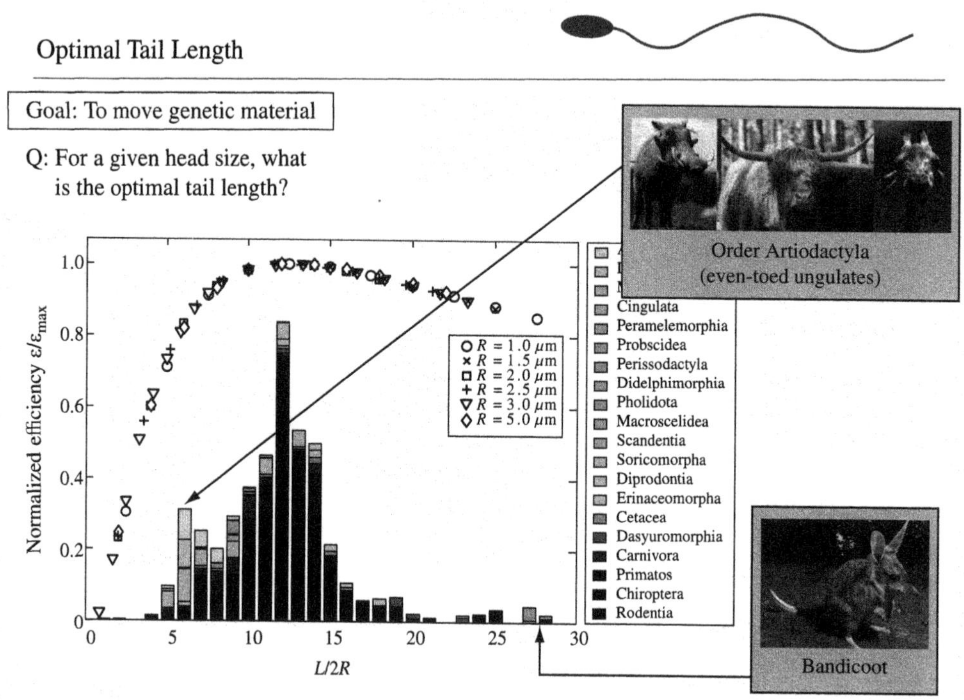

Fig. 1.6 The data points show the optimal efficiency as a function of head-to-tail length ratio, $L/2R$; the histogram represents the number of mammalian species that exhibit a given value of $L/2R$.

However, one could argue that the most interesting data points in the histogram are the ones that are *far* from optimal, as those points represent species that either are suboptimal or have been subject to unique evolutionary pressures and constraints. In our data set, the organisms at the far right (long-tailed sperm) correspond to several species of bandicoot and other mammals where the sperm have unusually thick tails. These organisms violate our assumption that the radius of the tail is set by the 9+2 microtubule structure discussed earlier, as the flagellum is encased in a thick sheath, which adds mechanical toughness. At the other end of the spectrum (short tails), the outliers correspond to monotremes, the egg-laying mammals. In these species, the head of the sperm has a helical shape, far from our spherical approximation, so again it is not surprising that they fall far from the computed optimum.

The set of outliers that is more difficult to explain corresponds to the peak in the histogram for the even-toed ungulates (pigs, sheep, goats, cows, ...). These points deviate consistently from the optimal value of tail length by about a factor of two. We have been unable to rationalize this discrepancy, and we leave it as a puzzle for the students. For further details, see Tam and Hosoi (2010).

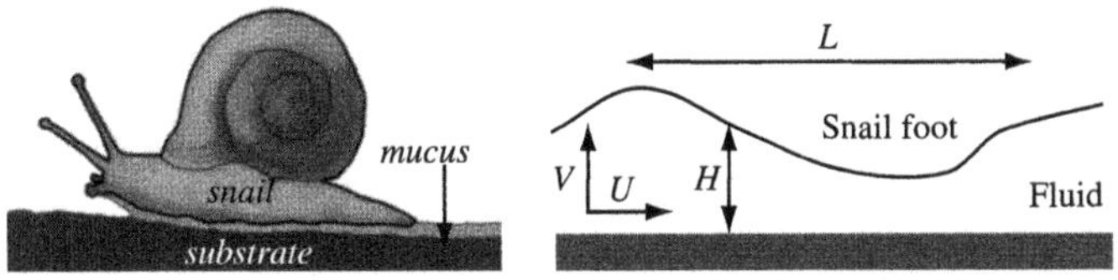

Fig. 1.7 Schematic illustration of snail and thin film. The characteristic velocities in the vertical and horizontal directions are denoted by V and U, respectively.

1.3 Lecture 2: Crawling

Our studies of crawling on fluid begin with snails. It is well known that snails crawl on top of a thin film of viscous fluid (pedal mucus), as shown in Fig. 1.7. The foot is not in direct contact with the substrate, and the only way that the snail's foot can interact with the substrate is by generating stresses within this thin film.

At this point we should ask: Does this qualify as low-Reynolds-number locomotion? Recall that I told you last time that in order to qualify as a low-Reynolds-number-swimmer, $L_{\text{organism}} \lesssim 1$ mm, and most snails are bigger than that (some, such as the giant African land snail, can grow to up to 30 cm in length!). However, we are saved because there are *two* characteristic length scales in a thin film, H and L (see Fig. 1.7). Comparing viscous and inertial forces, we find that we are allowed to neglect inertial terms in the equations of motion provided

$$\rho\frac{U^2}{L} \ll \mu\frac{U}{H^2} \;\Rightarrow\; \frac{\rho U H^2}{\mu L} \ll 1 \;\Rightarrow\; \left(\frac{H}{L}\right)^2 Re_L \ll 1.$$

In snails, $H \sim 10 - 20\,\mu$m and $L \sim 1$ mm, and hence $(H/L)^2 \sim 10^{-4}$. Combining this with the chart shown in Fig. 1.1, we find that even snails with sizes on the order of 10 cm still qualify as low-Reynolds-number locomotors.

1.3.1 Locomotion on thin films: Out-of-plane waves (see Chan *et al.* 2005)

In this lecture, we begin by considering a waving sheet near a wall (analogous to Taylor's sheet in the previous lecture). This is a naively simple model for a snail, but we will later see that even though snails do not use this method for propulsion, we can glean useful information from this simple approximation. Before we calculate anything, we can argue that we expect this configuration to generate propulsion. As fluid is squeezed beneath constrictions between the foot and the substrate, high-pressure regions develop in front of the wave (See Fig. 1.9 later). This high pressure acts on the tilted foot surface, resulting in a net propulsive force opposite to the direction in which the wave travels. We model the sheet as inextensible and assume that the wave amplitude is small, limited by the presence of the rigid substrate. Again, the goal is to prescribe the kinematics, i.e. the shape of the foot as a function of time $h(x,t)$, and find the crawling velocity V_s associated with a given deformation. As with Taylor's sheet, we will work in a frame moving with the wave to eliminate the time dependence in the problem. Rescaling the equations of motion using $\hat{u} = u\hat{V}_w$,

$\hat{v} = v(\hat{H}/\hat{L})\hat{V}_w$, $\hat{y} = y\hat{H}$, $\hat{x} = x\hat{L}$, and $\hat{p} = p\mu\hat{V}w\hat{L}/\hat{H}^2$, the dimensionless lubrication equations become to lowest order

$$\frac{\partial p}{\partial x} = \frac{\partial^2 u}{\partial y^2}, \quad \frac{\partial p}{\partial y} = 0.$$

Solving for the velocity with the boundary conditions given in Fig.1.8, we find

$$u = \frac{\partial p}{\partial x}\frac{1}{2}y(y-h) + V_s\left(\frac{y}{h} - 1\right) + 1. \tag{1.2}$$

If we consider steady-state crawling, the volumetric flux Q in the x direction must be a constant (i.e. not a function of x or t), and hence

$$Q = \int_0^h u\,dy = \text{constant}.$$

Combining this with the expression for u yields

$$\frac{\partial p}{\partial x} = \frac{12}{h^3}\left[h\left(1 - \frac{V_s}{2}\right) - Q\right]. \tag{1.3}$$

Next we assume periodic boundary conditions in x; in particular, $p(0) = p(1)$. This implies

$$\int_0^1 \frac{\partial p}{\partial x}dx = 0 = \int_0^1 \frac{12}{h^3}\left[h\left(1 - \frac{Vs}{2}\right) - Q\right]dx = \left(1 - \frac{Vs}{2}\right)\underbrace{\int_0^1 \frac{dx}{h^2}}_{I_2} - Q\underbrace{\int_0^1 \frac{dx}{h^3}}_{I_3}.$$

Solving for Q, we find

$$Q = \left(1 - \frac{Vs}{2}\right)\frac{I_2}{I_3},$$

where we have defined the integrals $I_n = \int_0^1 dx/h^n$.

So far we have invoked *conservation of momentum*, *conservation of mass*, and *periodicity*. The last step is to enforce *force balance*. The traction acting on the bottom of the foot of the snail is defined as

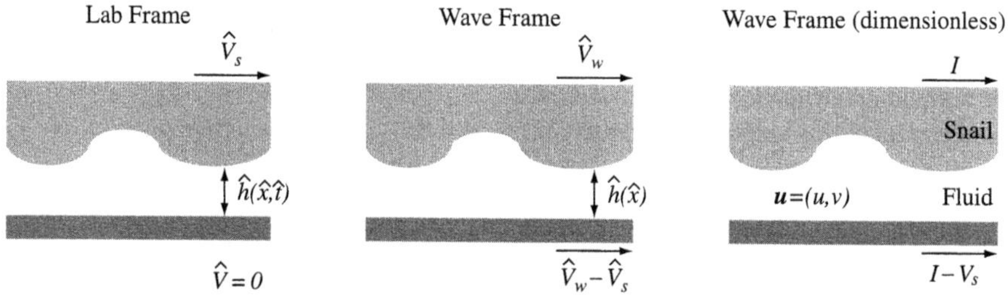

Fig. 1.8 Snail boundary conditions in stationary and moving frames.

$$\text{Force/length} = \text{traction} = \mathbf{f} = \sigma \cdot \mathbf{n}.$$

In the y-direction, the weight of the crawler is balanced by the pressure in the thin film (which doesn't tell us anything about the crawling velocity). Force balance in the x-direction is more useful:

$$f_x = p\frac{dh}{dx} + \frac{\partial u}{\partial y}.$$

Integrating over one wavelength, we find

$$F_x = \int_0^1 \left(p\frac{dh}{dx} + \frac{\partial u}{\partial y} \right)\bigg|_{y=h} dx = 0 \tag{1.4}$$

$$= ph|_0^1 - \int_0^1 h\frac{\partial p}{\partial x}\bigg|_{y=h} + \int_0^1 \frac{\partial u}{\partial y}\bigg|_{y=h} dx = 0. \tag{1.5}$$

Solving for V_s, we find

$$V_s = \frac{6(1-A)}{4-3A}, \text{where } A = \frac{I_2^2}{I_1 I_3} \text{ and } I_n = \int_0^1 \frac{dx}{I_n^n}, \tag{1.6}$$

which tells us how to find the crawling speed given the shape of the foot. A demonstration of the effectiveness of this crawling strategy is illustrated in Fig. 1.10, which shows a robotic crawler using out-of-plane waves.

We can now use eqn (1.6) to select optimal wave shapes for crawling. First, consider the properties of the shape function A. Using the Cauchy–Schwartz inequality, we can find upper and lower bounds for A:

$$I_2^2 = \left(\int h^{-1/2} h^{-2/3}\, dx \right)^2 \leq \int h^{-3}\, dx \int h^{-1}\, dx = I_3 I_1.$$

Hence $0 < A < 1$. Combining this with eqn (1.2), we find that the maximum crawling velocity (Fig. 1.9) is achieved when $A \to 0$, which corresponds to "sharp" profiles. This

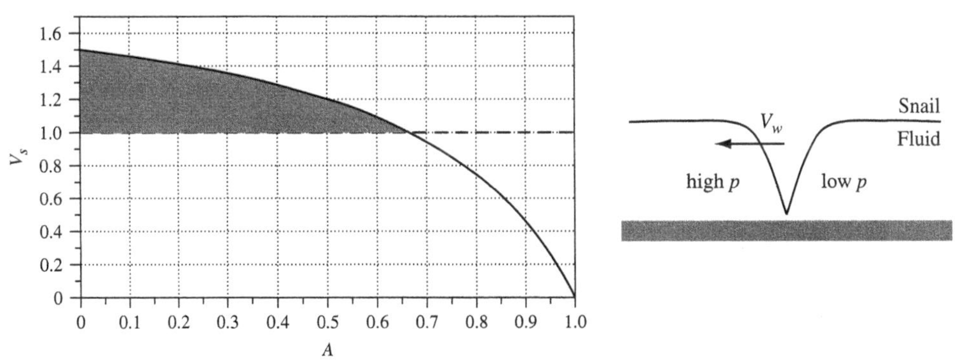

Fig. 1.9 Crawling velocity as a function of the shape function A.

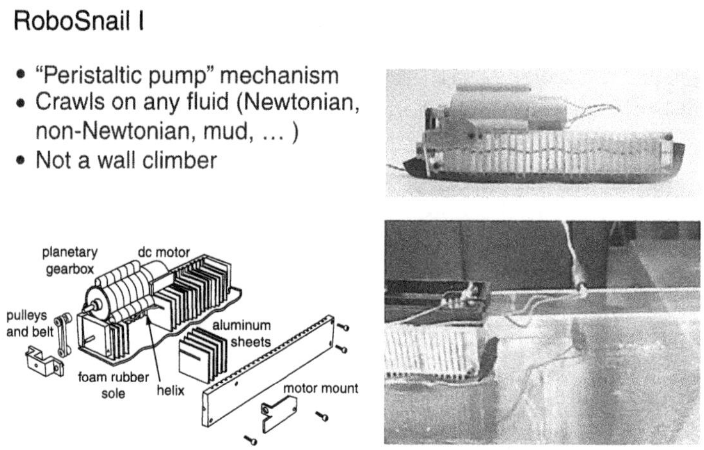

Fig. 1.10 Robotic crawler, built by Brian Chan, which crawls on a layer of glycerol using the mechanism described in the text.

is in line with what we would expect, as optimal profiles should have the following characteristics:

- They should include portions of the foot that nearly touch the substrate to maximize the pressure drop across these constrictions.
- The constrictions should be narrow to minimize viscous dissipation.
- The foot profile should be nearly vertical around these constrictions so that the force resulting from the high-pressure region acts primarily in the direction of motion.

Surprisingly, as $A \to 0$, $V_s \to 3/2$, which means that the snail moves *faster* than the wave (recall that V_s is the dimensionless crawling velocity, which has been normalized by the wave speed). Two-dimensional simulations by Shelley and Wilkening indicate that this phenomenon exists in full simulations as well and is not an artifact of lubrication theory breaking down in the neighborhood of sharp features. For further details of these 2D simulations and of the optimal profiles, see Wilkening and Hosoi (2008).

1.3.2 Locomotion on thin films: In-plane waves

Unfortunately, real snails don't move this way (although there is some evidence that limpets might). In live snails, there is very little out-of-plane motion of the wave and, instead, the animal sends waves of compression and expansion along the foot. A quick back-of-the-envelope calculation shows that the force required to locally change the film thickness H is much greater than the force required to slide the foot on top of the thin layer if $H/L \ll 1$. (We leave the details of this calculation as an exercise for the student.) Characteristic values for muscle strength are readily available in the literature, which reveal that the snail's muscle can produce enough force to slide but not enough to lift the foot.

To determine whether crawling is possible using waves of expansion and contraction on a film of constant thickness, we consider the following simple calculation based on

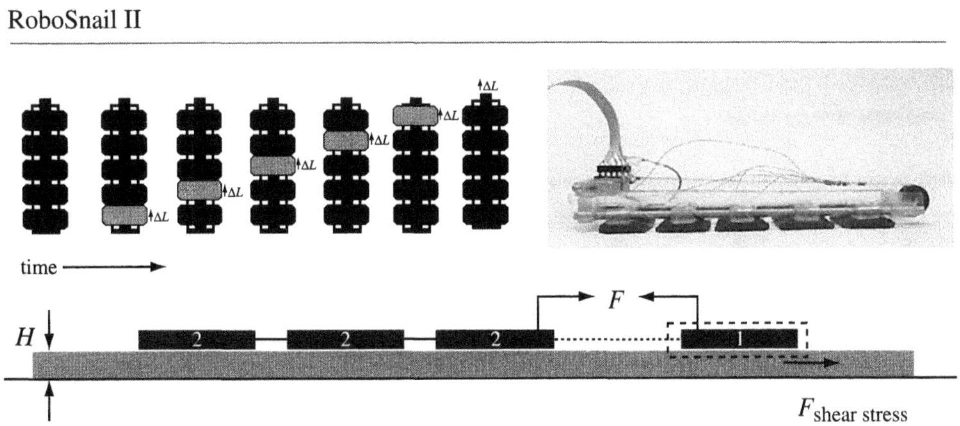

Fig. 1.11 Mechanical crawler RoboSnail II and schematic illustration of crawling.

the geometry of our second mechanical crawler, shown in Fig. 1.11. Consider the first "step" in crawling, in which the front pads are connected in a rigid block (labeled "2") and pad "1" slides forward. The actuator, or muscle, between blocks 1 and 2 applies a force F as indicated in the figure. On pad 1, F is balanced by the shear force generated in the thin film:

$$-F = A\frac{V_1\mu_1}{H},$$

where A is the area of the pad, μ_1 is the viscosity of the fluid under pad 1, and V_1 is the velocity of pad 1. Similarly, for the pads in the rigid block, 2,

$$F = (N-1)A\frac{V_2\mu_2}{H},$$

where N is the total number of pads on the crawler. To find the velocity of the center of mass of the crawler, we average over the velocities of all of the pads:

$$[V_1 + (N-1)V_2]/N = V_{\rm cm}.$$

Combining these three equations to solve for the velocity of the center of mass, we find

$$V_{\rm cm} = \left[-\frac{FH}{A\mu_1} + (N-1)\frac{FH}{A\mu_2(N-1)}\right]/N = \frac{FH}{AN}\left(\frac{1}{\mu_2} - \frac{1}{\mu_1}\right).$$

Therefore, the snail crawls only if $\mu_2 \neq \mu_1$ and will *not* crawl on a Newtonian fluid. In order to generate propulsion, there needs to be some type of nonlinearity in the constitutive relationship for the viscous thin film. Fortunately, the mechanical response of real snail mucus is extremely nonlinear, as first measured by M. Denny in the 1980s. Figure 1.12 shows the "design space" for snail mucus and other common yield stress fluids; snail mucus is indicated by stars. Stress-versus-strain data for mucus from a common garden snail is shown in the graph at the lower right. (In this representation,

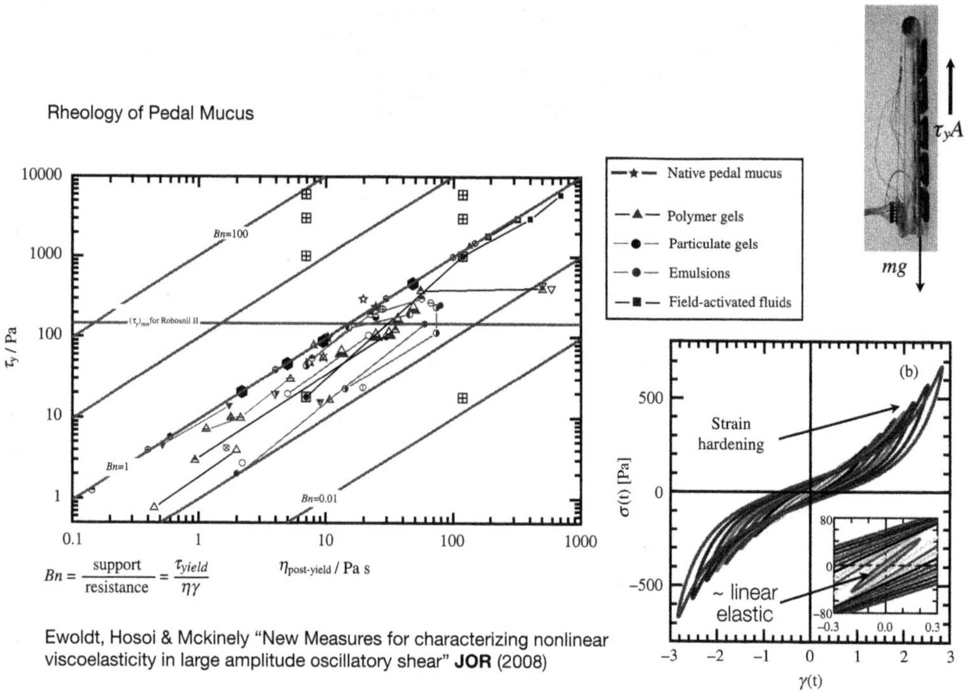

Fig. 1.12 Yield stress of common materials and measured stress versus strain for snail mucus (data taken by Randy Ewoldt).

a linear elastic material would appear as a straight line; a Newtonian viscous material would appear as an ellipse with its major and minor axes aligned with the axes for stress and strain.)

Finally, we can ask: What is the "best" rheology for crawling? Details of this calculation can be found in Lauga and Hosoi (2006). The calculation again applies lubrication theory, this time to a mucus rheology that is slightly non-Newtonian (i.e. the constitutive relationship is perturbed away from a Newtonian base state), and we find that the optimal rheology (not surprisingly) depends on our choice of the cost function. If we opt to minimize viscous dissipation in the thin film (which is roughly equivalent to minimizing the amount of work that the muscles need to do), we find that a shear-thickening rheology is preferable. However, if instead we opt to minimize mucus production (which is arguably more relevant for real snails which spend 80–90% of their energy budget on making mucus), a shear-thinning mucus is advantageous.

1.4 Lecture 3: Burrowing

In this final lecture, we will discuss moving through complex media at low Reynolds numbers. There has been quite a bit of excellent work on swimming in viscoelastic fluids (e.g. studies by Shelley, Powers, and Lauga), which will be discussed by other lecturers in this summer's program. Today we will focus instead on moving through *granular media, slurries,* and *suspensions.* In addition, we will limit our discussion to granular materials that are fully saturated with fluid.

Unlike the first two lectures, most of the research that I will discuss today represents work in progress. We first became interested in this topic in a collaboration with Bluefin Robotics, a company that makes small (15–20 ft long) autonomous underwater vehicles and is interested in developing new compact technologies for anchoring. To understand existing biological solutions to this challenge, we searched through the literature to find organisms that are good diggers and that live primarily in littoral zones. In this search, it was revealed that *clams* are surprisingly fast and efficient burrowers. In particular, we became interested in the razor clam, *Ensis directus* (which is so fast that my graduate student Amos Winter refers to it as "the Ferrari of bivalves").

First, we would like to determine the amount of force required to submerge a clam shell in a saturated granular medium. This can be measured directly by pushing the shell into a bucket of wet sand, as shown in Fig. 1.13. The data shows that the required force scales linearly with depth. This is not surprising, as soil and granular

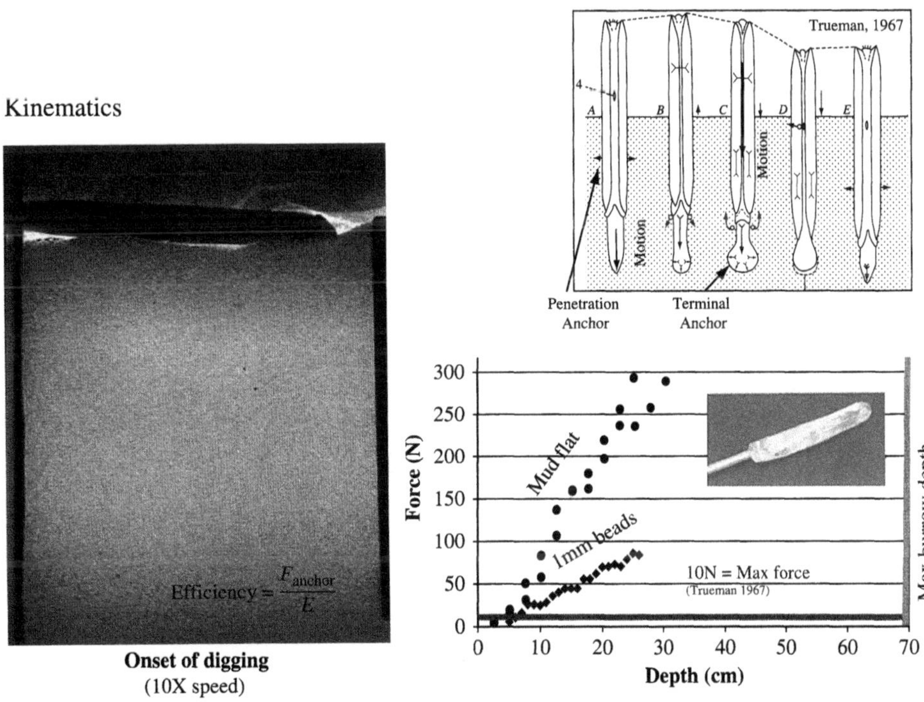

Fig. 1.13 (Left) Hele–Shaw clam visualization cell. (Top right) clam kinematics, reproduced from Trueman (1975). (Bottom right) Force required to submerge a clam shell in a saturated granular material.

media typically have a yield stress which (for sand) comes from the frictional contact between the grains:

$$\text{yield stress} = \sigma_y \sim \text{frictional force} \sim \text{normal force between grains} \sim \rho g d.$$

Hence $\sigma_y \sim d$, where d is the depth of the submerged object. The data in the figure also gives us some idea of how deep clams should be able to dig. It is known that a clam's muscle can exert a maximum force of approximately 10 N. Comparing this with our data, we find that a clam should be able to bury its shell to a depth of 5–10 cm. However, clams are known to burrow through approximately 1 m of sand, an order of magnitude deeper than one might expect!

To understand this discrepancy, we built a viewing tank (shown in Fig. 1.13) that allowed us to visualize the kinematics of the clam as it digs. Videos show that the clam does not move steadily downwards; rather, at the beginning of each cycle, it pushes *up*, executing a quick upward stroke before it moves down. This upward motion produces a pocket of fluidized material around the clam as illustrated in Fig. 1.14. To understand the gains associated with fluidization, we note that the energy required to dig scales as $E \sim F \cdot d$. The data in Fig. 1.13 shows that the force required to move through a packed granular system scales as $F_{\text{packed}} \sim d$, implying that $E_{\text{packed}} \sim d^2$. In contrast, the force required to move through a fluidized system is independent of d, and hence $E_{\text{fluidized}} \sim d$. Therefore, we expect that beyond a certain depth, it always pays to fluidize.

Fig. 1.14 also indicates two regions around a burrowing clam: the fluidized region and the packed granular medium, which are separated by a failure surface. In order to fully characterize the motion of the clam, we need to compute:

1. The location of the yield surface.
2. The coupling of the kinematics of the clam with motion through a dense suspension.
3. The flow through the porous medium.

In this lecture, we will address the first item: finding the location of the yield surface. The second item, coupling motion to kinematics as the clam moves through the effective viscous medium, involves calculations that are similar to those we have seen in the first two lectures. Again, the dominant force on the organism is drag, and the

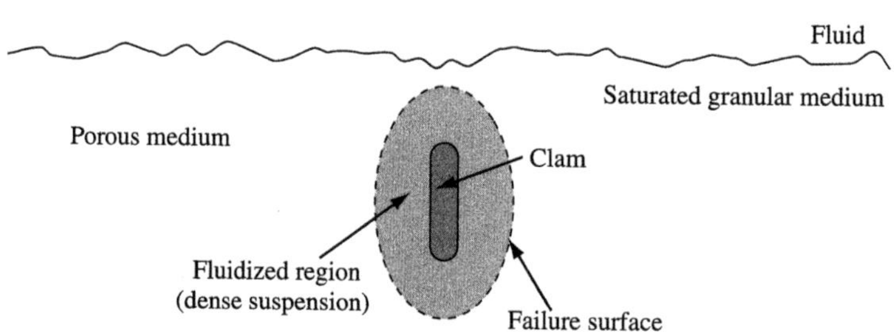

Fig. 1.14 Illustration of the various regimes around a digging clam.

digger must be force- and torque-free. The third item, flow through a porous medium, has been well studied in many contexts and will not be discussed here.

1.4.1 Finding the failure surface

In determining the failure zone, there are many empirical failure criteria to choose from. We begin with the simplest one, namely that the critical shear stress, at which the material yields is proportional to the normal stress. (i.e. if we squeeze the particles together, the material is harder to shear):

$$\tau_{\mathrm{crit}} = \alpha\sigma_{\mathrm{n}}.$$

This criterion is slightly complicated to apply, because stress is a tensor and it is not immediately obvious which direction is the "shear" direction and which is "normal." Consider the material elements shown in Fig. 1.15. The state of stress is the *same* in both elements but the coordinate system has been rotated by 45° in the case of the second. As a consequence, the state of stress in the first instance consists entirely of normal stresses and that in the second is entirely shear. Given this ambiguity, we take the "worst-case scenario" to apply our failure criterion. If any coordinate system exists in which $\tau_{\mathrm{crit}} > \alpha\tau_{\mathrm{n}}$, the soil fails locally. This can be checked graphically using Mohr's circle. To transform from the $x - y$ coordinate system (known) to the $x' - y'$ coordinate system (unknown) as shown in Fig. 1.16, we apply the following transformation relations:

$$\sigma_{x'x'} - \frac{\sigma_{xx} + \sigma_{yy}}{2} = \frac{\sigma_{xx} + \sigma_{yy}}{2}\cos 2\theta + \sigma x y \sin 2\theta, \tag{1.7}$$

$$\sigma_{x'x'} = \frac{-(\sigma_{xx} + \sigma_{yy})}{2}\sin 2\theta + \sigma_{xy}\cos 2\theta. \tag{1.8}$$

Squaring both equations and adding them together, we obtain

$$\Bigg(\underbrace{\sigma_{x'x'}}_{\sigma_{\mathrm{n}}} - \underbrace{\frac{\sigma_{xx} + \sigma_{yy}}{2}}_{\equiv \overline{\sigma}}\Bigg)^2 + \underbrace{\sigma^2_{x'y'}}_{\tau^2} = \underbrace{\left(\frac{\sigma_{xx} - \sigma_{yy}}{2}\right)^2 + \sigma^2_{xy}}_{\equiv R^2},$$

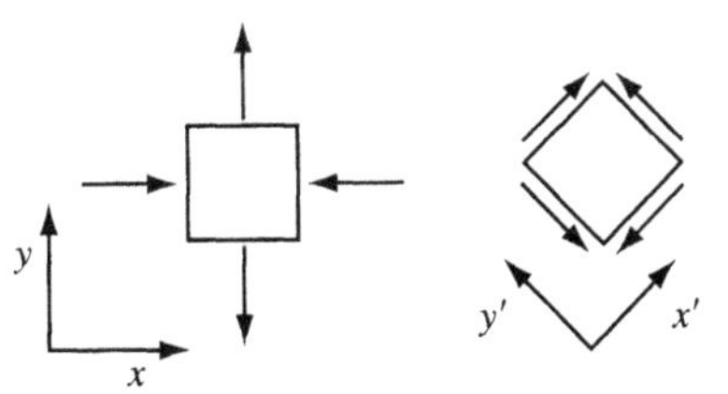

Fig. 1.15 Components of the stress tensor in two different coordinate systems.

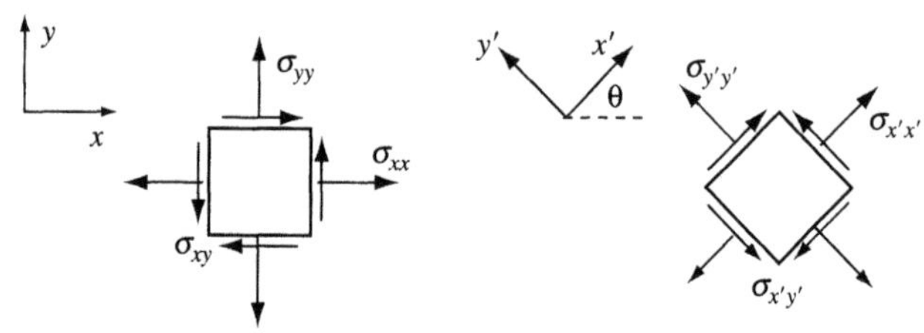

Fig. 1.16 Schematic of stresses in original $(x - y)$ and rotated $(x' - y')$ coordinate system.

which yields the equation for Mohr's circle:

$$(\sigma_n - \bar{\sigma})^2 + \tau^2 = R^2.$$

Mohr's circle represents a simple graphical way to transform the local state of stress. Points on the circle represent the state of stress in different rotated coordinate systems. This can now be used to graphically represent the failure criterion. (Note: from here on, we will use the sign convention commonly used in soil mechanics, in which compressive stresses are positive.)

As the clam closes its shell, the horizontal stress decreases around the clam, which moves the minimum principal stress on Mohr's circle to the left (for now, consider an infinitely long, cylindrically symmetric clam). As the circle grows (larger circles correspond to smaller r, indicating neighborhoods closer to the clam), it crosses the failure line (shown in Fig. 1.17 and defined by $\tau_{\text{crit}} = \alpha\sigma_{\text{n}}$) at some critical value of r and the soil within this critical radius fails. We can now use this to find the failure radius geometrically. From the diagram in Fig. 1.17, we see that

$$\sin\theta = \frac{R}{\sigma'_{\text{H}} + R}, \qquad R = \frac{\sigma'_{\text{V}} - \sigma'_{\text{H}}}{2},$$

which defines the failure surface as

$$\sigma'_{\text{V}} = \frac{1 + \sin\theta}{1 - \sin\theta}\sigma'_{\text{H}},$$

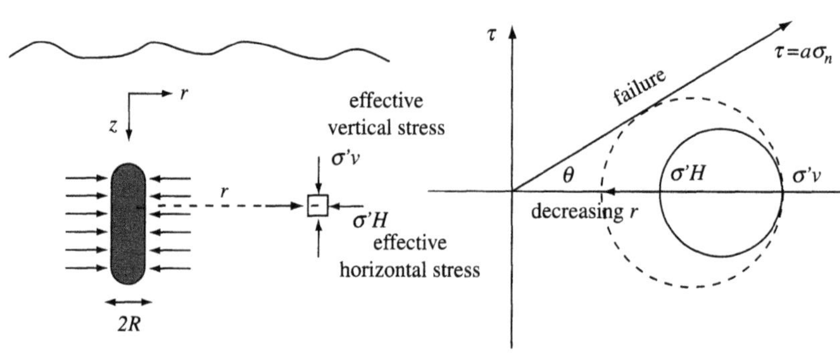

Fig. 1.17 Schematic of stress field surrounding the burrower and corresponding Mohr's circle.

where θ is the angle of repose of the granular medium and σ'_{H} and σ'_{V} are the effective horizontal and vertical stresses, respectively.

The final step is to write down expressions for σ'_{H} and σ'_{V} in terms of known quantities. The vertical stress is hydrostatic:

$$\sigma_{\mathrm{V}} = \rho_{\mathrm{tot}} g z,$$

where ρ_{tot} is the effective density of the fluid/particle mixture: $\rho_{\mathrm{tot}} = \rho_{\mathrm{p}}\phi + \rho_{\mathrm{f}}(1-\phi)$. Here ϕ is the volume fraction of particles, ρ_{p} is the particle density, and ρ_{f} is the fluid density. To find the effective vertical stress σ'_{V}, we need to subtract the pore pressure which arises from the lubrication forces between particles:

$$\sigma'_{\mathrm{V}} = \rho_{\mathrm{tot}} g z - \rho_{\mathrm{f}} g z = \phi\ \Delta\rho\ g z,$$

where $\Delta\rho \equiv \rho_{\mathrm{p}} - \rho_{\mathrm{f}}$.

The horizontal stress can be found by solving $\nabla \cdot \sigma = 0$, assuming axisymmetry, and applying the following boundary conditions (see e.g. Landau and Lifshitz 1986):

1. $\sigma_{\mathrm{H}} = p_0$ at $r = R$.
2. $\sigma_{\mathrm{H}} \to \sigma_{\infty}$ as $r \to \infty$.

This gives us an expression for σ_{H}:

$$\sigma_{\mathrm{H}} = \frac{R^2}{r^2}(p_0 - \sigma_{\infty}) + \sigma_{\infty}.$$

Subtracting the pore pressure again, we find the effective horizontal stress

$$\sigma'_{\mathrm{H}} = \frac{R^2}{r^2}(p_0 - \sigma_{\infty}) + \sigma_{\infty} - \rho_{\mathrm{f}} g z.$$

Here R is the radius of the clam and p_0 is set by the closing shell, so the only remaining unknown quantity is σ_{∞}, the horizontal stress in the far field, which can be measured. In general, soil engineers do not measure σ_{∞}; rather, they report K_0, the *coefficient of lateral earth pressure.* This is a of the soil property that depends both on the history of the soil and material properties of the granular medium. In unperturbed soil (i.e. no clam), $\sigma'_{\mathrm{H}} = K_0 \sigma'_{\mathrm{V}}$, which defines K_0.

Substituting this back into our failure criterion, we find that the failure radius is given by

$$\frac{r_{\mathrm{fail}}}{R} = C_0 \left[K_0 + \frac{\rho_{\mathrm{f}}}{\phi\Delta\rho}\left(1 - \frac{p_0}{\rho_{\mathrm{f}} g z}\right)\right]^{1/2},$$

as shown in Fig. 1.18, where $C_0 \equiv (K_0 - (1-\sin\theta)/(1+\sin\theta))^{-1/2}$ is a constant that depends on the soil properties. As $z \to \infty$, $r_{\mathrm{fail}}/R \to C_0\,(K_0 + \rho_{\mathrm{f}}/\phi\Delta\rho)^{1/2}$. Taking reasonable values for glass beads ($K_0 = 0.5$ and $\theta = 35°$), we find $C_0 \approx 2$, which gives $r_{\mathrm{fail}}/R \approx 3.3$. In fact, for a wide variety of soil parameters, we find that the failure radius varies between roughly 1 and 3 clam radii, implying that the fluidization is a fairly localized phenomenon (which is consistent with what we observe in our clam viewing tank).

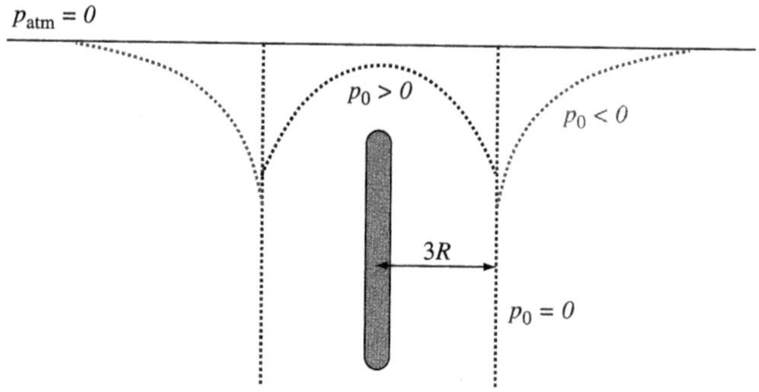

Fig. 1.18 Sketches of yield surfaces for various values of p_0.

Finally, I will end with an image of our mechanical digger, RoboClam, which burrows using kinematics inspired by live clams, reducing the energy required to dig (Fig. 1.19). Using a genetic optimization algorithm, we were able to find kinematics that corresponded to a digging-energy scaling of $E \sim d^{1.17}$, which is close to the optimal fluidized limit of $E \sim d$. (Again, I should emphasize that this is work in progress and our current results are likely to be a local minimum; as we continue to explore the phase space, we hope to achieve the optimal power of 1.)

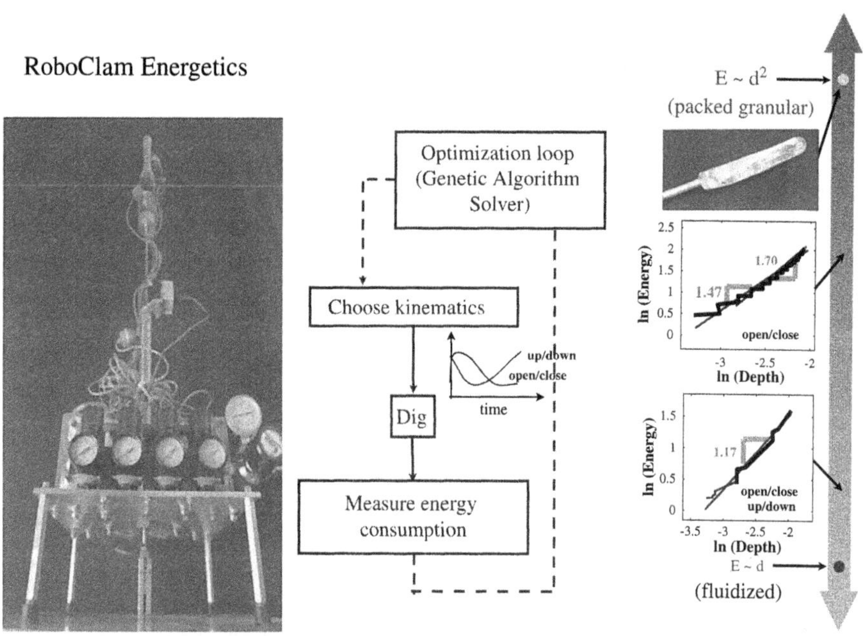

Fig. 1.19 Roboclam, a mechanical digger (built by Amos Winter) that uses kinematics inspired by live clams to locally fluidize the soil and increase digging efficiency.

References

L. E. Becker, S. A. Koehler, and H. A. Stone (2003). On self-propulsion of micro-machines at low Reynolds number: Purcell's three-link swimmer. *Journal of Fluid Mechanics*, **490**, 15–35.

B. Chan, N. J. Balmforth, and A. E. Hosoi (2005). Building a better snail: Lubrication and adhesive locomotion. *Physics of fluids*, **17**, 113101.

A Chwang and T. Wu (1974). Hydrodynamics of low-Reynolds-number flow. Part 2. Singularity method for Stokes flow. *Journal of Fluid Mechanics*, **67**, 787–815.

J. Happel and H. Brenner (1973). *Low Reynolds Number Hydrodynamics.* Noordhoff, Leiden.

J. B. Keller and S. I. Rubinow (1976). Slender-body theory for slow viscous flow. *Journal of Fluid Mechanics*, **75**, 705–714.

L. D. Landau and E. M. Lifshitz (1986). *Theory of Elasticity.* Course of Theoretical Physics, Vol. 7, 3rd edn. Butterworth-Heinemann, Oxford.

E. Lauga and D. Bartolo (2008). No many-scallop theorem: Collective locomotion of reciprocal swimmers. *Physical Review E*, **78**, 030901.

E. Lauga and A. E. Hosoi (2006). Tuning gastropod locomotion: Modeling the influence of mucus rheology on the cost of crawling. *Physics of Fluids*, **18**, 113102.

J. Lighthill (1975). *Mathematical Biofluiddynamics.* CBMS-NSF Regional Conference Series in Applied Mathematics. SIAM, Philadelphia.

J. Lighthill (1976). Flagellar hydrodynamics. SIAM *Review*, **18**(2), 161–230.

E. M. Purcell (1977). Life at low Reynolds number. *American Journal of physics*, **45**(1), 3–11.

D. Tam and A. E. Hosoi (2010). Optimal feeding and swimming gaits of biflagellated organisms. *Proceedings of the National academy of Sciences of the USA*, in press.

G. I. Taylor (1951). Analysis of the swimming of microscopic organisms. *Proceedings of the Royal Society A*, **209**, 447–461.

R. Trouilloud, T. S. Yu, A. E. Hosoi, and E. Lauga (2008). Soft swimming: Exploiting deformable interfaces for low Reynolds number locomotion. *Physical Review Letters*, **101**, 048102.

E. R. Trueman (1975). *The Locomotion of Soft-Bodied Animals.* Edward Arnold, London.

J. Wilkening and A. E. Hosoi (2008). Shape optimization of a sheet swimming over a thin liquid layer. *Journal of Fluid Mechanics*, **601**, 25–61.

2
Surface tension

John W. M. Bush

Department of Mathematics, MIT

2.1 Introduction

These lecture notes have been drawn from many sources, including textbooks, journal articles, and lecture notes from courses taken by the author as a student. They are not intended as a complete discussion of the subject, or as a scholarly work in which all relevant references are cited. Rather, they are intended as an introduction that will hopefully motivate the interested student to learn more about the subject. Topics have been chosen according to their perceived value in developing the physical insight of the students.

2.2 The definition and scaling of surface tension

2.2.1 Surface tension: A working definition

Discussions of the molecular origins of surface or interfacial tension may be found elsewhere; ours follows that of De Gennes *et al.* (2004).

Molecules in a fluid feel a mutual attraction. When this attractive force is overcome by thermal agitation, the molecules pass into a gaseous phase. Let us first consider a free surface, for example that between air and water. A water molecule in the fluid bulk is surrounded by attractive neighbors, while a molecule at the surface is attracted by a reduced number of neighbors and so is in an energetically unfavorable state. The creation of new surface is thus energetically costly, and a fluid system will act to minimize surface areas. It is thus that small fluid bodies tend to evolve into spheres; for example, a thin fluid jet emerging from the top in your kitchen sink will generally pinch off into spherical drops in order to minimize the total surface area (see Section 2.6).

If U is the total cohesive energy per molecule, then a molecule at a flat surface will lose $U/2$. Surface tension is a direct measure of this energy loss per unit area of surface. If the characteristic molecular dimension is R and its area is thus R^2, then the surface tension is $\sigma \sim U/(2R^2)$. Note that surface tension increases as the intermolecular attraction increases and the molecular size decreases. For most oils, $\sigma \sim 20$ dyn/cm, while for water, $\sigma \sim 70$ dyn/cm. The highest surface tensions are for liquid metals; for example, liquid mercury has $\sigma \sim 500$ dyn/cm.

The origins of interfacial tension are analogous. Interfacial tension is a material property of a fluid–fluid interface whose origins lie in the different attractive intermolecular forces that act in the two fluid phases. The result is an interfacial energy per area that acts to resist the creation of new interface, and that is equivalent to a line tension acting in all directions parallel to the interface. Fluids between which no interfacial tension arises are said to be miscible. For example, salt molecules will diffuse freely across a boundary between fresh and salt water; consequently, these fluids are miscible, and there is no interfacial tension between them. Our discussion will be confined to immiscible fluid–fluid interfaces (or fluid surfaces), at which an effective interfacial (or surface) tension acts.

Surface tension σ has the units of force/length or, equivalently, energy/area, and so may be thought of as a negative surface pressure. Pressure is generally an isotropic force per area that acts throughout the bulk of a fluid: a small surface element dS will feel a total force $p(\mathbf{x})\, dS$ owing to the local pressure field $p(\mathbf{x})$. If the surface S

is closed, and the pressure uniform, the net pressure force acting on S is zero and the fluid remains static. Pressure gradients correspond to body forces (with units of force per unit volume) within a fluid, and so appear explicitly in the Navier–Stokes equations. Surface tension has the units of force per length, and its action is confined to the free surface. Consider, for the sake of simplicity, a perfectly flat interface. A surface line element $d\ell$ will feel a total force $\sigma\, d\ell$ owing to the local surface tension $\sigma(\mathbf{x})$. If the surface line element is a closed loop C and the surface tension is uniform, the net surface tension force acting on C is zero, and the fluid remains static. If surface tension gradients arise, there may be a net force on the surface element that acts to distort it through driving flow.

2.2.2 Governing equations

The motion of a fluid of uniform density ρ and viscosity μ is governed by the Navier–Stokes equations, which represent a continuum statement of Newton's laws:

$$\rho\left(\frac{\partial \mathbf{u}}{\partial t} + \mathbf{u}\cdot\nabla\mathbf{u}\right) = -\nabla p + \mathbf{F} + \mu\nabla^2\mathbf{u}, \tag{2.1}$$

$$\nabla\cdot\mathbf{u} = 0. \tag{2.2}$$

This represents a system of four equations in four unknowns (the fluid pressure p and the three components of the velocity field $\mathbf{u}$). Here $\mathbf{F}$ represents any body force acting on the fluid; for example, in the presence of a gravitational field, $\mathbf{F} = \rho\mathbf{g}$, where $\mathbf{g}$ is the acceleration due to gravity.

Surface tension acts only at the free surface; consequently, it does not appear in the Navier–Stokes equations, but rather enters through the boundary conditions. The boundary conditions appropriate at a fluid–fluid interface are formally developed in Section 2.3. Here, we simply state them for the simple case of a free surface (such as between air and water, in which one of the fluids is not dynamically significant) in order to get a feeling for the scaling of surface tension.

The normal stress balance at a free surface must be balanced by the curvature force associated with the surface tension:

$$\mathbf{n}\cdot\mathbf{T}\cdot\mathbf{n} = \sigma\,(\nabla\cdot\mathbf{n}), \tag{2.3}$$

where $T = -p\mathbf{I} + \mu[\nabla\mathbf{u} + (\nabla\mathbf{u})^{\mathrm{T}}] = -p\mathbf{I} + 2\mu\mathbf{E}$ is the stress tensor, $\mathbf{E} = 1/2\ [\nabla\mathbf{u} + (\nabla\mathbf{u})^T]$ is the deviatoric stress tensor, and $\mathbf{n}$ is the unit normal to the surface. The tangential stress at a free surface must balance the local surface tension gradient:

$$\mathbf{n}\cdot\mathbf{T}\cdot\mathbf{t} = \nabla\sigma\cdot\mathbf{t}, \tag{2.4}$$

where $\mathbf{t}$ is the unit tangent to the interface.

2.2.3 The scaling of surface tension

We consider a fluid of density ρ and viscosity $\mu = \rho\nu$ with a free surface characterized by a surface tension σ. The flow is marked by characteristic length and velocity scales of a and U, respectively, and evolves in the presence of a gravitational field $\mathbf{g} = -g\hat{z}$. We thus have a physical system defined in terms of six physical variables $(\rho, \nu, \sigma, a, U, g)$ that may be expressed in terms of three fundamental units: mass, length, and time. Buckingham's Theorem thus indicates that the system may be uniquely prescribed in terms of three dimensionless groups. We choose

$$Re = \frac{Ua}{\nu} = \frac{\text{INERTIA}}{\text{VISCOSITY}} = \text{Reynolds number}, \tag{2.5}$$

$$Fr = \frac{U^2}{ga} = \frac{\text{INERTIA}}{\text{GRAVITY}} = \text{Froude number}, \tag{2.6}$$

$$Bo = \frac{\rho g a^2}{\sigma} = \frac{\text{GRAVITY}}{\text{CURVATURE}} = \text{Bond number}. \tag{2.7}$$

The Reynolds number prescribes the relative magnitudes of inertial and viscous forces in the system, while the Froude number prescribes those of inertial and gravity forces. The Bond number indicates the relative importance of forces induced by gravity and surface tension. Note that these two forces are comparable when $Bo = 1$, which arises on a length scale corresponding to the capillary length $\ell_c = (\sigma/(\rho g))^{1/2}$. For an air–water surface, for example, $\sigma \approx 70$ dyn/cm, $\rho = 1\,\text{g/cm}^3$, and $g = 980$ cm/s^2, so that $\ell_c \approx 2\,\text{mm}$. Bodies of water in air are dominated by the influence of surface tension provided they are smaller than the capillary length. Roughly speaking, the capillary length prescribes the maximum size of pendant drops that may hang inverted from a ceiling, and the maximum size of water-walking insects and raindrops. Note that as a fluid system becomes progressively smaller, the relative importance of surface tension and gravity increases; it is thus that surface tension effects are dominant in microscale engineering processes.

Finally, we note that other frequently arising dimensionless groups may be formed from products of B, Re, and Fr:

$$We = \frac{\rho U^2 a}{\sigma} = \frac{\text{INERTIA}}{\text{CURVATURE}} = \text{Weber number}, \tag{2.8}$$

$$Ca = \frac{\rho \nu U}{\sigma} = \frac{\text{VISCOUS}}{\text{CURVATURE}} = \text{capillary number}. \tag{2.9}$$

The Weber number indicates the relative magnitudes of inertial and curvature forces within a fluid, and the capillary number those of viscous and curvature forces. Finally, we note that if the flow is marked by a Marangoni stress of characteristic magnitude $\Delta\sigma/L$, then an additional dimensionless group arises that characterizes the relative magnitude of the Marangoni and curvature stresses: $a\,\Delta\sigma/(L\sigma)$.

We now demonstrate how these dimensionless groups arise naturally from the nondimensionalization of the Navier–Stokes equations and the surface boundary conditions. We first introduce a dynamic pressure $p_d = p - \rho\mathbf{g}\cdot\mathbf{x}$, so that gravity appears only in the boundary conditions. We consider the special case of high-Reynolds-number flow, for which the characteristic dynamic pressure is ρU^2. We define dimensionless primed variables according to

$$\mathbf{u} = U\mathbf{u}', \quad p_{\rm d} = \rho U^2 p_{\rm d}', \quad \mathbf{x} = a\mathbf{x}', \quad t = \frac{a}{U}t'. \tag{2.10}$$

Nondimensionalizing the Navier–Stokes equations and appropriate boundary conditions yield the following system:

$$\left(\frac{\partial \mathbf{u}'}{\partial t'} + \mathbf{u}' \cdot \nabla' \mathbf{u}'\right) = -\nabla p_{\rm d}' + \frac{1}{Re}\nabla'^2 \mathbf{u}', \quad \nabla' \cdot \mathbf{u}' = 0. \tag{2.11}$$

The normal-stress condition assumes the dimensionless form

$$-p_{\rm d}' + \frac{1}{Fr}z' + \frac{2}{Re}\,\mathbf{n} \cdot \mathbf{E}' \cdot \mathbf{n} = \frac{1}{We}\,\nabla' \cdot \mathbf{n}. \tag{2.12}$$

The importance of surface tension relative to gravity and viscous stresses is prescribed by the relative magnitudes of the Weber, Froude, and Reynolds numbers. In the high-Re limit of interest, the normal force balance requires that the dynamic pressure be balanced by either gravitational or curvature stresses, the relative magnitudes of which are prescribed by the Bond number.

The nondimensionalization scheme will depend on the physical system of interest. Our purpose here was simply to illustrate the manner in which the dimensionless groups arise in the theoretical formulation of the problem. Moreover, we see that those involving surface tension enter exclusively through the boundary conditions.

2.3 Stress conditions at a fluid–fluid interface

We proceed by deriving the normal- and tangential-stress boundary conditions appropriate at a fluid–fluid interface characterized by an interfacial tension σ.

Consider an interfacial surface S bounded by a closed contour C (Figs. 2.1–2.3). One may think of there being a force per unit length of magnitude σ in the s-direction at every point along C that acts to flatten the surface S. Perform a force balance on a volume element V enclosing the interfacial surface S defined by the contour C:

$$\int_V \rho \frac{D\mathbf{u}}{Dt}\, dV = \int_V \mathbf{f}\, dV + \int_S [\mathbf{t}(\mathbf{n}) + \hat{\mathbf{t}}(\hat{\mathbf{n}})]\, dS + \int_C \sigma \mathbf{s}\, d\ell.$$

Here ℓ indicates an arc length, and so $d\ell$ indicates a length increment along the curve C. $\mathbf{t}(\mathbf{n}) = \mathbf{n} \cdot \mathbf{T}$ is the stress vector, the force/area exerted by the upper fluid (+) on the interface. The stress tensor is defined in terms of the local fluid pressure and velocity field as $\mathbf{T} = -p\mathbf{I} + \mu[\nabla \mathbf{u} + (\nabla \mathbf{u})^{\rm T}]$. Similarly, the stress exerted on the interface by the lower fluid – is $\hat{\mathbf{t}}(\hat{\mathbf{n}}) = \hat{\mathbf{n}} \cdot \hat{\mathbf{T}} = -\mathbf{n} \cdot \hat{\mathbf{T}}$, where $\hat{\mathbf{T}}, = -\hat{p}\mathbf{I} + \hat{\mu}[\nabla \hat{\mathbf{u}} + (\nabla \hat{\mathbf{u}})^{\rm T}]$.

2.3.1 Physical interpretation of terms

$\int_V \rho \frac{D\mathbf{u}}{Dt}\, dV$: inertial force associated with acceleration of fluid within V.

$\int_V \mathbf{f}\, dV$: body forces acting on fluid within V.

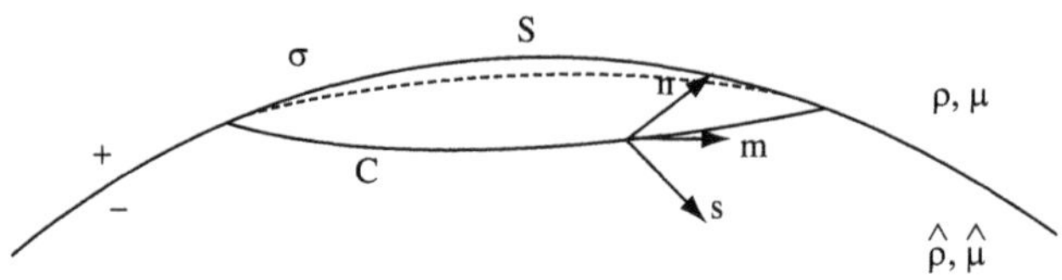

Fig. 2.1 A surface S and bounding contour C on an interface between two fluids. The upper fluid (+) has density ρ and viscosity μ; the lower fluid – has corresponding values $\hat{\rho}$ and $\hat{\mu}$. $\mathbf{n}$ represents the unit outward normal to the surface, and $\hat{\mathbf{n}} = -\mathbf{n}$ the unit inward normal. $\mathbf{m}$ represents the unit tangent to the contour C, and $\mathbf{s}$ the unit vector normal to C but tangent to S.

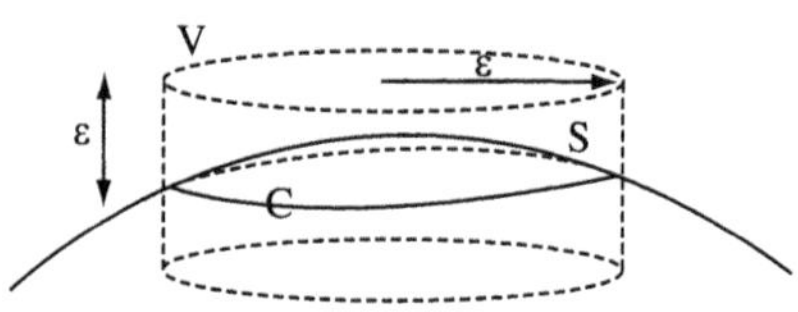

Fig. 2.2 A Gaussian fluid pillbox of height and radius ϵ spanning the interface evolves under the combined influence of volume and surface forces.

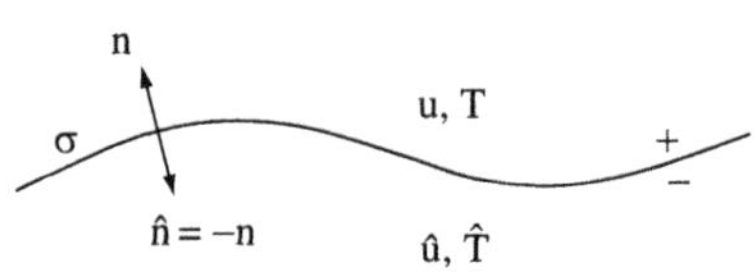

Fig. 2.3 A definitional sketch of a fluid–fluid interface. Carets denote variables in the lower fluid.

$\int_S \mathbf{t}(\mathbf{n})\, dS$: hydrodynamic force exerted at interface by fluid +.

$\int_S \hat{\mathbf{t}}(\hat{\mathbf{n}})\, dS$: hydrodynamic force exerted at interface by fluid –.

$\int_C \sigma \mathbf{s}\, d\ell$: surface tension force exerted along perimeter C.

Now, if ϵ is the typical length scale of the element V, then the acceleration and body forces will scale as ϵ^3, but the surface forces will scale as ϵ^2. Hence, in the limit of $\epsilon \to 0$, we have that the surface forces must balance:

$$\int_S [\mathbf{t}(\mathbf{n}) + \hat{\mathbf{t}}(\hat{\mathbf{n}})]\, dS + \int_C \sigma\, \mathbf{s}\, d\ell = 0.$$

Now we have that

$$\mathbf{t}(\mathbf{n}) = \mathbf{n} \cdot \mathbf{T}, \qquad \hat{\mathbf{t}}(\mathbf{n}) = \hat{\mathbf{n}} \cdot \hat{\mathbf{T}} = -\mathbf{n} \cdot \hat{\mathbf{T}}.$$

Moreover, the application of Stokes' Theorem (see Appendix A) allows us to write

$$\int_C \sigma \mathbf{s}\, d\ell \;=\; \int_S \nabla_s \sigma \;-\; \sigma \mathbf{n}\,(\nabla_s \cdot \mathbf{n})\; dS,$$

where the tangential gradient operator, defined by

$$\nabla_s \;=\; [\mathbf{I} \;-\; \mathbf{nn}] \cdot \nabla \;=\; \nabla \;-\; \mathbf{n}\frac{\partial}{\partial n},$$

appears because σ and $\mathbf{n}$ are defined only on the surface. We proceed by dropping the subscript s on ∇, with this understanding.

The surface force balance thus becomes

$$\int_S [\mathbf{n} \cdot \mathbf{T} \;-\; \mathbf{n} \cdot \hat{\mathbf{T}}]\; dS \;=\; \int_S \sigma \mathbf{n}\,(\nabla \cdot \mathbf{n}) \;-\; \nabla \sigma\; dS. \tag{2.13}$$

Now, since the surface element is arbitrary, the integrand must vanish identically. One thus obtains the interfacial stress balance equation.

2.3.2 Stress balance equation

$$\mathbf{n} \cdot \mathbf{T} \;-\; \mathbf{n} \cdot \hat{\mathbf{T}} \;=\; \sigma \mathbf{n}\,(\nabla \cdot \mathbf{n}) \;-\; \nabla \sigma. \tag{2.14}$$

2.3.3 Interpretation of terms

$\mathbf{n} \cdot \mathbf{T}$: stress (force/area) exerted by + on − (will generally have both normal and tangential components).

$\mathbf{n} \cdot \hat{\mathbf{T}}$: stress (force/area) exerted by − on + (will generally have both normal and tangential components).

$\sigma \mathbf{n}(\nabla \cdot \mathbf{n})$: normal curvature force per unit area associated with local curvature of interface, $\nabla \cdot \mathbf{n}$.

$\nabla \sigma$: tangential stress associated with gradients of surface tension.

Both the normal and the tangential stress must be balanced at the interface. We consider each component in turn.

2.3.4 Normal-stress balance

Taking $\mathbf{n} \cdot$ eqn(2.13) yields the normal-stress balance at the interface:

$$\mathbf{n} \cdot \mathbf{T} \cdot \mathbf{n} \;-\; \mathbf{n} \cdot \hat{\mathbf{T}} \cdot \mathbf{n} \;=\; \sigma\,(\nabla \cdot \mathbf{n}). \tag{2.15}$$

The jump in normal stress across the interface must balance the curvature force per unit area. We note that a surface with nonzero curvature ($\nabla \cdot \mathbf{n} \neq 0$) reflects a jump in normal stress across the interface.

2.3.5 Tangential-stress balance

Taking $\mathbf{t} \cdot eqn(2.13)$, where $\mathbf{t}$ is any unit vector tangent to the interface, yields the tangential-stress balance at the interface:

$$\mathbf{n} \cdot \mathbf{T} \cdot \mathbf{t} \;-\; \mathbf{n} \cdot \hat{\mathbf{T}} \cdot \mathbf{t} \;.=\; \nabla\sigma \cdot \mathbf{t}. \tag{2.16}$$

The physical interpretation of this is that:

- the LHS represents the jump in the tangential components of the hydrodynamic stress at the interface;
- the RHS represents the tangential stress associated with gradients in σ, as may result from gradients in temperature or chemical composition at the interface;
- the LHS contains only velocity gradients, not pressure; therefore, a nonzero $\nabla\sigma$ at a fluid interface must *always* drive motion.

2.4 Fluid statics

We begin by considering static fluid configurations, for which the stress tensor reduces to the form $\mathbf{T} = -p\mathbf{I}$, so that $\mathbf{n} \cdot \mathbf{T} \cdot \mathbf{n} = -p$, and the normal-stress balance assumes the form

$$\hat{p} - p \;=\; \sigma\, \nabla \cdot \mathbf{n}. \tag{2.17}$$

The pressure jump across the interface is balanced by the curvature force at the interface. Now, since $\mathbf{n} \cdot \mathbf{T} \cdot \mathbf{s} = 0$ for a static system, the tangential-stress balance equation indicates that $0 = \nabla\sigma$. This leads us to the following important conclusion.

There cannot be a static system in the presence of surface tension gradients. While pressure jumps can sustain normal-stress jumps across a fluid interface, they do not contribute to the tangential-stress jump. Consequently, tangential surface stresses can only be balanced by viscous stresses associated with fluid motion.

We proceed by applying eqn (2.17) to a number of static situations.

2.4.1 Stationary bubble

We consider a spherical bubble of radius R submerged in a static fluid. What is the pressure drop across the bubble surface?

The curvature of the spherical surface is computed simply:

$$\nabla \cdot \mathbf{n} \;=\; \nabla \cdot \mathbf{r} \;=\; \frac{1}{r^2}\frac{\partial}{\partial r}(r^2) \;=\; \frac{2}{R}. \tag{2.18}$$

So, the normal-stress jump (eqn(2.17)) indicates that

$$\hat{p} - p \;=\; \frac{2\sigma}{R}. \tag{2.19}$$

The pressure within the bubble is higher than that outside by an amount proportional to the surface tension, and inversely proportional to the bubble size. It is thus that small bubbles are louder than large ones when they burst at a free surface: champagne is louder than beer. We note that soap bubbles in air have two surfaces, which define the inner and outer surfaces of the soap film; consequently, the pressure differential is twice that across a single interface.

2.4.2 Static meniscus

Consider a situation where the pressure within a static fluid varies owing to the presence of a gravitational field: $p = p_0 + \rho g z$, where p_0 is the constant ambient pressure, and $\vec{g} = -g\hat{z}$ is the gravitational acceleration. The normal-stress balance thus requires that the interface satisfy the *Young–Laplace equation*,

$$\rho g z \;=\; \sigma \, \nabla \cdot \mathbf{n}. \tag{2.20}$$

The vertical gradient in fluid pressure must be balanced by the curvature pressure; as the gradient is constant, the curvature must likewise increase linearly with z. Such a situation arises in a static meniscus (see Fig. 2.4).

The shape of the meniscus is prescribed by two factors: the contact angle between the air–water interface and the wall, and the balance between hydrostatic pressure and curvature pressure. We treat the contact angle θ as given; it depends on the physics of the log–water–air interaction. The normal-force balance is expressed by the Young–Laplace equation, where now $\rho = \rho_{\rm w} - \rho_{\rm a} \approx \rho_{\rm w}$ is the density difference between water and air.

We define the free surface by $z = \eta(x)$; equivalently, we define a functional $f(x, z) = z - \eta(x)$ that vanishes on the surface. The normal to the surface is thus

$$\mathbf{n} \;=\; \frac{\nabla f}{|\nabla f|} \;=\; \frac{\hat{z} - \eta'(x)\mathbf{x}}{(1 + \eta'(x)^2)^{1/2}}. \tag{2.21}$$

As deduced in Appendix C, the curvature of the free surface, $\nabla \cdot \hat{n}$, may be expressed as

$$\nabla \cdot \hat{n} = \frac{\eta_{xx}}{(1 + \eta_x^2)^{3/2}} \approx \eta_{xx}. \tag{2.22}$$

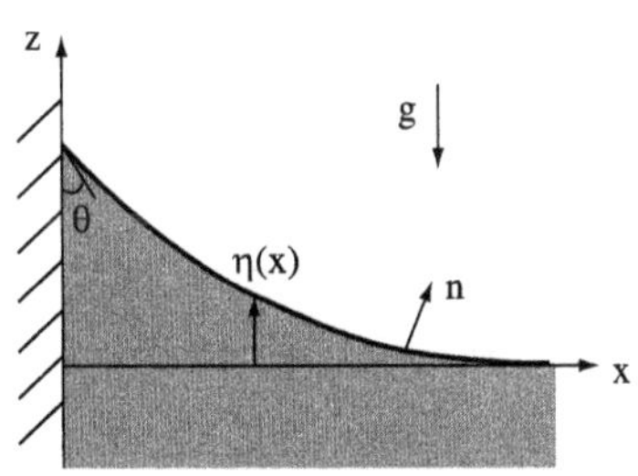

Fig. 2.4 A definitional sketch of a planar meniscus at an air–water interface. The free surface is defined by $z = \eta(x)$, varying from its maximum elevation at its point of contact with the wall ($x = 0$) to zero at large x. The shape is prescribed by the Young–Laplace equation.

Assuming that the slope of the meniscus remains small, i.e. $\eta_x \gg 1$, allows one to linearize eqn (2.22) so that eqn (2.20) assumes the form

$$\rho g \eta \;=\; \sigma \eta_{xx}. \tag{2.23}$$

Applying the boundary condition $\eta(\infty) = 0$ and the contact condition $\eta_x(0) = -\cot\theta$ and solving eqn (2.23) thus yields

$$\eta(x) = \ell_c \cot(\theta) e^{-x/\ell_c}, \tag{2.24}$$

where $\ell_c = \sqrt{\sigma/\rho g}$ is the capillary length. The meniscus formed by a log in water is exponential, dying off on the scale of ℓ_c.

2.4.3 Radial force on a circular hydraulic jump

Hydraulic jumps may be generated when a vertical jet strikes a flat plate. The jet spreads radially, giving rise to a fluid layer that generally thins with radius until reaching a critical radius, at which it increases dramatically (see Fig. 2.5). Here, we calculate the total radial force acting on the jump surface owing to the curvature of the jump between points A and B, located at radii R_1 and R_2, respectively. Assume that the points A and B are the points nearest the jump, upstream and downstream, respectively, at which the slope of the surface vanishes identically.

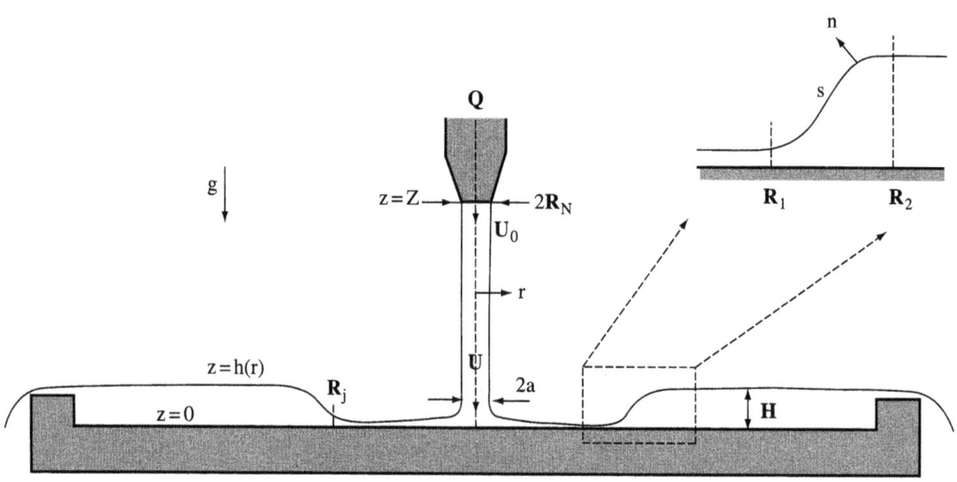

Fig. 2.5 Schematic illustration of the geometry of a circular hydraulic jump. A jet of radius a impacts a reservoir of outer depth H at a speed U. The curvature force associated with the surface tension σ depends only on the geometry of the jump; specifically, on the radial distance $\Delta R = R_2 - R_1$ and the arc length s between the two nearest points upstream and downstream of the jump at which the surface $z = h(r)$ has vanishing slope.

We know that the curvature force per unit area is $\sigma(\nabla \cdot \mathbf{n})\mathbf{n}$. We must integrate the radial component of this force over the jump surface:

$$F_c = \sigma \int\int_S \nabla \cdot \mathbf{n}\, (\mathbf{n} \cdot \hat{\mathbf{r}})\, dS. \tag{2.25}$$

Now, defining the surface as $z = h(r)$ allows us to express the area element as

$$dS = r\, d\theta (dr^2 + dz^2)^{1/2} = r\, d\theta\, (1 + h_r)^{1/2}\, dr, \tag{2.26}$$

so that our radial force assumes the form

$$F_c = \sigma \int_0^{2\pi} \int_{R_1}^{R_2} \nabla \cdot \mathbf{n}\, (\mathbf{n} \cdot \hat{\mathbf{r}})\, r\, (1 + h_r)^{1/2}\, dr\, d\theta. \tag{2.27}$$

Now we use the appropriate forms for $\mathbf{n}$ and $\nabla \cdot \mathbf{n}$ from Appendix C:

$$\mathbf{n} = \frac{\hat{z} - h_r \hat{r}}{(1 + h_r^2)^{1/2}}, \qquad \nabla \cdot \mathbf{n} = -\frac{1}{r}\frac{d}{dr}\frac{r h_r}{(1 + h_r^2)^{1/2}}, \tag{2.28}$$

so that

$$(\nabla \cdot \mathbf{n})(\mathbf{n} \cdot \hat{\mathbf{r}}) = \frac{h_r}{(1 + h_r^2)^{1/2}} \frac{1}{r}\frac{d}{dr}\frac{r h_r}{(1 + h_r^2)^{1/2}}, \tag{2.29}$$

and the radial curvature force becomes

$$F_c = 2\pi\sigma \int_{R_1}^{R_2} \frac{h_r}{(1 + h_r^2)^{1/2}} \frac{1}{r}\frac{d}{dr}\left(\frac{r h_r}{(1 + h_r^2)^{1/2}}\right) r\, (1 + h_r)^{1/2}\, dr \tag{2.30}$$

$$= 2\pi\sigma \int_{R_1}^{R_2} h_r \frac{d}{dr}\left(\frac{r h_r}{(1 + h_r^2)^{1/2}}\right) dr. \tag{2.31}$$

Integrating by parts, i.e. $\int_a^b u\, dv = uv|_a^b - \int_a^b v\, du$ with $u = h_r$, $v = r h_r/(1 + h_r^2)^{1/2}$, yields

$$F_c = 2\pi\sigma \left[r h_r^2 \frac{1}{(1 + h_r^2)^{1/2}} \Big|_{R_1}^{R_2} - \int_{R_1}^{R_2} \frac{r h_r h_r r}{(1 + h_r^2)^{1/2}}\, dr \right] \tag{2.32}$$

$$= -2\pi\sigma \int_{R_1}^{R_2} \frac{r h_r h_r r}{(1 + h_r^2)^{1/2}}\, dr. \tag{2.33}$$

Integrating by parts again, with $u = r$, $v = (1 + h_r^2)^{1/2}$, yields

$$F_c = 2\pi\sigma \left[r(1 + h_r^2)^{1/2} \Big|_{R_1}^{R_2} - \int_{R_1}^{R_2} (1 + h_r^2)^{1/2} dr \right] \tag{2.34}$$

$$= -2\pi\sigma \left[(R_2 - R_1) - \int_{R_1}^{R_2} (1 + h_r^2)^{1/2}\, dr \right], \tag{2.35}$$

since $h_r = 0$ at $r = R_1, R_2$ by assumption. We note also that

$$\int_{R_1}^{R_2} (1 + h_r^2)^{1/2} \, dr \;=\; \int_A^B (dr^2 + dz^2)^{1/2} \;=\; \int_A^B d\ell \;=\; S, \qquad (2.36)$$

where S is defined as the total arc length of the surface between points A and B. We define $\Delta R = R_1 - R_2$ in order to obtain the simple result

$$F_c \;=\; 2\pi\sigma(S - \Delta R). \qquad (2.37)$$

We note that this relation yields reasonable results in two limits of interest. First, for a flat interface, $S = \Delta R$, so that $F_c = 0$. Second, for an abrupt jump in height of ΔH, $\Delta R = 0$ and $S = \Delta H$, so that $F_c = 2\pi\sigma \, \Delta H$. This result is commensurate with the total force exerted by a cylindrical annulus of radius R and height ΔH, which is deduced from the product of σ, the area $2\pi R \, \Delta H$, and the curvature $1/R$:

$$F_c \;=\; \sigma \left(2\pi R \, \Delta H \, \frac{1}{R} \right) \;=\; 2\pi\sigma\Delta H. \qquad (2.38)$$

The influence of this force has been found to be significant for small hydraulic jumps. Its importance relative to the hydrostatic pressure in containing the jump is given by

$$\frac{\text{Curvature}}{\text{Gravity}} \;=\; \frac{\sigma/R}{\rho g \, \Delta H} \;=\; \frac{H}{R} Bo^{-1}, \qquad (2.39)$$

where the Bond number is defined here as

$$Bo \;=\; \frac{\rho g H^2}{\sigma}. \qquad (2.40)$$

The simple result (2.37) was derived by Bush and Aristoff (2003), who also presented the results of an experimental study of circular hydraulic jumps. Their experiments indicated that the curvature force becomes appreciable for jumps with a characteristic radius of less than 3 cm.

While the influence of surface tension on the radius of a circular hydraulic jump is generally small, it may have a qualitative influence on the shape of the jump. In particular, it may prompt the axisymmetry-breaking instability responsible for the polygonal hydraulic jump structures discovered by Ellegaard *et al.* (1998), and more recently examined by Bush *et al.* (2006). See also http://www-math.mit.edu/bush/jump.htm.

2.4.4 Floating bodies

Floating bodies must be supported by some combination of buoyancy and curvature forces. Specifically, since the fluid pressure beneath the interface is related to the atmospheric pressure P_0 above the interface by

$$p \;=\; P_0 \;+\; \rho g z \;+\; \sigma \nabla \cdot \mathbf{n},$$

one may express the vertical force balance as

$$Mg = \mathbf{z} \cdot \int_C -p\, \mathbf{n}\, d\ell = F_b + F_c. \tag{2.41}$$

The buoyancy force

$$F_b = \mathbf{z} \cdot \int_C \rho g z\, \mathbf{n}\, d\ell = \rho g V_b \tag{2.42}$$

is thus simply the weight of the fluid displaced above the object and inside the line of tangency (Fig. 2.6). A simple expression for the curvature force may be deduced using the first of the Frenet–Serret equations (see Appendix B),

$$F_c = \mathbf{z} \cdot \int_C \sigma(\nabla \cdot \mathbf{n})\mathbf{n}\, d\ell = \sigma\mathbf{z} \cdot \int_C \frac{d\mathbf{t}}{d\ell}\, d\ell = \sigma\mathbf{z} \cdot (\mathbf{t}_1 - \mathbf{t}_2) = 2\sigma \sin\theta. \tag{2.43}$$

At the interface, the buoyancy and curvature forces must balance precisely, so the Young–Laplace relation is satisfied:

$$0 = \rho g z + \sigma \nabla \cdot \mathbf{n}. \tag{2.44}$$

Integrating the fluid pressure over the meniscus yields the vertical force balance:

$$F_b^m + F_c^m = 0, \tag{2.45}$$

where

$$F_b^m = \mathbf{z} \cdot \int_{C_m} \rho g z \mathbf{n}\, d\ell = \rho g V_m, \tag{2.46}$$

$$F_c^m = \mathbf{z} \cdot \int_{C_m} \sigma(\nabla \cdot \mathbf{n})\mathbf{n}\, d\ell = \sigma\mathbf{z} \cdot \int_{C_m} \frac{d\mathbf{t}}{d\ell}\, d\ell = \sigma\mathbf{z} \cdot (\mathbf{t}_x - \mathbf{t}_2) = -2\sigma \sin\theta, \tag{2.47}$$

where we have again used the Frenet–Serret equations to evaluate the curvature force.

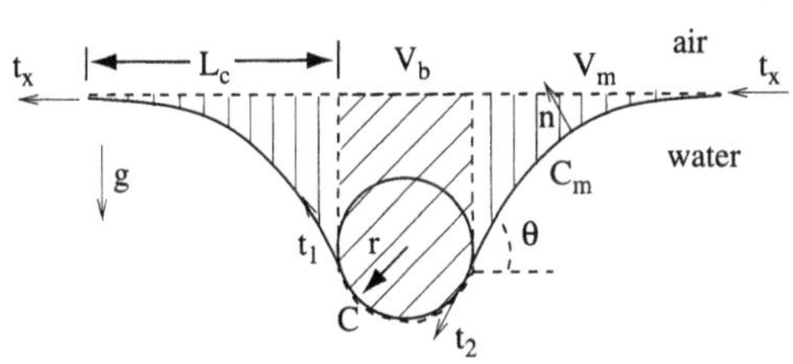

Fig. 2.6 A nonwetting two-dimensional body of radius r and mass M floats on a free surface with surface tension σ. In general, its weight Mg must be supported by some combination of curvature and buoyancy forces. V_b and V_m denote the fluid volumes displaced, respectively, inside and outside the line of tangency.

Equations (2.44)–(2.47) thus indicate that the curvature force acting on the floating body is expressible in terms of the fluid volume displaced *outside* the line of tangency:

$$F_c = \rho g V_m. \tag{2.48}$$

The relative magnitudes of the buoyancy and curvature forces supporting a floating, nonwetting body are thus prescribed by the relative magnitudes of the volumes of the fluid displaced inside and outside the line of tangency:

$$\frac{F_b}{F_c} = \frac{V_b}{V_m}. \tag{2.49}$$

For 2D bodies, we note that since the meniscus will have a length comparable to the capillary length $\ell_c = (\sigma/(\rho g))^{1/2}$, the relative magnitudes of the buoyancy and curvature forces,

$$\frac{F_b}{F_c} \approx \frac{r}{\ell_c}, \tag{2.50}$$

are prescribed by the relative magnitudes of the body size and capillary length. Very small floating objects ($r \ll \ell_c$) are supported principally by curvature rather than buoyancy forces. This result has been extended to three-dimensional floating objects by Keller (1998).

2.4.5 Water-walking insects (Figs. 2.7 and 2.8)

Small objects such as paper clips, pins, or insects may reside at rest on a free surface provided the curvature force induced by their deflection of the free surface is sufficient to bear their weight. For example, for a body of contact length L and total mass M, static equilibrium on the free surface requires that

$$\frac{Mg}{2\sigma L \sin\theta} < 1, \tag{2.51}$$

where θ is the angle of tangency of the floating body.

Fig. 2.7 Water-walking insects deflect the free surface, thus generating curvature forces that bear their weight. The water strider has characteristic length of 1 cm and weight of 1–10 mg. See http://www-math.mit.edu/dhu/Striderweb/striderweb.html.

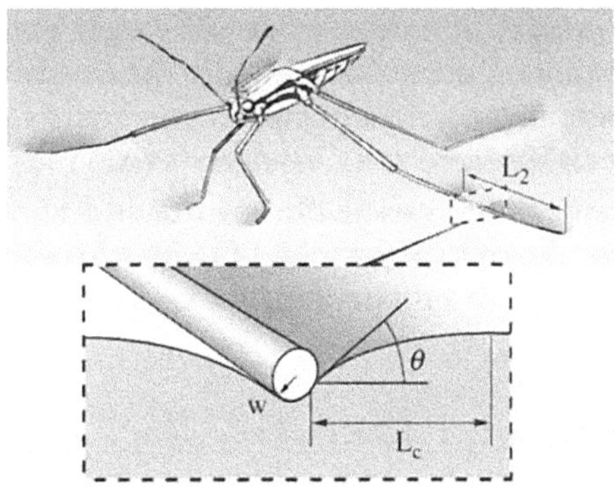

Fig. 2.8 The legs of a water strider are covered with hair, rendering them effectively nonwetting. The tarsal segments of its legs rest on the free surface. The free surface makes an angle θ with the horizontal, resulting in an upward curvature force per unit length $2\sigma \sin\theta$ that bears the insect's weight.

This simple criterion is an important geometric constraint on water-walking insects. Figure 2.9 indicates the dependence of contact length on body weight for over 300 species of water striders, the most common water-walking insect. Note that the solid line corresponds to the requirement (2.51) for static equilibrium. Smaller insects

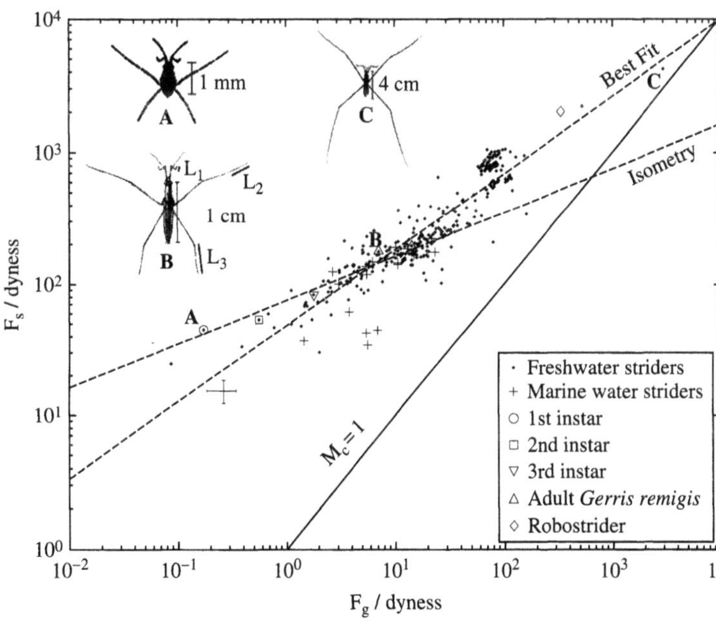

Fig. 2.9 The relation between the maximum curvature force $F_s = \sigma P$ and body weight $F_g = Mg$ for 342 species of water striders. $P = 2(L_1 + L_2 + L_3)$ is the combined length of the tarsal segments (see strider B). From Hu et al. (2003).

maintain a considerable margin of safety, while the larger striders live close to the edge. The size of water-walking insects is limited by the constraint (2.51).

If body proportions were independent of size L, one would expect the body weight to scale as L^3 and the curvature force as L. Isometry would thus suggest a dependence of the form $F_c \sim F_g^{1/3}$, represented as the dashed line in Fig. 2.9. The fact that the best-fit line has a slope considerably larger than 1/3 indicates a variance from isometry: the legs of large water striders are proportionally longer.

2.5 Marangoni flows

Marangoni flows are flows driven by surface tension gradients. In general, the surface tension σ depends on both the temperature and the chemical composition at the interface; consequently, Marangoni flows may be generated by gradients in either temperature or chemical concentration at an interface.

Recall the tangential-stress balance at a free surface,

$$\mathbf{n} \cdot \mathbf{T} \cdot \mathbf{t} = -\mathbf{t} \cdot \nabla\sigma, \tag{2.52}$$

where $\mathbf{n}$ is the unit outward normal to the surface, and $\mathbf{t}$ is any unit tangent vector. The tangential component of the hydrodynamic stress at the surface must balance the tangential stress associated with gradients in σ. Such Marangoni stresses may result from gradients in temperature or chemical composition at the interface. For a static system, since $\mathbf{n} \cdot \mathbf{T} \cdot \mathbf{t} = 0$, the tangential-stress balance equation indicates that $0 = \nabla\sigma$. As seen in Section 2.4, a static situation cannot arise in the presence of Marangoni stresses.

2.5.1 Tears of wine

The first Marangoni flow considered was the tears-of-wine phenomenon (Thomson 1885); Thomson's treatment actually predates Marangoni's first published work on the subject by a decade. The tears-of-wine phenomenon is readily observed in a glass of any but the weakest wines following the establishment of a thin layer of wine on the walls of the wine glass.

The "tears" or "legs" of wine are taken by sommeliers to be an indicator of the quality of wine. An illustration of the tears-of-wine phenomenon is presented in Fig. 2.10 (see also http://www-math.mit.edu/bush/tears.html). Evaporation of alcohol occurs everywhere on the free surface. The alcohol concentration in the thin layer is thus reduced relative to that in the bulk owing to the enhanced surface-area-to-volume ratio. As the surface tension decreases with alcohol concentration, the surface tension is higher in the thin film than in the bulk; the associated Marangoni stress drives upflow throughout the thin film. The wine climbs until it reaches the top of the film, where it accumulates in a band of fluid that thickens until eventually it becomes gravitationally unstable and releases the tears of wine. The tears or "legs" roll back to replenish the bulk reservoir, but with fluid that is depleted in alcohol.

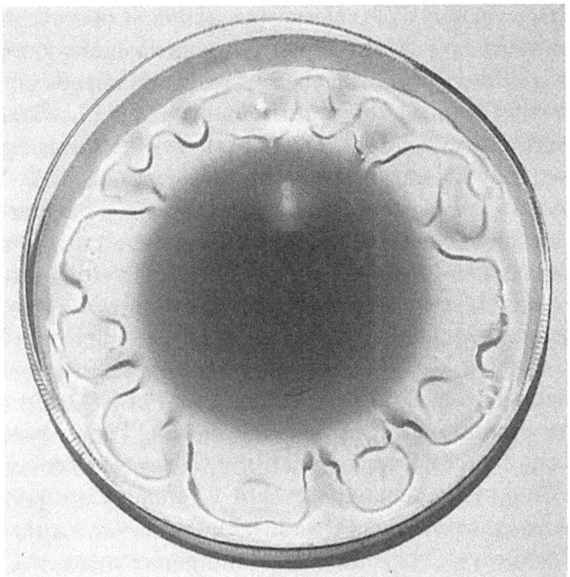

Fig. 2.10 The tears of wine. Fluid is drawn from the bulk up the thin film on the walls of the glass by Marangoni stresses induced by evaporation of alcohol from the free surface.

The flow relies on the transfer of chemical potential energy to kinetic and ultimately gravitational potential energy. The process continues until the alcohol is completely evaporated.

2.5.2 Surfactants

Surfactants are molecules that have an affinity for interfaces; common examples include soap and oils. Owing to their molecular structure (often a hydrophilic head and hydrophobic tail), they find it energetically favorable to reside at the free surface. Their presence reduces the surface tension; consequently, gradients in the surfactant concentration Γ result in surface tension gradients. Surfactants thus generate a special class of Marangoni flows. There are many different types of surfactants, some of which are insoluble (and so remain on the surface), and others which are soluble in the suspending fluid and so diffuse into the bulk. For a wide range of common surfactants, the surface tension is a monotonically decreasing function of Γ until a critical concentration is achieved, beyond which σ remains constant.

The concentration of surfactant Γ on a free surface evolves according to

$$\frac{\partial \Gamma}{\partial t} + \nabla_s \cdot (\Gamma \mathbf{u}_s) + \Gamma(\nabla_s \cdot \mathbf{n})(\mathbf{u} \cdot \mathbf{n}) = J(\Gamma, C_s) + D_s \nabla_s^2 \Gamma, \tag{2.53}$$

where $\mathbf{u}_s$ is the surface velocity, ∇_s is the surface gradient operator, and D_s is the surface diffusivity of the surfactant (Stone 1990). J is a surfactant source term associated with adsorption onto or desorption from the surface, and depends on both the surface surfactant concentration Γ and the concentration in the bulk C_s. Tracing the evolution of a contaminated free surface requires the use of the Navier–Stokes equations, relevant boundary conditions, and the surfactant evolution equation (2.53). The dependence of the surface tension on the surfactant concentration, $\sigma(\Gamma)$, requires the coupling of the flow field and surfactant field. In certain special cases, the system may be made more tractable. For example, for insoluble surfactants, $J = 0$, and many surfactants have a sufficiently small D_s that the surface diffusivity may be safely neglected.

The principal dynamical influence of surfactants is to impart an effective elasticity to the interface. Specifically, the presence of surfactants will serve to alter not only the normal-stress balance (through the reduction of σ) but also the tangential-stress balance (through the generation of Marangoni stresses). The presence of surfactants will act to suppress any fluid motion characterized by nonzero surface divergence. For example, consider a fluid motion characterized by a radially divergent surface motion. The presence of a surfactant results in the redistribution of the surfactant: Γ is reduced near the point of divergence. The resulting Marangoni stresses act to suppress the surface motion, resisting it through an effective surface elasticity. Similarly, if the flow is characterized by a radial convergence, the resulting accumulation of surfactant in the region of convergence will result in Marangoni stresses that serve to resist it. It is this effective elasticity that gives soap films their longevity: the divergent motions that would cause a pure liquid film to rupture are suppressed by the surfactant layer on the soap film surface.

The ability of surfactants to suppress flows with nonzero surface divergence is evident throughout the natural world (Lucassen-Reynders and Lucassen 1969). It was first remarked upon by Pliny the Elder, who rationalized that the absence of capillary waves in the wake of ships is due to the ships stirring up surfactant. This phenomenon was also widely known to spear-fishermen, who poured oil on the water to increase their ability to see their prey, and to sailors, who would do similarly in an attempt to calm troubled seas. Finally, the suppression of capillary waves by surfactants is responsible for the "footprints of whales" (see Fig. 2.11). In the wake of whales, even in turbulent roiling seas, one sees patches on the sea surface (of characteristic width 5–10 m) that are perfectly flat. These are generally acknowledged to result from the whales sweeping biomaterial to the surface with their tails; this biomaterial serves as a surfactant that suppresses capillary waves.

2.5.3 The soap boat

Consider a floating body with perimeter C in contact with the free surface, which we assume for the sake of simplicity to be flat. Recalling that σ may be thought of as a force per length in a direction tangent to the surface, we see that the total surface tension force acting on the body is

$$F_c = \int_C \sigma \mathbf{s} \, d\ell, \tag{2.54}$$

Fig. 2.11 The "footprint" of a whale, caused by the whale sweeping biomaterial to the sea surface. The biomaterial acts as a surfactant in locally suppressing the capillary waves evident elsewhere on the sea surface. Observed in the wake of a whale on a whale watch from Boston Harbor.

where $\mathbf{s}$ is the unit vector tangent to the free surface and normal to C, and $d\ell$ is an incremental arc length along C. If σ is everywhere constant, then this line integral vanishes identically by the Divergence Theorem. However, if $\sigma = \sigma(\mathbf{x})$, then it may result in a net propulsive force.

A "soap boat" may be simply generated by coating one end of a toothpick with soap, which acts to reduce surface tension (see Fig. 2.12). The concomitant gradient in surface tension results in a net propulsive force that drives the boat away from the soap. We note that an analogous Marangoni propulsion technique arises in the natural world: certain water-walking insects eject surfactant and use the resulting surface tension gradients as an emergency mechanism for avoiding predation. Moreover, when a pine needle falls into a lake or pond, it is propelled across the surface in an analogous fashion owing to the influence of the resin at its base decreasing the local surface tension.

2.5.4 Bubble motion

Theoretical predictions of the rise speed of small drops or bubbles do not adequately describe the observations. Specifically, air bubbles rise at low Reynolds number at

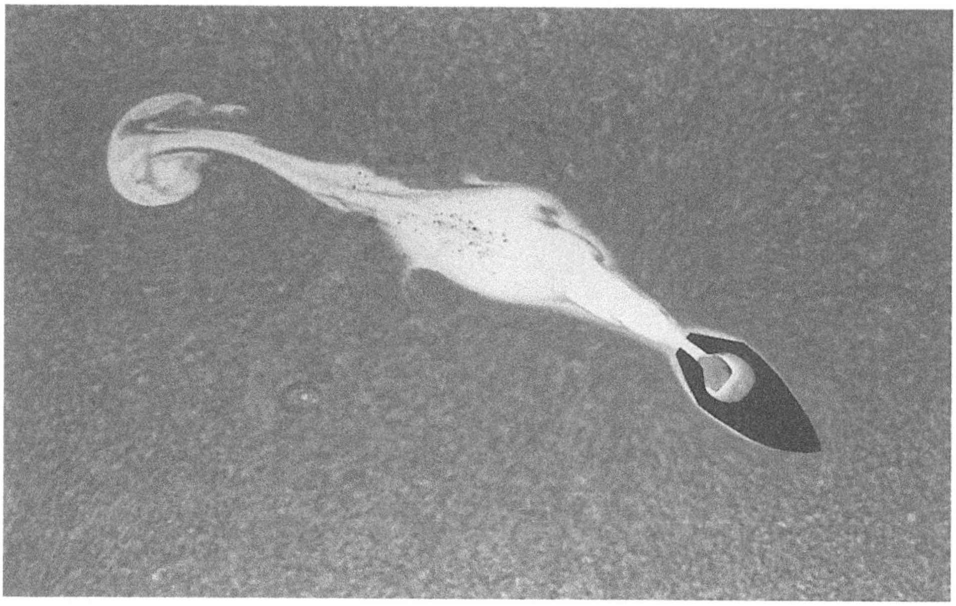

Fig. 2.12 The soap boat. A floating body (length 2.5 cm) contains a small volume of soap, which serves as its fuel for propelling it across the free surface. The soap exits the rear of the boat, decreasing the local surface tension. The resulting fore-to-aft surface tension gradient propels the boat forward. The water surface was covered with thymol blue, which parts owing to the presence of the soap, which is visible as a white streak.

rates appropriate to for rigid spheres with equivalent buoyancy in all but the most carefully cleaned fluids. This discrepancy may be rationalized through consideration of the influence of surfactants on the surface dynamics. The flow generated by a clean spherical bubble of radius a rising at low $Re = Ua/\nu$ is intuitively obvious. The interior flow is toroidal, while the surface motion is characterized by divergence and convergence at the leading and trailing surfaces, respectively. The presence of surface contamination changes the flow qualitatively.

The effective surface elasticity imparted by a surfactant acts to suppress the surface motion. Surfactant is generally swept to the trailing edge of the bubble, where it accumulates, giving rise to a local decrease in surface tension. The resulting fore-to-aft surface tension gradient results in a Marangoni stress that resists surface motion, and so rigidifies the bubble surface. The air bubble thus moves as if its surface were stagnant, and it is thus that its rise speed is commensurate with that predicted for a rigid sphere: the no-slip boundary condition is more appropriate than the free-slip condition. Finally, we note that the characteristic Marangoni stress $\Delta\sigma/a$ is most pronounced for small bubbles. It is thus that the influence of surfactants is greatest on small bubbles.

2.6 Fluid jets

We consider here the form and stability of fluid jets falling under the influence of gravity.

2.6.1 The shape of a falling fluid jet

Consider a circular orifice of radius a ejecting a flux Q of fluid of density ρ and kinematic viscosity ν (Fig. 2.13). The resulting jet is shot downwards, and accelerates under the influence of gravity $-g\hat{z}$. We assume that the Reynolds number $Re = Q/(a\nu)$ of the jet is sufficiently high that the influence of viscosity is negligible; furthermore, we assume that the speed of the jet is independent of radius, and so is adequately described by $U(z)$. We proceed by deducing the shape $r(z)$ and speed $U(z)$ of the evolving jet.

Applying Bernoulli's Theorem at the points A and B,

$$\frac{1}{2}\rho U_0^2 + \rho g z + P_{\mathrm{A}} = \frac{1}{2}\rho U^2(z) + P_{\mathrm{B}}. \tag{2.55}$$

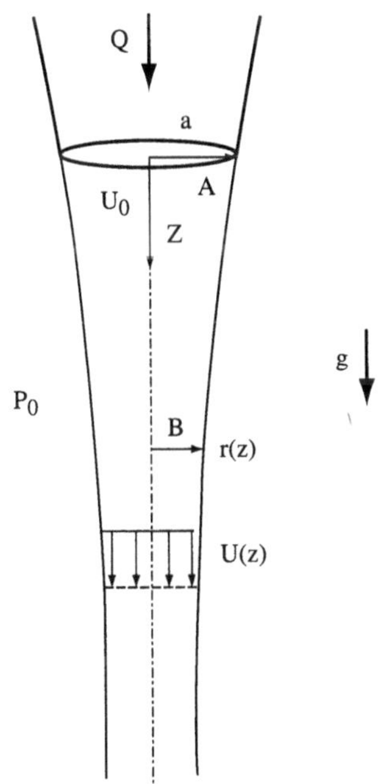

Fig. 2.13 A fluid jet extruded from an orifice of radius a accelerates under the influence of gravity. Its shape is influenced both by the gravitational acceleration $\mathbf{g}$ and by the surface tension σ.

The local curvature of a slender thread may be expressed in terms of the two principal radii of curvature, R_1 and R_2:

$$\nabla \cdot \mathbf{n} = \frac{1}{R_1} + \frac{1}{R_2} \approx \frac{1}{r}.$$

Thus, the fluid pressures within the jet at the points A and B may be simply related to that of the ambient, P_0:

$$P_{\mathrm{A}} \approx P_0 + \frac{\sigma}{a}, \qquad P_{\mathrm{B}} \approx P_0 + \frac{\sigma}{r}. \tag{2.56}$$

Substituting into eqn (2.55) thus yields

$$\frac{1}{2}\rho U_0^2 + \rho g z + P_0 + \frac{\sigma}{a} = \frac{1}{2}\rho U^2(z) + P_0 + \frac{\sigma}{r}, \tag{2.57}$$

from which one finds

$$\frac{U(z)}{U_0} = \left[1 + \frac{2}{Fr}\frac{z}{a} + \frac{2}{We}\left(1 - \frac{a}{r}\right)\right]^{1/2}, \tag{2.58}$$

where we define the dimensionless groups

$$Fr = \frac{U_0^2}{ga} = \frac{\text{INERTIA}}{\text{GRAVITY}} = \text{Froude number}, \tag{2.59}$$

$$We = \frac{\rho U_0^2 a}{\sigma} = \frac{\text{INERTIA}}{\text{CURVATURE}} = \text{Weber number}. \tag{2.60}$$

Now flux conservation requires that

$$Q = 2\pi \int_0^r U(z) r(z)\, dr = \pi a^2 U_0 = \pi\, r^2 U(z), \tag{2.61}$$

from which one obtains

$$\frac{r(z)}{a} = \left(\frac{U_0}{U(z)}\right)^{1/2} = \left[1 + \frac{2}{Fr}\frac{z}{a} + \frac{2}{We}\left(1 - \frac{a}{r}\right)\right]^{-1/4}. \tag{2.62}$$

This may be solved algebraically to yield the thread shape $r(z)/a$, and then this result may be substituted into eqn (2.61) to deduce the velocity profile $U(z)$. In the limit of $We \to \infty$, one obtains

$$\frac{r}{a} = \left(1 + \frac{2gz}{U_0^2}\right)^{-1/4}, \qquad \frac{U(z)}{U_0} = \left(1 + \frac{2gz}{U_0^2}\right)^{1/2}.$$

2.6.2 The Plateau–Rayleigh instability

Here, we summarize the work of Plateau and Rayleigh on the instability of cylindrical fluid jets bound by surface tension. It is precisely this Rayleigh–Plateau instability

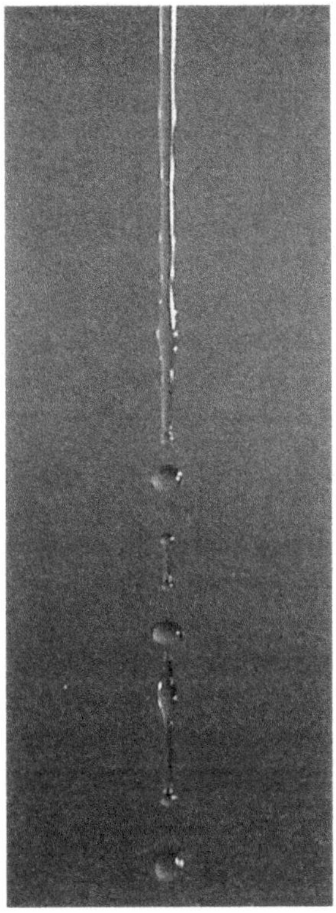

Fig. 2.14 The capillary-driven instability of a water thread falling under the influence of gravity. The initial jet diameter is approximately 3 mm.

that is responsible for the pinch-off of thin water jets emerging from kitchen taps (see Fig. 2.14).

The equilibrium base state consists of an infinitely long, quiescent, cylindrical, inviscid fluid column of radius R_0, density ρ, and surface tension σ (Fig. 2.15). The influence of gravity is neglected. The pressure p_0 is constant inside the column and may be calculated by balancing the normal stresses with surface tension at the boundary. Assuming zero external pressure yields

$$p_0 = \sigma \nabla \cdot \mathbf{n} \quad \Rightarrow \quad p_0 = \frac{\sigma}{R_0}. \tag{2.63}$$

We consider the evolution of infinitesimal varicose perturbations on the interface, which enables us to linearize the governing equations. The perturbed columnar surface takes the form

$$\widetilde{R} = R_0 + \epsilon e^{\omega t + ikz}, \tag{2.64}$$

where the perturbation amplitude $\epsilon \ll R_0$, ω is the growth rate of the instability, and k is the wavenumber of the disturbance in the z-direction. The corresponding

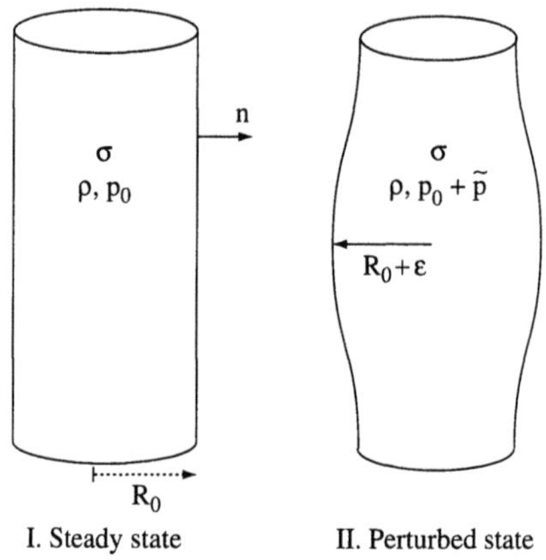

Fig. 2.15 A cylindrical column of initial radius R_0, comprising an inviscid fluid of density ρ and bound by surface tension σ.

wavelength of the varicose perturbations is necessarily $2\pi/k$. We denote the radial component of the perturbation velocity by $\widetilde{u}_r$ the axial component by $\widetilde{u}_z$, and the perturbation pressure by $\widetilde{p}$. Substituting these perturbation fields into the Navier–Stokes equations and retaining terms only to order ϵ yields

$$\frac{\partial \widetilde{u}_r}{\partial t} = -\frac{1}{\rho}\frac{\partial \widetilde{p}}{\partial r}, \tag{2.65}$$

$$\frac{\partial \widetilde{u}_z}{\partial t} = -\frac{1}{\rho}\frac{\partial \widetilde{p}}{\partial z}. \tag{2.66}$$

The linearized continuity equation becomes

$$\frac{\partial \widetilde{u}_r}{\partial r} + \frac{\widetilde{u}_r}{r} + \frac{\partial \widetilde{u}_z}{\partial z} = 0. \tag{2.67}$$

We anticipate that the disturbances in velocity and pressure will have the same form as the surface disturbance (2.64), and so write the perturbation velocities and pressure as

$$\widetilde{u_r} = R(r)e^{\omega t + ikz}, \quad \widetilde{u_z} = Z(r)e^{\omega t + ikz}, \text{ and } \widetilde{p} = P(r)e^{\omega t + ikz}. \tag{2.68}$$

Substituting eqn (2.68) into eqns (2.65) –(2.67) yields the linearized equations governing the perturbation fields:

Momentum equations:

$$\omega R = -\frac{1}{\rho}\frac{dP}{dr}, \tag{2.69}$$

$$\omega Z = -\frac{ik}{\rho}P; \tag{2.70}$$

Continuity:

$$\frac{dR}{dr} + \frac{R}{r} + ikZ = 0. \tag{2.71}$$

Eliminating $Z(r)$ and $P(r)$ yields a differential equation for $R(r)$:

$$r^2 \frac{d^2 R}{dr^2} + r\frac{dR}{dr} - \left(1 + (kr)^2\right) R = 0. \tag{2.72}$$

This corresponds to modified Bessel equation of order 1, whose solutions may be written in terms of the modified Bessel functions of the first and second kind, $I_1(kr)$ and $K_1(kr)$, respectively. We note that $K_1(kr) \to \infty$ as $r \to 0$; therefore, the well-behavedness of our solution requires that $R(r)$ take the form

$$R(r) = CI_1\,(kr), \tag{2.73}$$

where C is an as yet unspecified constant to be determined later by application of appropriate boundary conditions.

The pressure may be obtained from eqns (2.73) and (2.69) and by using the Bessel function identity $I_0'(\xi) = I_1(\xi)$:

$$P(r) = -\frac{\omega\rho C}{k} I_0(kr). \tag{2.74}$$

We proceed by applying appropriate boundary conditions. The first is the kinematic condition on the free surface:

$$\frac{\partial \widetilde{R}}{\partial t} = \underline{\widetilde{u}} \cdot \underline{n} \simeq \widetilde{u}_r. \tag{2.75}$$

Substitution of eqn (2.73) into this condition yields

$$C = \frac{\epsilon\omega}{I_1\,(kR_0)}. \tag{2.76}$$

Second, we require a normal-stress balance on the free surface:

$$p_0 + \widetilde{p} = \sigma\nabla \cdot \underline{n}. \tag{2.77}$$

We write the curvature as $\sigma\nabla \cdot \underline{n} = (1/R_1 + 1/R_2)$, where R_1 and R_2 are the principal radii of curvature of the jet surface:

$$\frac{1}{R_1} = \frac{1}{R_0 + \epsilon e^{\omega t + ikz}} \simeq \frac{1}{R_0} - \frac{\epsilon}{R_0^2} e^{\omega t + ikz}, \tag{2.78}$$

$$\frac{1}{R_2} = \epsilon k^2 e^{\omega t + ikz}. \tag{2.79}$$

Substitution of eqns (2.78) and (2.79) into eqn (2.77) yields

$$p_0 + \widetilde{p} = \frac{\sigma}{R_0} - \frac{\epsilon\sigma}{R_0^2}\left(1 - k^2 R_0^2\right) e^{\omega t + ikz}. \tag{2.80}$$

Cancellation via eqn (2.63) yields an equation for $\widetilde{p}$ accurate to order ϵ:

$$\widetilde{p} = -\frac{\epsilon\sigma}{R_0^2}\left(1 - k^2 R_0^2\right) e^{\omega t + ikz}. \tag{2.81}$$

Combining eqns (2.74), (2.76), and (2.81) yields the dispersion relation, which indicates the dependence of the growth rate ω on the wavenumber k:

$$\boxed{\omega^2 = \frac{\sigma}{\rho R_0^3} kR_0 \frac{I_1(kR_0)}{I_0(kR_0)}\left(1 - k^2 R_0^2\right)}. \tag{2.82}$$

We note first that unstable modes are possible only when

$$kR_0 < 1. \tag{2.83}$$

The column is thus unstable to disturbances whose wavelengths exceed the circumference of the cylinder. A plot of the dispersion relation is shown in Fig. 2.16.

The fastest-growing mode occurs for $kR_0 = 0.697$, i.e. when the wavelength of the disturbance is

$$\lambda_{\text{max}} \simeq 9.02 R_0. \tag{2.84}$$

By inverting the maximum growth rate ω_{max}, one may estimate the characteristic breakup time:

$$t_{\text{breakup}} \simeq 2.91\sqrt{\frac{\rho R_0^3}{\sigma}}. \tag{2.85}$$

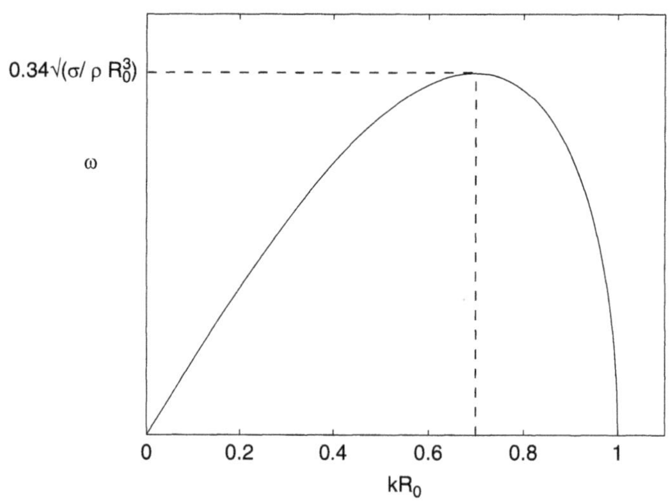

Fig. 2.16 The dependence of the growth rate ω on the wavenumber k for the Rayleigh–Plateau instability.

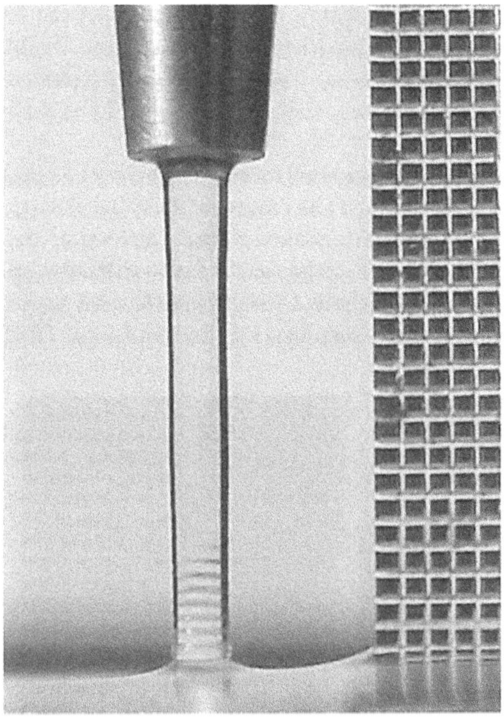

Fig. 2.17 The field of stationary capillary waves excited on the base of a water jet impinging on a horizontal water reservoir. The grid at the right is millimetric.

A water jet of diameter 1 cm has a characteristic breakup time of about 1/8 s, which is consistent with casual observation of jet break-up in a kitchen sink.

When a vertical water jet impinges on a horizontal reservoir of water, a field of standing waves may be excited on the base of the jet (see Fig. 2.17). The wavelength is determined by the requirement that the wave speed correspond to the local jet speed: $U = -\omega/k$. Using our dispersion relation (2.82) thus yields

$$U^2 = \frac{\omega^2}{k^2} = \frac{\sigma}{\rho k R_0^2} \frac{I_1(kR_0)}{I_0(kR_0)} \left(1 - k^2 R_0^2\right). \tag{2.86}$$

Provided the jet speed U is known, this equation may be solved in order to deduce the wavelength of the waves that will travel at U and so appear to be stationary in the lab frame. For jets falling from a nozzle, the result (2.86) may be used to deduce the local jet speed.

2.6.3 Fluid pipes

The following system may be readily observed in a kitchen sink. When the volume flux exiting the tap is such that the falling stream has a diameter of 2–3 mm, obstructing the stream with a finger at a distance of several centimeters from the tap gives rise to a stationary field of varicose capillary waves upstream of the finger. If the finger is dipped in a liquid detergent (or soap) before insertion into the stream, the capillary waves begin at some critical distance above the finger, below which the stream is cylindrical. Closer inspection reveals that the surface of the jet's cylindrical base is quiescent.

An analogous phenomenon arises when a vertical fluid jet impinges on a deep water reservoir (Figs. 2.17 and 2.18). When the reservoir is contaminated by a surfactant, the surface tension of the reservoir is diminished relative to that of the jet. The associated surface tension gradient draws the surfactant a finite distance up the jet, prompting two salient alterations in the jet surface. First, the surfactant suppresses surface waves, so that the base of the jet surface assumes a cylindrical form (Fig. 2.18). Second, the

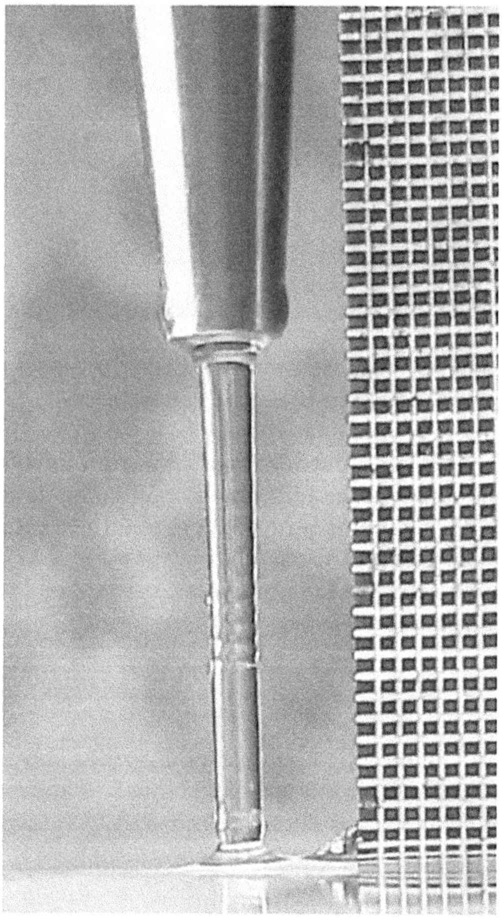

Fig. 2.18 A fluid pipe generated by a falling water jet impinging on a contaminated water reservoir. The field of stationary capillary waves is excited above the fluid pipe. The grid at the right is millimetric.

jet surface becomes stagnant at its base: the Marangoni stresses associated with the surfactant gradient are balanced by the viscous stresses generated within the jet. The quiescence of the jet surface may be simply demonstrated by sprinkling a small amount of talc or lycopodium powder onto the jet. The fluid jet thus enters a contaminated reservoir as if through a rigid pipe.

A detailed theoretical description of the fluid pipe has been given by Hancock and Bush (2002). We present here a simple scaling that yields the dependence of the vertical extent H of the fluid pipe on the governing system parameters. We assume that, once the jet enters the fluid pipe, a boundary layer develops on its outer wall owing to the no-slip boundary condition appropriate there. Balancing viscous and Marangoni stresses on the pipe surface yields

$$\rho\nu\frac{V}{\delta_H} \sim \frac{\Delta\sigma}{H}, \tag{2.87}$$

where $\Delta\sigma$ is the surface tension differential between the jet and the reservoir, V is the jet speed at the top of the fluid pipe, and δ_H is the boundary layer thickness at the base of the fluid pipe. We assume that the boundary layer thickness increases with the distance z from the inlet according to classical boundary layer scaling:

$$\frac{\delta}{a} \sim \left(\frac{\nu z}{a^2 V}\right)^{1/2}. \tag{2.88}$$

Substituting for $\delta(H)$ from eqn(2.88) into eqn(2.87) yields

$$H \sim \frac{(\Delta\sigma)^2}{\rho\mu V^3}. \tag{2.89}$$

The pipe height increases with the surface tension differential and pipe radius, and decreases with fluid viscosity and jet speed.

2.7 Fluid sheets

The dynamics of high-speed fluid sheets was first considered by Savart after his early work on electromagnetism with Biot, and was subsequently examined in a series of papers by Taylor (1959). Fluid sheets have recently received a great deal of attention owing to their relevance in a number of spray atomization processes. Such sheets may be generated from a variety of source conditions, for example the collision of jets with rigid impactors, and jet–jet collisions.

There is generally a curvature force acting on the sheet edge which acts to contain the fluid sheet. For a 2D (planar) sheet, the magnitude of this curvature force is given by

$$\mathbf{F}_c = \int_C \sigma(\nabla\cdot\mathbf{n})\,\mathbf{n}\,d\ell. \tag{2.90}$$

Using the first Frenet–Serret equation (Appendix B),

$$(\nabla \cdot \mathbf{n})\,\mathbf{n} = \frac{d\mathbf{t}}{d\ell}, \tag{2.91}$$

this yields

$$\mathbf{F}_c = \int_C \sigma \frac{d\mathbf{t}}{d\ell}\, d\ell = \sigma(\mathbf{t}_1 - \mathbf{t}_2) = 2\sigma\mathbf{x}. \tag{2.92}$$

There is thus an effective force per unit length 2σ along the length of the sheet rim acting to contain the rim.

We now consider how this result may be applied to compute sheet shapes for three distinct cases: (i) a circular sheet, (ii) a lenticular sheet with an unstable rim, and (iii) a lenticular sheet with a stable rim.

2.7.1 Circular sheet

We consider the geometry considered in Savart's original experiment. A vertical fluid jet strikes a small, horizontal circular impactor. If the flow rate is sufficiently high that gravity does not influence the shape of the sheet, the fluid is ejected radially, giving rise to a circular, free fluid sheet (Fig. 2.19). The sheet radius is prescribed by a balance of radial forces; specifically, the inertial force must balance the curvature force:

$$\rho u^2 h = 2\sigma. \tag{2.93}$$

Fig. 2.19 A circular fluid sheet generated by the impact of a water jet on a circular impactor (bottom). The impacting circle has a diameter of 1 cm.

Continuity requires that the sheet thickness h depend on the speed u, the jet flux Q, and the radius r as

$$h = \frac{Q}{2\pi r u}. \tag{2.94}$$

Experiments (specifically, tracking of particles suspended within the sheet) indicate that the sheet speed u is independent of radius; consequently, the sheet thickness decreases as $1/r$. Substituting the form (2.94) for h into the force balance (2.93) yields the sheet radius, or Taylor radius,

$$R_{\mathrm{T}} = \frac{\rho Q u}{4\pi\sigma}. \tag{2.95}$$

The sheet radius increases with source flux and sheet speed, but decreases with surface tension. We note that the fluid proceeds radially to the sheet edge, where it accumulates until it goes unstable via a modified Rayleigh–Plateau instability, often referred to as the Rayleigh–Plateau–Savart instability, as it was first observed on a sheet edge by Savart.

2.7.2 Lenticular sheet with an unstable rim

We now consider a nonaxisymmetric fluid sheet, such as may be formed by the oblique collision of water jets (see Fig. 2.20), a geometry originally considered by Taylor in 1960. Fluid is ejected radially from the origin into a sheet with a flux distribution given by $Q(\theta)$, so that the volume flux flowing into the sector between θ and $\theta + d\theta$ is $Q(\theta)\, d\theta$. As in the previous case of the circular sheet, the sheet rim is unstable, and

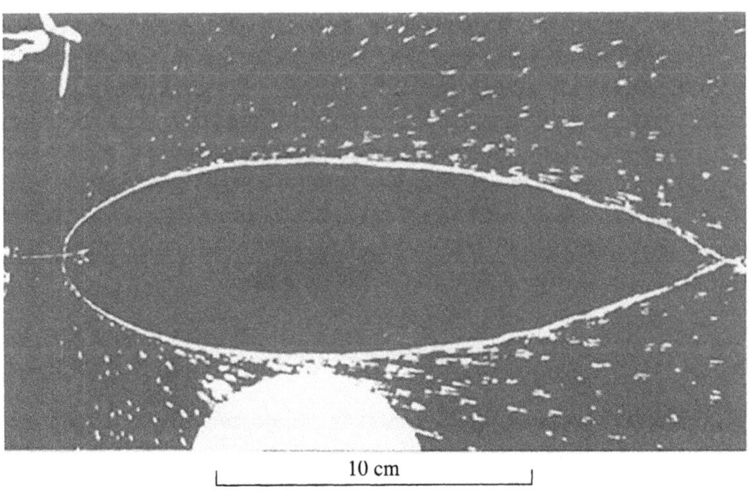

Fig. 2.20 A water sheet generated by collision of water jets at the left. The fluid streams radially outward in a thinning sheet; once the fluid reaches the sheet rim, it is ejected radially in the form of droplets. From Taylor (1960).

fluid drops are continuously ejected from it. The sheet shape is computed in a similar manner, but now depends explicitly on the flux distribution within the sheet, $Q(\theta)$. The normal-force balance on the sheet edge now depends on the normal component of the sheet speed, $u_{\rm n}$:

$$\rho u_{\rm n}^2 h = 2\sigma. \tag{2.96}$$

The sheet thickness is again prescribed by eqn (2.94), but now $Q = Q(\theta)$, so the sheet radius $R(\theta)$ is given by

$$R(\theta) = \frac{\rho u_{\rm n}^2 Q(\theta)}{4\pi\sigma u}. \tag{2.97}$$

Computing sheet shapes thus relies on measurement of the flux distribution $Q(\theta)$ within the sheet.

2.7.3 Lenticular sheet with a stable rim

In a certain region of parameter space, specifically, for fluids more viscous than water, one may encounter fluid sheets with a stable rims (see http://www-math.mit.edu/bush/bones.html). The force balance describing the sheet shape must change accordingly. When the rim is stable, fluid entering the rim proceeds along the rim. As a result, there is a centripetal force normal to the fluid rim associated with flow along the curved rim that must be considered in order to correctly predict the sheet shape.

The relevant geometry is presented in Fig. 2.21. $r(\theta)$ is defined to be the distance from the origin to the rim centerline, and $u_{\rm n}(\theta)$ and $u_{\rm t}(\theta)$ are the normal and tangential components of the fluid velocity in the sheet where it contacts the rim. $v(\theta)$ is defined to be the velocity of flow in the rim, $R(\theta)$ is the rim radius, and $\psi(\theta)$ is the angle between the position vector $\mathbf{r}$ and the local tangent to the rim centerline. Finally, $r_{\rm c}(\theta)$ is defined to be the radius of curvature of the rim centerline, and s is the arc length along the rim centerline. The differential equations governing the shape of a stable fluid rim bounding a fluid sheet may be deduced by consideration of conservation of mass in the rim and the local normal and tangential force balances at the rim.

For a steady sheet, continuity requires that the volume flux from the sheet balances the tangential gradient of the volume flux along the rim:

$$0 = u_{\rm n} h - \frac{\partial}{\partial s}\left(v\pi R^2\right). \tag{2.98}$$

The normal-force balance requires that the curvature force associated with the rim's surface tension balances the force resulting from the normal flow into the rim from the fluid sheet and the centripetal force resulting from the flow along the curved rim:

$$\rho u_{\rm n}^2 h + \frac{\rho\pi R^2 v^2}{r_c} = 2\sigma. \tag{2.99}$$

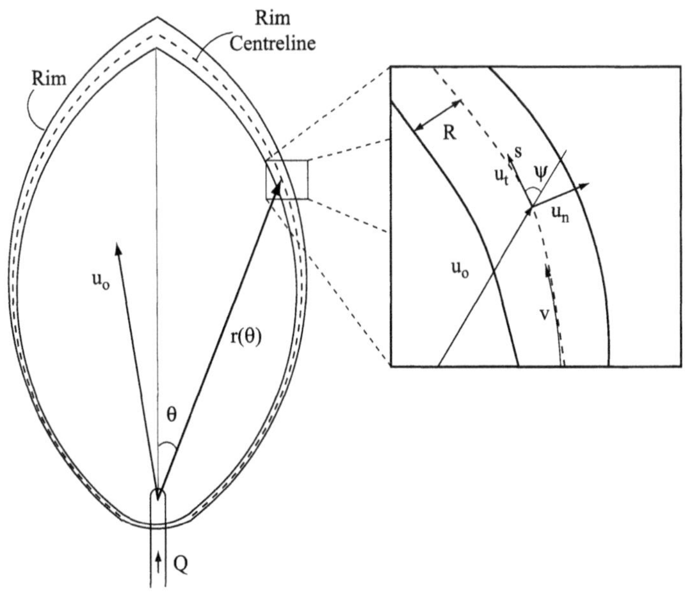

Fig. 2.21 Schematic illustration of a fluid sheet bounded by a stable rim.

Note that the force balance (2.96) is augmented here by the centripetal force. The tangential-force balance at the rim requires a balance between tangential gradients of tangential momentum flux, tangential gradients of curvature pressure, viscous resistance to stretching of the rim, and the tangential momentum flux arriving from the sheet. For most applications involving high-speed sheets, the Reynolds number characterizing the rim dynamics is sufficiently large that viscous resistance may be safely neglected. Moreover, the curvature term $\nabla \cdot \hat{n}$ generally depends on θ; however, with an accuracy to $O(R/r_c)$, we may use $\nabla \cdot \hat{n} = 1/R$. One thus obtains

$$\frac{\partial}{\partial s}(\pi R^2 v^2) = h u_t u_n - \frac{\pi R^2 \sigma}{\rho} \frac{\partial}{\partial s}\left(\frac{1}{R}\right). \tag{2.100}$$

Equations (2.98)–(2.100) must be supplemented by the continuity relation,

$$h(r, \theta) = \frac{Q(\theta)}{u_0 r}, \tag{2.101}$$

in addition to a number of relations that follow directly from the system geometry:

$$u_n = u_0 \sin \Psi, \; u_t = u_0 \cos \Psi, \tag{2.102}$$

$$\frac{1}{r_c} = \frac{\sin \Psi}{r}\left(\frac{\partial \Psi}{\partial \theta} + 1\right). \tag{2.103}$$

This resulting system of equations may be nondimensionalized, and reduces to a set of coupled ordinary equations in the four variables $r(\theta)$, $v(\theta)$, $R(\theta)$, and $\Psi(\theta)$. Given a flux distribution $Q(\theta)$, the system may be integrated to deduce the sheet shape.

2.7.4 Water bells

All of the fluid sheets considered thus far have been confined to a plane. In Section 2.7.1, we considered a circular sheet generated from a vertical jet striking a horizontal impactor. The sheet remains planar only if the flow is sufficiently fast that the fluid reaches its Taylor radius before sagging substantially under the influence of gravity. Decreasing the flow rate will make this sagging more pronounced, and the sheet will no longer be planar. While one might expect the sheet to fall along a parabolic trajectory, the toroidal curvature of the bell induces curvature pressures that act to close the sheet. Consequently, the sheet may actually close upon itself, giving rise to a water bell, as illustrated in Fig. 2.22. We proceed by outlining the theory required to compute the shapes of water bells. See Clanet (2007).

We consider a fluid sheet extruded radially at a speed u_0 and subsequently sagging under the influence of a gravitational field $\mathbf{g} = -g\hat{z}$. The inner and outer sheet surfaces are characterized by a constant surface tension σ. The sheet has a constant density ρ and a thickness $t(r, z)$, and is assumed to be inviscid. Q is the total volume flux in the sheet.

We define the origin to be the center of the impact plate; r and z are the radial and vertical distances, respectively, from the origin. u is the sheet speed, and ϕ is the angle between the sheet and the vertical. r_c is the local radius of curvature of a meridional

Fig. 2.22 A water bell produced by the impact of a descending water jet and a solid impactor. The impactor radius is 1 cm. Fluid is splayed radially by the impact, and then sags under the influence of gravity. The sheet may close on itself owing to the azimuthal curvature of the bell.

line, and s is the arc length along a meridional line measured from the origin. Finally, ΔP is the pressure difference between the outside and inside of the bell.

Flux conservation requires that

$$Q = 2\pi r t u, \tag{2.104}$$

while Bernoulli's Theorem indicates that

$$u^2 = u_0^2 + 2gz. \tag{2.105}$$

The total curvature force acting normal to the bell surface is given by

$$2\sigma \, \nabla \cdot \mathbf{n} = 2\sigma \left(\frac{1}{r_c} + \frac{\cos \phi}{r} \right). \tag{2.106}$$

The factor of two results from there being two free surfaces. Therefore, the force balance normal to the sheet takes the form

$$\frac{2\sigma}{r_c} + \frac{2\sigma \cos \phi}{r} - \Delta P + \rho g t \sin \phi - \frac{\rho t u^2}{r_c} = 0. \tag{2.107}$$

Equations (2.104), (2.105), and (2.107) may be appropriately nondimensionalized and integrated to determine the shape of the bell.

2.8 Appendix A

Recall Stokes' Theorem,

$$\int_C \mathbf{F} \cdot \vec{d\ell} = \int_S \mathbf{n} \cdot (\nabla \wedge \mathbf{F}) \, dS.$$

Along the contour C, $\vec{d\ell} = \mathbf{m} \, d\ell$, so that we have

$$\int_C \mathbf{F} \cdot \mathbf{m} \, d\ell = \int_S \mathbf{n} \cdot (\nabla \wedge \mathbf{F}) \, dS.$$

Now let $\mathbf{F} = \mathbf{f} \wedge \mathbf{b}$, where $\mathbf{b}$ is an arbitrary *constant* vector. We thus have

$$\int_C (\mathbf{f} \wedge \mathbf{b}) \cdot \mathbf{m} \, d\ell = \int_S \mathbf{n} \cdot (\nabla \wedge (\mathbf{f} \wedge \mathbf{b})) \, dS.$$

We now use standard vector identities to see that

$$\begin{aligned}
(\mathbf{f} \wedge \mathbf{b}) \cdot \mathbf{m} &= -\mathbf{b} \cdot (\mathbf{f} \wedge \mathbf{m}), \\
\nabla \wedge (\mathbf{f} \wedge \mathbf{b}) &= \mathbf{f}(\nabla \cdot \mathbf{b}) - \mathbf{b}(\nabla \cdot \mathbf{f}) + \mathbf{b} \cdot \nabla \mathbf{f} - \mathbf{f} \cdot \nabla \mathbf{b} \\
&= \mathbf{b}(\nabla \cdot \mathbf{f}) + \mathbf{b} \cdot \nabla \mathbf{f},
\end{aligned}$$

since $\mathbf{b}$ is a constant vector. We thus have

$$\mathbf{b}\cdot\int_C (\mathbf{f}\wedge\mathbf{m})\, d\ell \;=\; \mathbf{b}\cdot\int_S \left[\mathbf{n}(\nabla\cdot\mathbf{f}) - (\nabla\mathbf{f})\cdot\mathbf{n}\right] dS.$$

Since $\mathbf{b}$ is arbitrary, we thus have

$$\int_C (\mathbf{f}\wedge\mathbf{m})\, d\ell \;=\; \int_S \left[\mathbf{n}(\nabla\cdot\mathbf{f}) - (\nabla\mathbf{f})\cdot\mathbf{n}\right] dS.$$

We now choose $\mathbf{f} = \sigma\mathbf{n}$, and recall that $\mathbf{n}\wedge\mathbf{m} = -\mathbf{s}$. We thus obtain

$$\begin{aligned} -\int_C \sigma\mathbf{s}\, d\ell &= \int_S \left[\mathbf{n}\nabla\cdot(\sigma\mathbf{n}) \;-\; \nabla(\sigma\mathbf{n})\cdot\mathbf{n}\right] dS. \\ &= \int_S \left[\mathbf{n}\nabla\sigma\cdot\mathbf{n} + \sigma\mathbf{n}(\nabla\cdot\mathbf{n}) - \nabla\sigma - \sigma(\nabla\mathbf{n})\cdot\mathbf{n}\,\right] dS. \end{aligned}$$

We note that $\nabla\sigma\cdot\mathbf{n} = 0$ since $\nabla\sigma$ must be tangent to the surface S; moreover, $(\nabla\mathbf{n})\cdot\mathbf{n} \;=\; \frac{1}{2}\nabla(\mathbf{n}\cdot\mathbf{n}) \;=\; \frac{1}{2}\nabla(1) \;= 0$. We thus obtain the desired result:

$$\int_C \sigma\mathbf{s}\, d\ell \;=\; \int_S [\nabla\sigma \;-\; \sigma\mathbf{n}\,(\nabla\cdot\mathbf{n})]\, dS.$$

2.9 Appendix B: The Frenet–Serret equations

Differential geometry yields relations that are often useful in computing curvature forces on 2D interfaces:

$$(\nabla\cdot\mathbf{n})\;\mathbf{n} = \frac{d\mathbf{t}}{d\ell}, \tag{2.108}$$

$$-(\nabla\cdot\mathbf{n})\;\mathbf{t} = \frac{d\mathbf{n}}{d\ell}. \tag{2.109}$$

Note that the LHS of eqn (2.108) is proportional to the curvature pressure acting on an interface.

2.10 Appendix C: Computing curvatures

We see the appearance of the divergence of the surface normal, $\nabla\cdot\mathbf{n}$, in the normal-stress balance. We proceed by briefly reviewing how to formulate this curvature term in two common geometries.

In Cartesian coordinates (x, y, z), we consider a surface defined by $z = h(x, y)$. We define a functional $f(x, y, z) = z - h(x, y)$ that necessarily vanishes on the surface. The normal to the surface is defined by

$$\mathbf{n} \;=\; \frac{\nabla f}{|\nabla f|} \;=\; \frac{\hat{z} - h_x\hat{x} - h_y\hat{y}}{(1 + h_x^2 + h_y^2)^{1/2}}, \tag{2.110}$$

and the local curvature may thus be computed:

$$\nabla \cdot \mathbf{n} = \frac{-(h_{xx} + h_{yy}) - (h_{xx}h_y^2 + h_{yy}h_x^2) + 2h_x h_y h_{xy}}{(1 + h_x^2 + h_y^2)^{3/2}}. \qquad (2.111)$$

In the simple case of a 2D interface $z = h(x)$, these results assume the simple forms

$$\mathbf{n} = \frac{\hat{z} - h_x \hat{x}}{(1 + h_x^2)^{1/2}}, \quad \nabla \cdot \mathbf{n} = \frac{-h_{xx}}{(1 + h_x^2)^{3/2}}. \qquad (2.112)$$

Note that $\mathbf{n}$ is dimensionless, while $\nabla \cdot \mathbf{n}$ has the units of $1/L$.

In 3D polar coordinates (r, θ, z), we consider a surface defined by $z = h(r, \theta)$. We define a functional $g(r, \theta, z) = z - h(r, \theta)$ that vanishes on the surface, and compute the normal:

$$\mathbf{n} = \frac{\nabla g}{|\nabla g|} = \frac{\hat{z} - h_r \hat{r} - (1/r) h_\theta \hat{\theta}}{(1 + h_r^2 + 1/r^2 h_\theta^2)^{1/2}}, \qquad (2.113)$$

from which the local curvature is computed:

$$\nabla \cdot \mathbf{n} = \frac{-h_{\theta\theta} - h_r^2 h_{\theta\theta} + h_r h_\theta - r h_r - (2/r) h_r h_\theta^2 - r^2 h_{rr} - h_{rr} h_\theta^2 + h_r h_\theta h_{r\theta}}{r^2 (1 + h_r^2 + (1/r^2) h_\theta^2)^{3/2}}. \qquad (2.114)$$

In the case of an axisymmetric interface $z = h(r)$, these results reduce to

$$\mathbf{n} = \frac{\hat{z} - h_r \hat{r}}{(1 + h_r^2)}, \quad \nabla \cdot \mathbf{n} = \frac{-r h_r - r^2 h_{rr}}{r^2 (1 + h_r^2)^{3/2}}. \qquad (2.115)$$

References

Bush, J. W. M. and Aristoff, J. M. (2003). The influence of surface tension on the circular hydraulic jump. *Journal of Fluid Mechanics*, **489**, 229–238.

Bush, J. W. M., Aristoff, J. M., and Hosoi, A. E. (2006). An experimental investigation of the stability of the circular hydraulic jump. *Journal of Fluid Mechanics*, **558**, 33–52.

Clanet, C. (2007). Waterbells and liquid sheets. *Annual Review of Fluid Mechanics*, **39**, 469–496.

De Gennes, P. G., Brochard-Wyart, F., and Quéré, D. (2004). *Capillarity and Wetting Phenomena: Drops, Bubbles, Pearls, Waves.* Springer.

Ellegaard, C., Hansen, A. E., Haaning, A., Hansen, K., Marcussen, A., Bohr, T., Hansen, J. L., and Watanabe, S. (1998). Creating corners in kitchen sinks. *Nature*, **392**(6678), 767–768.

Hancock, M. J. and Bush, J. W. M. (2002). Fluid pipes. *Journal of Fluid Mechanics*, **466**, 285–304.

Hu, D. L., Chan, B., and Bush, J. W. M. (2003). The hydrodynamics of water strider locomotion. *Nature*, **424**(6949), 663–666.

Keller, J. B. (1998). Surface tension force on a partly submerged body. *Physics of Fluids*, **10**, 3009.

Lucassen-Reynders, E. H. and Lucassen, J. (1969). Properties of capillary waves. *Adv. Colloid Interface Sci.*, **2**, 347–395.

Stone, H. A. (1990). A simple derivation of the time-dependent convective-diffusion equation for surfactant transport along a deforming interface. *Physics of Fluids*, **2**(1), 111–112.

Taylor, G. (1959). The dynamics of thin sheets of fluid. III. Disintegration of fluid sheets. *Proceedings of the Royal Society of London, Series A, Mathematical and Physical Sciences*, **253**, 313–321.

Taylor, G. I. (1960). Formation of thin flat sheets of water. *Proceedings of the Royal Society of London, Series A*, **259**, 1–17.

Thomson, J. (1885). On certain curious motions observable at the surfaces of wine and other alcoholic liquors. *Philosophical Magazine*, **10**, 330.

3
Dynamics of complex biofluids

Christel HOHENEGGER and Michael J. SHELLEY

Courant Institute of Mathematical Sciences, New York University

3.1 Introduction

Complex fluids, such as polymer solutions, particulate suspensions, and many biological fluids, form a broad class of liquids whose mechanical and dynamical properties must be described on multiple length scales. In recent years, biological flow phenomena involving complex fluids, such as peristaltic pumping and sperm motility in the reproductive tracts, have received much attention. It has also been fruitful to consider systems such as bacterial baths as complex fluids themselves when describing them at the macroscopic scale. One of the most challenging issues when characterizing the transport properties of these systems is capturing the interactions between the fluid and the suspended microstructures (e.g. polymer coils, colloidal particles, flexible and rigid fibers, and "active" particles such as bacteria). This is important, as it is these couplings that lead to very complicated dynamical structures and large-scale flow associated with mixing or enhanced swimming efficiency. In many cases, numerical simulations are as challenging as model development, since complex fluid systems can have many degrees of freedom.

The general field of complex fluids presents many interesting flow phenomena and important applications, such as elastic turbulence and low-Reynolds-number mixing (Groisman and Steinberg 2001, 2004), microscopic to macroscopic instabilities such as coil-stretch (Arratia *et al.* 2006; Thomases and Shelley 2009) and stretch-coil (Young and Shelley 2007), and the use of viscoelastic nonlinearities to perform logical operations in microfluidic chips (Groisman and Quake 2004). Of particular interest here are transport phenomena in active suspensions and complex fluids. This includes mixing, pumping, and swimming, both of single organisms and collectively. The most classical problem of swimming, in a biological fluid is that of sperm motility. In studies of collective dynamics, Dombrowski *et al.* (2004) have observed the emergence of large-scale vortices and jets in suspensions of swimming *B. subtilis.* The possible utility of the transport and mixing in these systems has driven the development of artificial analogues, such as the chemically driven synthetic microswimmers fabricated by Paxton *et al.* (2004), which mimic the motile behavior of swimming bacteria.

The general mathematical framework for the derivation of many complex fluid models can be summarized as follows. Consider a small fluid volume moving with the fluid, filled with many "particles", and over which the macroscopic velocity field and its gradient can be assumed uniform. Given the current configuration of the particles, approximate the "extra" stress that these particles induce in the surrounding fluid. The evolution of the particle configuration (location, orientation, distension, etc.), from which one approximates this extra stress, is described by a Smoluchowski equation (or conservation of probability). The inclusion of the extra stress as a source of force in the macroscopic momentum balance equations for a continuous medium often closes the system. In Section 3.2 we develop these ideas, which from the basics of non-Newtonian fluid mechanics. We then focus on the two main examples, of rod-shaped and dumbbell-shaped immersed particles. The latter is a necessary element in the derivation of the Oldroyd-B model and is described in Section 3.3. In order to find the extra stress due to a suspension of rod-like bodies, the Kirkwood formula is recalled in Section 3.2.3 and its application to an actively swimming rod is presented. The last section,

Section 3.4, consists of an overview of recent work on two important applications, pumping and swimming. In-depth details are given about the continuum model describing a suspension of active rods and its stability behavior around a state of uniformity and isotropy.

3.2 Basics of non-Newtonian fluid mechanics

3.2.1 Conservation of probability

We consider a particle whose initial configuration (e.g. orientation, center of mass, length, and volume) is given by $\mathbf{z}$ and which evolves according to

$$\frac{d\mathbf{Z}}{dt} = \mathbf{V}(\mathbf{Z}, t), \qquad \mathbf{Z}(\mathbf{z}, 0) = \mathbf{z}.$$

Example 1 (dumbbell). The configuration variables are the position of the center of mass $\mathbf{X}_c$ and the end-to-end vector $\mathbf{R}$.

Example 2 (rod). The configuration variables are the position of the center of mass $\mathbf{X}_c$ and the orientation vector $\mathbf{p}$, with $|\mathbf{p}| = 1$.

Let $\Psi(\mathbf{Z}, t)$ be the probability density for realizing a set of configuration variables $\mathbf{Z}$ at time t, and let $\Omega(t) \subset \mathbb{R}^n$ be an arbitrary volume that tracks a set of particle trajectories from the initial configuration $\mathbf{z}$ through configuration space.

The "number" of particles in this evolving set, defined as $N = \int_{\Omega(t)} \Psi(\mathbf{Z}, t)\, dV_Z$, is taken to be conserved. Diffusional processes will be added later. To express N in the Lagrangian frame, we define the flow map $\mathbf{J} = \nabla_z \mathbf{Z}$. The derivation of the conservation of probability follows the same argument as for the conservation of mass in a continuum medium. First, by a change of variable, we express N in the reference configuration $\Omega(0)$. Next, using Liouville's formula $\partial_t \mathbf{J} = (\nabla_Z \cdot \mathbf{V})J$ ($J = \det \mathbf{J}$) and the fact that N is conserved, we have

$$0 = \frac{dN}{dt} = \int_{\Omega(0)} \left[\frac{\partial \Psi}{\partial t} + \mathbf{V} \cdot \nabla_Z \Psi + (\nabla_Z \cdot \mathbf{V})\Psi \right] J\, dV_Z.$$

Transforming back to the arbitrary $\Omega(t)$, we arrive at the Smoluchowski equation,

$$\frac{\partial \Psi}{\partial t} + \nabla_Z \cdot (\mathbf{V}\Psi) = 0. \tag{3.1}$$

In order to close eqn (3.1), we need boundary and initial conditions, as well as equations for the particle fluxes $\mathbf{V}$.

3.2.2 Kinematics of rods and dumbbells

We now develop kinematic equations for immersed rod and dumbbell- shaped particles.

3.2.2.1 Rods

First, consider a slender object immersed in a Newtonian Stokesian fluid, described by its centerline $\mathbf{X}(s,t)$, with s being its arc length (Batchelor 1970*b*; Keller and Rubinow 1976; Johnson 1980). Here, we make use of the local slender-body-theory approximation to the centerline velocity $\mathbf{V}(s,t)$,

$$8\pi\mu\left[\mathbf{V}(s,t) - \mathbf{U}(\mathbf{X},t)\right] = -c(\mathbf{I} + \mathbf{X}_s\mathbf{X}_s^{\mathrm{T}})\mathbf{f}, \tag{3.2}$$

where $c = \ln(e\varepsilon^2) < 0$, with $\varepsilon \ll 1$ being the particle aspect ratio, $\mathbf{U}(\mathbf{X},t)$ is the background velocity field, and $\mathbf{f}$ is the force per unit length exerted by the rod upon the fluid. For brevity, we define $\eta = 8\pi\mu/(-c) > 0$.

We will study three examples in detail: a straight rod in a linear background flow, and a straight rod that locomotes through a prescribed surface stress, considered both in the absence of a background flow and in a linear background flow.

Example 3 (straight rod in a linear background flow). Consider a rigid rod of fixed length l described by its center of mass and orientation: $\mathbf{X}(s,t) = \mathbf{X}_{\mathrm{c}}(t) + s\mathbf{p}(t)$ for $-l/2 \le s \le l/2$ and $|\mathbf{p}| = 1$. The background flow is taken as linear: $\mathbf{U}(\mathbf{X}) = \mathbf{A}\mathbf{X}$, with $\mathrm{tr}(\mathbf{A}) = 0$. We determine $\mathbf{f}$ and the evolution equations for $\mathbf{X}_{\mathrm{c}}$ and $\mathbf{p}$ under the conditions of zero net force and torque. Integrating eqn (3.2) with respect to s and using the zero-force condition yields

$$\dot{\mathbf{X}}_{\mathrm{c}} = \mathbf{A}\mathbf{X}_{\mathrm{c}}, \qquad \mathbf{f} = s\mathbf{f}_1, \tag{3.3}$$

where $\dot{\mathbf{X}}_{\mathrm{c}} = \partial\mathbf{X}/\partial t$. From the zero-torque condition it follows that $\mathbf{f}_1 = \alpha\mathbf{p}$. Substituting eqn (3.3) into eqn (3.2) gives $\eta(\dot{\mathbf{p}} - \mathbf{A}\mathbf{p}) = 2\alpha\mathbf{p}$. Since $\mathbf{p}\cdot\dot{\mathbf{p}} = 0$, α is found by taking the dot product of the last equality with $\mathbf{p}$. Combining everything, we obtain Jeffery's equation (Jeffery 1922),

$$\dot{\mathbf{p}} = (\mathbf{I} - \mathbf{p}\mathbf{p}^{\mathrm{T}})\mathbf{A}\mathbf{p}. \tag{3.4}$$

The force per unit length is therefore $\mathbf{f} = -(\eta s/2)(\mathbf{p}^T\mathbf{A}\mathbf{p})\mathbf{p}$.

Example 4 (propulsive rod with no background flow). Consider a rod where a constant propulsive tangential stress $-f_{\|}\mathbf{p} = -2\pi a g\mathbf{p}$, where a is the particle radius and g is the magnitude of the stress, is imposed on one half of its length (without loss of generality, for $s < 0$) and a no-slip condition is imposed on the other half ($s > 0$) (Saintillan and Shelley 2007). This model is an idealization of a "Pusher", such as *B. subtilis*, which generates a thrust (via rotary flagellar motion) through its trailing flagellar bundle (schematically illustrated in Fig. 3.1). In this case, eqn (3.2) becomes

$$\eta\left[U + u_{\|}(s)\right]\mathbf{p} = (\mathbf{I} + \mathbf{p}\mathbf{p}^{\mathrm{T}})\mathbf{f}_1, \quad -\frac{l}{2} \le s \le 0, \tag{3.5}$$

$$\eta U\mathbf{p} = (\mathbf{I} + \mathbf{p}\mathbf{p}^{\mathrm{T}})\mathbf{f}_2, \quad 0 \le s \le \frac{l}{2}, \tag{3.6}$$

where $\mathbf{f}_1 = -f_{\|}\mathbf{p} = -2\pi a g\mathbf{p}$ is the motive force per unit length exerted by the rod on the fluid, $\dot{\mathbf{X}}_{\mathrm{c}} = U\mathbf{p}$ is the translational velocity of the rod, $\mathbf{u}_s = u_s(s)\mathbf{p}$ is the

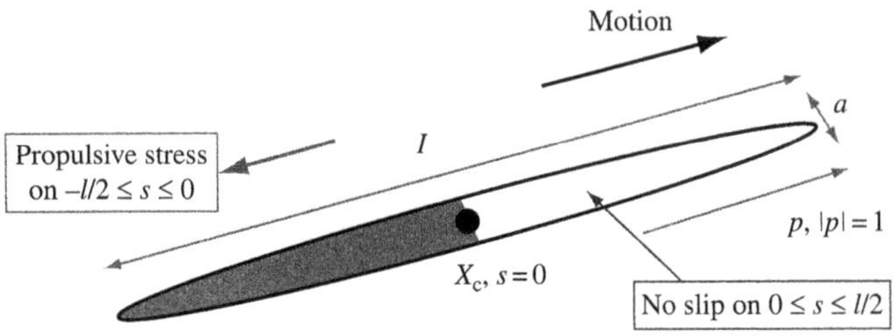

Fig. 3.1 An ellipsoidal rod with propulsive stress on the posterior half of its body.

surface slip velocity, and $\mathbf{f}_2(s) = f_1\mathbf{p}$ is the drag force. The zero-force condition yields $f_2 = f_{\|}$, and thus we have $U = (2/\eta)f_{\|}$ and $u_s = -(4/\eta)f_{\|}$. Using the fact that $\eta = 8\pi\mu/|\ln(e\varepsilon^2)|$, we rewrite U as

$$U = \frac{\varepsilon|\ln(\varepsilon^2 e)|}{2}\frac{lg}{\mu} = \kappa_2\frac{lg}{\mu}, \tag{3.7}$$

where $\kappa_2 = \varepsilon|\ln(\varepsilon^2 e)|/2$ is a purely geometric constant. We will use this expression later as the individual swimming speed in a suspension of many rods.

Example 5 (propulsive rod in linear background flow). We now look at the same rod model swimming within a linear background flow $\mathbf{U} = \mathbf{AX}$. Equation (3.2) becomes

$$\eta\left[\dot{\mathbf{X}}_c + \mathbf{u}_s\mathbf{p} + s\dot{\mathbf{p}} - \mathbf{A}(\mathbf{X}_c + s\mathbf{p})\right] = (\mathbf{I} + \mathbf{pp}^{\mathrm{T}})\mathbf{f}_1(s), \quad -\frac{l}{2} \le s \le 0, \tag{3.8}$$

$$\eta\left[\dot{\mathbf{X}}_c + s\dot{\mathbf{p}} - \mathbf{A}(\mathbf{X}_c + s\mathbf{p})\right] = (\mathbf{I} + \mathbf{pp}^{\mathrm{T}})\mathbf{f}_2(s), \quad 0 \le s \le \frac{l}{2}. \tag{3.9}$$

From the zero-torque condition, both $\mathbf{f}_1$ and $\mathbf{f}_2$ are in the $\mathbf{p}$-direction. Therefore, we set $\mathbf{f}_1 = -f_{\|}(s)\mathbf{p} = -(f_0 + f_1 s)\mathbf{p}$ and $\mathbf{f}_2 = (f_2 + f_3 s)\mathbf{p}$. As in Example 4, we consider only a constant propulsive force $f_0 = 2\pi a g$ and thus a slip velocity $u_s(s)$ independent of s. Upon integration of eqns (3.8) and (3.9) with respect to arc length, we obtain the kinematic equation of the center of mass,

$$\dot{\mathbf{X}}_c = \mathbf{AX}_c - \frac{u_s}{2}\mathbf{p}. \tag{3.10}$$

Substituting into eqns (3.8) and (3.9), taking the dot product with $\mathbf{p}$, and comparing coefficients in s allows us to find all the remaining unknowns: $u_s = -4/\eta f_0$, $f_2 = f_0$, $f_1 = (\eta/2)\,\mathbf{p}^{\mathrm{T}}\mathbf{Ap}$, and $f_3 = -(\eta/2)\,\mathbf{p}^{\mathrm{T}}\mathbf{Ap}$. From these expressions, we obtain again Jeffery's equation (3.4), and $\dot{\mathbf{X}}_c = \mathbf{AX} + (2/\eta)\,f_0\mathbf{p}$.

Summarizing the kinematic equations for a general disturbance flow whose rate-of-strain tensor is constant, we denote by $\mathbf{u}$ the linearized velocity field around the body and denote its gradient by $\nabla_x\mathbf{u}$. Again, we impose a constant propulsive stress $f_{\|} = f_0 = 2\pi a g$ on the posterior half of the body, and we let $U = \kappa_2 lg/\mu$ be the swimming speed of an individual rod of length l. We have

$$\dot{\mathbf{X}}_c = \mathbf{u} + U\mathbf{p} - d_x \nabla_x \ln \Psi, \tag{3.11}$$

$$\dot{\mathbf{p}} = (\mathbf{I} - \mathbf{p}\mathbf{p}^T)\nabla_x \mathbf{u}\mathbf{p} - d_p \nabla_p \ln \Psi, \tag{3.12}$$

where we have introduced translational and rotational diffusion processes with phase space (not particle) diffusion coefficients d_x and d_p, respectively (Doi and Edwards 1986). Equations 3.11 and 3.12 give the particle fluxes of the Smoluchowski equation.

3.2.2.2 Dumbbells

We now consider a dumbbell, or bead–spring, model. This model arises when a long polymer chain is modeled as a succession of beads and springs as in Fig. 3.2. The change in conformation of the chain is represented by its end-to-end displacement vector $\mathbf{R}$, and its response to distension is represented through the entropic spring force

$$\mathbf{F}_s = 2k_B T \beta^2 \mathbf{R}, \qquad \beta^2 = \frac{3}{2Nb^2}, \tag{3.13}$$

where k_B is the Boltzmann constant, T is the temperature, N is the number of links, and b is the length of a link (Doi and Edwards 1986; Larson 1995). Henceforth, the end-to-end vector will be considered as that of a simple dumbbell. Assuming that the beads are spherical and neglecting hydrodynamic effects between them, the drag force is $\mathbf{F}^d = (1/2)\,\zeta\left(\dot{\mathbf{R}} - \nabla \mathbf{u}\,\mathbf{R}\right)$, where $\zeta = 6\pi\mu a$ is the Stokes drag and $\nabla \mathbf{u}\,\mathbf{R}$ is the rate of stretching of a fluid element containing the beads. The Brownian force is given by $\mathbf{F}^B = k_B T \nabla_\mathbf{R} \ln \Psi$. The balance of forces on the dumbbell gives $\dot{\mathbf{R}}$:

$$\dot{\mathbf{R}} = \nabla \mathbf{u}\,\mathbf{R} - \frac{2k_B T}{\zeta}\left[2\beta^2 \mathbf{R} + \nabla_\mathbf{R} \ln \Psi\right]. \tag{3.14}$$

Equation (3.11) and eqn (3.12) or (3.14) define the particle fluxes in the Smoluchowski equation (3.1). However, this requires knowledge of $\mathbf{u}$ (and hence $\nabla_x \mathbf{u}$), which

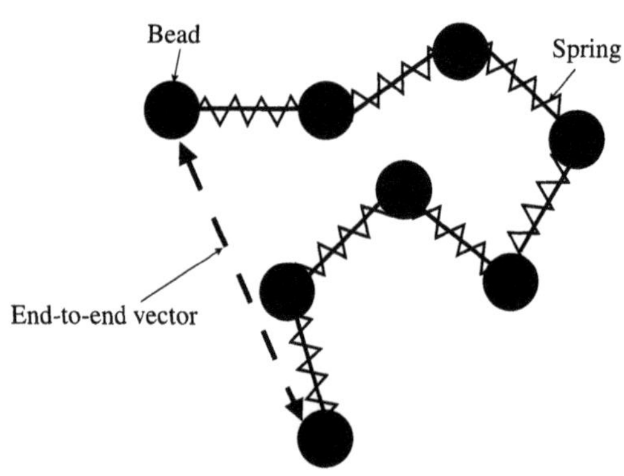

Fig. 3.2 Polymer chain modeled with beads and springs.

is obtained by including both the particle stress produced by the suspension and the Newtonian solvent stress in the macroscopic momentum balance equation:

$$\rho\frac{D\mathbf{u}}{Dt} - \mu\,\Delta_x \mathbf{u} + \nabla_x q = \nabla_x \cdot \boldsymbol{\sigma}^{(p)}, \tag{3.15}$$

where q is the pressure, μ is the Newtonian solvent viscosity, and ρ is the density. Incompressibility is usually assumed, i.e. $\nabla_x \cdot \mathbf{u} = 0$. The extra stress $\boldsymbol{\sigma}^{(p)}$ is given by the Kirkwood formula, which is the subject of the next section.

3.2.3 Kirkwood formula

The Kirkwood formula provides a means for determining the additional or "extra" stress created by particles (rods as dumbbells) suspended in a Newtonian solvent. Details can be found in Doi and Edwards (1986), Larson (1995), and Bird *et al.* (1987).

Example 6 (dumbbells). One basic construction of the Kirkwood formula assumes that the suspension is composed of pairs of beads, with a connecting vector $\mathbf{R}_m$, between which there is a "nonhydrodynamic" force $\mathbf{F}_m$. In particular, given a control volume V, the Kirkwood formula for a suspension of N bead pairs is

$$\boldsymbol{\sigma}^{(p)} = -\frac{1}{V}\sum_{m=1}^{N}\langle \mathbf{F}_m \mathbf{R}_m^{\mathrm{T}}\rangle. \tag{3.16}$$

As in statistical physics, for large N it is useful to replace the average $(1/N)\sum_{n=1}^{N} f_n$ by $\langle f\rangle$, the distributional average of f with respect to the probability density function Ψ.
Therefore, by defining the particle density as $n = N/V$, eqn (3.16) becomes

$$\boldsymbol{\sigma}^{(p)} = -n\langle \mathbf{F}\mathbf{R}^{\mathrm{T}}\rangle. \tag{3.17}$$

The Kirkwood formula as given by eqn (3.17) is particularly suitable for the dumbbell model. Application of eqn (3.17) to the dumbbell model, where $\mathbf{F}$ is the entropic spring force given by eqn (3.13), gives

$$\boldsymbol{\sigma}^{(p)} = 2nk_{\mathrm{B}}T\beta^2\langle \mathbf{R}\mathbf{R}^{\mathrm{T}}\rangle. \tag{3.18}$$

Example 7 (rods). Another form of the extra stress in a suspension, with slender bodies in mind, was proposed by Batchelor (1970*a*, 1977). Let $\boldsymbol{\sigma}$ be the microscopic stress evaluated on the surfaces ∂B_m of N bodies. Then in a volume V, Batchelor's formula is

$$\boldsymbol{\sigma}^{(p)} = \frac{1}{V}\sum_{m=1}^{N}\int_{\partial B_m} dA\,[\boldsymbol{\sigma}\hat{\mathbf{n}}\mathbf{X}_{\mathrm{c}}^{\mathrm{T}} - \mu(\mathbf{u}\hat{\mathbf{n}}^{\mathrm{T}} + \hat{\mathbf{n}}\mathbf{u}^{\mathrm{T}})]. \tag{3.19}$$

Here $\hat{\mathbf{n}}$ is the unit outward normal to the surface and $\mathbf{u}$ is the microscopic velocity on the surface. For slender bodies, the surface velocity is only a function of the arc length along the centerline, and therefore the surface integral containing $\mathbf{u} = \mathbf{u}(s)$ vanishes.

Denoting by $\mathbf{f} = -\boldsymbol{\sigma}\hat{\mathbf{n}}$ the force the body exerts on the fluid, eqn (3.19) therefore reduces to

$$\boldsymbol{\sigma}^{(p)} = -\frac{1}{V}\sum_{p=1}^{N}\int ds\, \mathbf{f}\mathbf{X}^{\mathrm{T}}. \tag{3.20}$$

The Kirkwood formula given by eqn (3.20) leads to the calculation of the extra stress generated by a single rod and then by a suspension of such rods.

Applying eqn (3.20) to a propulsive rod as described in Example 4 gives the contribution to the stress from a single swimmer stress,

$$\mathbf{S} = -\kappa_1 l^3 g \mathbf{p}\mathbf{p}^{\mathrm{T}} = -\sigma_0 \mathbf{p}\mathbf{p}^{\mathrm{T}}, \tag{3.21}$$

where $\kappa_1 = (\pi\varepsilon)/2$ is another geometric constant. We remark that the units of $\sigma_0 = \kappa_1 l^3 g$ are force times length; note these are the units of the strength of a force dipole or stresslet.

Next, we combine eqns (3.20) and (3.21) to find the volume-averaged extra stress in a box of volume L^3 containing N such swimmers. We assume that there are M swimmers in a smaller control volume L_M^3 and that the rate-of-strain tensor is constant over this smaller volume. Further, we assume a separation of scales $l \leq L_M \leq L$. After some manipulation, we find

$$\boldsymbol{\sigma}^{(p)} = -nC\sigma_0\frac{1}{M}\sum_{m=1}^{M}\mathbf{p}\mathbf{p}^{\mathrm{T}}, \tag{3.22}$$

where $n = N/L^3$ is the number density and $C = (M/L_M^3)/(N/L^3)$ is the local concentration.

For large M, the weighted sum converges to the configurational average with respect to Ψ_M, the probability density function for finding a rod with a given center-of-mass position and orientation in the small volume. In passing from the local distribution function and concentration to the macroscopic distribution function, we write $\Psi_M = \Psi/C$ and the extra stress becomes $\boldsymbol{\sigma}^{(p)} = -n\sigma_0\langle\mathbf{p}\mathbf{p}^{\mathrm{T}}\rangle$. Redefining Ψ as $n\Psi$, we obtain the extra stress generated by a suspension of rear-activated swimmers,

$$\boldsymbol{\sigma}^{(p)} = -\sigma_0\langle\mathbf{p}\mathbf{p}^{\mathrm{T}}\rangle. \tag{3.23}$$

As an aside, we note that the normalization of Ψ is $\int dV_x \int dS_p\, \Psi = nL^3$.

3.3 Viscoelastic fluid

One of the simplest models of a viscoelastic fluid is given by the Oldroyd-B equations. There, the polymer chains are modeled as dumbbells (see Fig. 3.2) that have a linear force response to distension. The Smoluchowski equation, eqn (3.1), gives the evolution of a probability density function whose configuration variables are the end-to-end displacement vector $\mathbf{R}$ and the center-of-mass location $\mathbf{X}_c$:

$$\Psi_t + \nabla_{\mathbf{x}} \cdot (\dot{\mathbf{X}}_c \Psi) + \nabla_{\mathbf{R}} \cdot (\dot{\mathbf{R}} \Psi) = 0. \tag{3.24}$$

Again, the Oldroyd-B model assumes that the force exerted by the dumbbell is that of a linear spring given by eqn (3.13). The particle fluxes are given by $\dot{\mathbf{X}}_c = \mathbf{u}$ and eqn (3.14) for $\dot{\mathbf{R}}$ (here, the effect of center-of-mass diffusion has been neglected). The macroscopic velocity $\mathbf{u}$ satisfies the momentum conservation equation (3.15), which balances viscous stress against the extra stress given by the Kirkwood formula, eqn (3.18).

Because of the simple Hookean response in the Oldroyd-B model, a macroscopic evolution equation for $\boldsymbol{\sigma}^{(p)}$ can be found directly, without recourse to any "closure approximation". Substituting the expressions for the particle fluxes into eqn (3.24) together with incompressibility yields

$$\Psi_t + \mathbf{u} \cdot \nabla_x \Psi + \nabla_{\mathbf{R}} \cdot \left[\Psi \nabla_x \mathbf{u} \mathbf{R} - \frac{2k_{\mathrm{B}}T}{\zeta} \left(2\beta^2 \mathbf{R} \Psi + \nabla_{\mathbf{R}} \Psi \right) \right] = 0. \tag{3.25}$$

Since $\boldsymbol{\sigma}^{(p)}$ is proportional to $\langle \mathbf{R}\mathbf{R}^{\mathrm{T}} \rangle$, the evolution equation for $\boldsymbol{\sigma}^{(p)}$ is found by multiplying eqn (3.25) by $\mathbf{R}\mathbf{R}^{\mathrm{T}}$ and integrating over the $\mathbf{R}$ volume. The derivation requires the following tensor equalities (Larson 1995):

$$\int dV_{\mathrm{R}}\, \mathbf{R}\mathbf{R}^{\mathrm{T}} \Psi_t = \frac{\partial}{\partial t} \langle \mathbf{R}\mathbf{R}^{\mathrm{T}} \rangle,$$

$$\int dV_{\mathrm{R}}\, \mathbf{R}\mathbf{R}^{\mathrm{T}} \nabla_{\mathrm{R}} \cdot (\Psi \nabla \mathbf{u}\, \mathbf{R}) = -\left[\nabla \mathbf{u} \langle \mathbf{R}\mathbf{R}^{\mathrm{T}} \rangle + \langle \mathbf{R}\mathbf{R}^{\mathrm{T}} \rangle \nabla \mathbf{u}^{\mathrm{T}} \right],$$

$$\int dV_{\mathrm{R}}\, \mathbf{R}\mathbf{R}^{\mathrm{T}} \nabla_{\mathrm{R}} \cdot (\Psi \mathbf{R}) = -2 \langle \mathbf{R}\mathbf{R}^{\mathrm{T}} \rangle,$$

$$\int dV_{\mathrm{R}}\, \mathbf{R}\mathbf{R}^{\mathrm{T}} \nabla_{\mathrm{R}} \cdot (\nabla_{\mathrm{R}} \Psi) = 2\mathbf{I}.$$

Applying these identities, we find

$$\frac{D}{Dt} \boldsymbol{\sigma}^{(p)} - \left(\nabla_x \mathbf{u}\, \boldsymbol{\sigma}^{(p)} + \boldsymbol{\sigma}^{(p)} \nabla_x \mathbf{u}^{\mathrm{T}} \right) + \tau^{-1} (\boldsymbol{\sigma}^{(p)} - G\mathbf{I}) = 0, \tag{3.26}$$

where $D/Dt = \partial/\partial t + \mathbf{u} \cdot \nabla$ is the material derivative, $\tau = \zeta/(8k_{\mathrm{B}}T\beta^2)$, and $G = \nu k_{\mathrm{B}}T$. Defining the upper convective derivative $\overset{\nabla}{\boldsymbol{\sigma}} = D\boldsymbol{\sigma}/Dt - (\nabla_x \mathbf{u}\, \boldsymbol{\sigma} + \boldsymbol{\sigma} \nabla_x \mathbf{u}^{\mathrm{T}})$, eqn (3.26) takes the form

$$\tau \overset{\nabla}{\boldsymbol{\sigma}}^{(p)} + (\boldsymbol{\sigma}^{(p)} - G\mathbf{I}) = 0. \tag{3.27}$$

In a state of rest with no flow, the extra-stress tensor is isotropic: $\boldsymbol{\sigma}^{(p)} = G\mathbf{I}$. Equation (3.27) is called the upper convected Maxwell equation or the Oldroyd-B equation, and is a simple model for a so-called Boger fluid.

Phenomenologically, viscoelastic fluids are characterized by a relaxation time for stress fluctuations. In the Oldroyd-B equations, this timescale is provided by τ. To see this, we rewrite eqn (3.27) in the Lagrangian frame. We denote by $\mathbf{X}$ the Lagrangian variable which gives initial data to the Lagrangian flow map $\mathbf{x} = \chi(\mathbf{X}, t)$, where $\mathbf{x}$ is the Eulerian spatial variable. Let $F_{ij}(t) = \partial x_i/\partial X_j$ be the deformation tensor and $\mathbf{C}(t) = \mathbf{F}^{\mathrm{T}}\mathbf{F}$ be the right Cauchy–Green tensor. It is straightforward to show that $\dot{\mathbf{F}} = \nabla_x \mathbf{u}\mathbf{F}$. Multiplying eqn (3.26) by $\mathbf{F}^{-1}$ on the left and by $\mathbf{F}^{-\mathrm{T}}$ on the right, we obtain

$$\left(\mathbf{F}^{-1}\boldsymbol{\sigma}^{(p)}\mathbf{F}^{-\mathrm{T}}\right)_t + \frac{1}{\tau}\left(\mathbf{F}^{(-1)}\boldsymbol{\sigma}^{(p)}\mathbf{F}^{-\mathrm{T}}\right) = \frac{G}{\tau}\mathbf{C}^{-1}. \tag{3.28}$$

Integrating with respect to time then gives

$$\boldsymbol{\sigma}^{(p)}(t) = e^{-t/\tau}\mathbf{F}\boldsymbol{\sigma}^{(p)}(0)\mathbf{F}^{\mathrm{T}} + \frac{G}{\tau}\int_0^t ds\, e^{-(t-s)/\tau}\mathbf{F}(t)\mathbf{C}^{-1}(s)\mathbf{F}^{\mathrm{T}}(t), \tag{3.29}$$

where $\mathbf{F}(0) = \mathbf{I}$. Equation (3.29) shows that memory of the current stress state is lost on a timescale of $O(\tau)$.

To nondimensionalize the Oldroyd-B equation, eqn (3.27), let τ_{f} be some timescale characterizing the fluid flow; for example, given a system length scale L and an external force of size F, this gives $\tau_{\mathrm{f}} = \mu/(\rho L F)$, and we can define the dimensionless relaxation time, the Deborah number, as $De = \tau/\tau_{\mathrm{f}}$. For a shear flow with characteristic time $\dot{\gamma}^{-1} = L/U$ (where L is the channel height and U is the velocity), the Deborah number is replaced by the Weissenberg number $Wi = \tau\dot{\gamma}$.

Rescaling time by τ_{f}, length by L, extra stress by G, and velocity by L/τ_{f}, eqn (3.15) becomes

$$Re\frac{D\mathbf{u}}{Dt} = -\nabla_x p + \Delta_x \mathbf{u} + \beta\nabla_x \cdot \boldsymbol{\sigma}^{(p)},$$

where $\beta = G\tau_{\mathrm{f}}/\mu$ measures the dimensionless size of the extra-stress-to-overall-momentum balance, and $Re = \rho U L/\mu$ is the Reynolds number. Equation (3.27) becomes

$$De\,\overset{\triangledown}{\boldsymbol{\sigma}}{}^{(p)} = -(\boldsymbol{\sigma}^{(p)} - \mathbf{I}).$$

Note that $De\,\beta = \tau G/\mu_{\mathrm{s}}$ is the ratio of the polymer viscosity to the solvent viscosity, and is a material constant. Moreover, in the limit of large Weissenberg number, the Navier–Stokes/Oldroyd-B equations describe an incompressible neo-Hookean solid.

In the next section (Section 3.4), we discuss applications to pumping and to swimming of small organisms, which is characterized by a low Reynolds number but usually not by a low Weissenberg number. Therefore, we replace the Navier–Stokes equations by the Stokes equations and obtain the Stokes–Oldroyd-B (Stokes-OB) equations

$$-\nabla_x p + \Delta_x \mathbf{u} = -\beta\nabla_x \cdot \boldsymbol{\sigma} - \mathbf{f}, \qquad \nabla_x \cdot \mathbf{u} = 0, \tag{3.30}$$

$$Wi\,\overset{\triangledown}{\boldsymbol{\sigma}} = -(\boldsymbol{\sigma} - \mathbf{I}), \tag{3.31}$$

where $\mathbf{f}$ is some external force. Here we have dropped the subscript (p) and are using a Weissenberg number rather than a Deborah number.

Before moving on to applications, we discuss some properties of eqns (3.30) and (3.31). First, we note that the existence of global solutions is unknown, even for two-dimensional flows. A good measure of the stress fluctuations in the system is the relative strain energy $\mathcal{E} = \frac{1}{2}\int dV_x \ \mathrm{tr}(\boldsymbol{\sigma}^{(p)} - \mathbf{I})$. The first property gives an equation for the time evolution of $\mathcal{E}$, while the other two are more general properties and pertain to the mathematical structure of the Stokes-OB equations.

3.3.1 Properties

1. *If* $\mathbf{E}$ *is the symmetric rate-of-strain tensor, then*

$$\dot{\mathcal{E}} + Wi^{-1}\,\mathcal{E} = \beta^{-1}\left(\int \mathbf{u}\cdot\mathbf{f} - 2\int \mathbf{E}:\mathbf{E}\right).$$

   Physically, $\int \mathbf{u}\cdot\mathbf{f}$ *is the input power from the background force and* $\int \mathbf{E}:\mathbf{E}$ *is the rate of viscous dissipation.*
2. *Equations* (3.30) *and* (3.31) *lack scale-dependent dissipation.*
3. *Unlike the Newtonian Stokes equations, eqns* (3.30) *and* (3.31) *are not time-reversible.*

Proof

1. By the transport theorem, we have $\dot{\mathcal{E}} = (1/2)\,\mathrm{tr}\int D\boldsymbol{\sigma}/Dt$ and, with $D\boldsymbol{\sigma}/Dt$ given by eqn (3.31),

$$\dot{\mathcal{E}} = \frac{1}{2}\mathrm{tr}\int \left[\nabla_x \mathbf{u}\boldsymbol{\sigma} + \boldsymbol{\sigma}\nabla_x \mathbf{u}^{\mathrm{T}}\right] - Wi^{-1}\,\mathcal{E}.$$

   This equation can be further simplified by integrating by parts, assuming either periodic or no-slip boundary conditions, to give

$$\dot{\mathcal{E}} = -\frac{1}{2}\int \left[u_i(\nabla_x\boldsymbol{\sigma})_i + u_i(\nabla_x\boldsymbol{\sigma})_i\right] - Wi^{-1}\,\mathcal{E} = -\int \mathbf{u}\cdot(\nabla_x\boldsymbol{\sigma}) - Wi^{-1}\,\mathcal{E}.$$

   Substituting eqn (3.30) for $\boldsymbol{\sigma}$, the claim follows by another integration by parts.
2. To understand dissipation in the Stokes-OB equations, we study the dynamics under a small background force perturbation $\mathbf{f} = \varepsilon\mathbf{g}$. Setting $\boldsymbol{\sigma} = \mathbf{I} + \varepsilon\mathbf{T}$ and $p = \varepsilon q$, eqns (3.30) and (3.31) are easily linearized. The linearized polymer extra-stress equation is

$$Wi\left(\frac{\partial \mathbf{T}}{\partial t} - (\nabla_x\mathbf{v} + \nabla_x\mathbf{v}^{\mathrm{T}})\right) = -\mathbf{T}. \tag{3.32}$$

   Assuming a periodicity of 2π, we transform the linearized equations using a spatial Fourier transform with wave vector $\mathbf{k}$. After manipulation of the Fourier transform of the Stokes equations, we obtain the following equation for the $k-$component of the Fourier transform of $\mathbf{T}$:

$$Wi\,\frac{\partial \mathbf{T}_k}{\partial t} = -\mathbf{T}_k - \beta\,Wi\,\mathbf{Q}\left(\frac{\mathbf{k}}{|\mathbf{k}|}\right)\mathbf{T}_k - i\frac{Wi}{k}\mathbf{P}\left(\frac{\mathbf{k}}{|\mathbf{k}|}\right)\mathbf{g}_k,$$

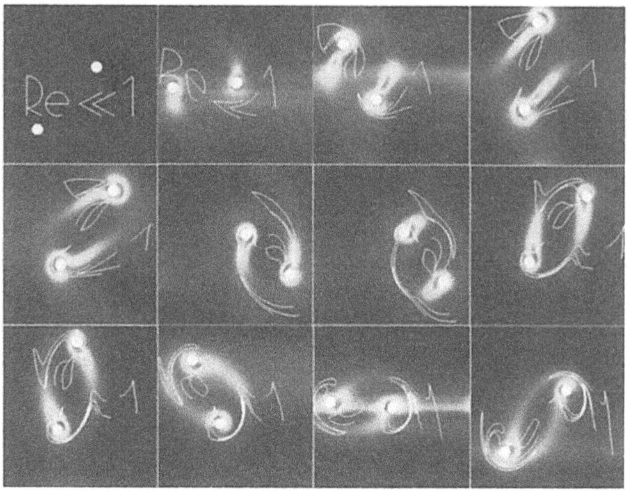

Fig. 3.3 Cylindrical pegs used to mix and then unmix a Stokes-OB fluid, showing the time irreversibility of the equations. The gray level represents the normal stress in the fluid. Reproduced with permission from Teran *et al.* (2008).

where $\mathbf{Q}$ is a rank-four tensor and $\mathbf{P}$ is a rank-three tensor. Since $\mathbf{Q}$ depends only upon the direction of the wave vector and not its amplitude, there is no scale-dependent damping.

3. This is immediately apparent from the energy dissipation law of Property 1, and follows by replacing t with $-t$ in eqn (3.31). A striking illustration of this irreversibility is provided by Fig. 3.3, from Teran *et al.* (2008). This figure shows the simulated motion of material points in an Oldroyd-B fluid as a pair of mixing pegs undergo a displacement with time-reversal symmetry (the pegs depart from and arrive back at their initial positions along the same spatio-temporal path). If the dynamics were time reversible, then the labeling "$Re \ll 1$" in the figure, transported as material lines by the flow, would reappear undeformed at the final time. The evolution of the polymer stress is shown by the gray level of the background.

We conclude with two simple rheological flows for the Stokes-OB model that illustrate the model's properties and limitations.

Example 8 (shear flow). We consider a shear flow $\mathbf{u} = (\dot{\gamma}y, 0, 0)$, where $\dot{\gamma}$ is the shear rate between two plates at $z = 0$ and $z = L$, and seek the tensor $\boldsymbol{\sigma}^{(p)}$ in the steady state. From eqn (3.27), we have

$$\boldsymbol{\sigma}^{(p)} = G \begin{pmatrix} 2\,Wi^2 + 1 & Wi & 0 \\ Wi & 1 & 0 \\ 0 & 0 & 1 \end{pmatrix}.$$

The first normal-stress difference is $N_1 = \sigma_{11} - \sigma_{22} = 2G\,Wi^2 > 0$, showing that this system develops normal stresses, as is typically found in viscoelastic fluids. This is unlike the case of a Newtonian fluid, for which $N_1 = 0$. The shear viscosity is likewise calculated as $\mu = \sigma_{12}/\dot{\gamma} = \mu_s + \tau G$, showing that μ is independent of $\dot{\gamma}$, and hence the system does not capture shear thinning. In this sense, the Oldroyd-B equations model a Boger fluid (Larson 1995).

Example 9 (extensional flow). Consider an extensional flow $\mathbf{u} = (\dot{\epsilon}x, -\dot{\epsilon}y/2, -\dot{\epsilon}z/2)$, where $\dot{\epsilon}$ is the elongation rate. The steady-state solution for $\boldsymbol{\sigma}^{(p)}$ is given by

$$\boldsymbol{\sigma}^{(p)} = G \begin{pmatrix} 1/(1-2\dot{\epsilon}\tau) & 0 & 0 \\ 0 & 1/(1+\dot{\epsilon}\tau) & 0 \\ 0 & 0 & 1/(1+\dot{\epsilon}\tau) \end{pmatrix}.$$

The extensional viscosity is $\mu_e = (\sigma_{11} - \sigma_{22})/\dot{\epsilon}$ and is divergent at $\dot{\epsilon} = 1/(2\tau)$. This divergence of viscosity at a finite strain rate is a consequence of the assumption of a linear force response to distension in the microscopic model. That is, there is no limit to the stretching of polymer coils in the Oldroyd-B model.

That said, recent numerical studies by Thomases and Shelley (2007) suggest that such divergences are only realized exponentially in time, and their singular character is strongly dependent upon the Weissenberg number and the flow geometry. Further, flows associated with polymer stretching in hyperbolic stagnation can undergo symmetry-breaking instabilities at critical values of the Weissenberg number of the flow, and these instabilities can lead to new dynamical states associated with coherent structures and fluid mixing (Arratia *et al.* 2006; Thomases and Shelley 2009).

The Oldroyd-B model is one of the simplest models of a viscoelastic fluid that is derived from microscopic principles. It has the advantage of being closed at the macroscopic level, but it is limited by the absence of shear thinning and, more notably, by the lack of a limit on polymer distension. FENE (finitely extensible nonlinear elastic spring) models overcome the latter shortcoming by using a nonlinear spring law that diverges at a finite distension length. However, FENE models do not generally close at the macroscopic level, requiring the evolution of a Smoluchowski equation at every point in the domain. A commonly used closure approximation is the so-called FENE-P model, which replaces $\mathrm{tr}(\mathbf{R}\mathbf{R}^{\mathrm{T}})$ in the nonlinear force law by $\mathrm{tr}\langle\mathbf{R}\mathbf{R}^{\mathrm{T}}\rangle$. Other models with different transport operators and nonlinearities include the Johnson–Segalman model and the Giesekus model (Larson 1995).

3.4 Applications

Owing to the wide range of elastic moduli (0.1–10Pa), viscosities (0.1–10Pa s), and relaxation times (1–10 s) (Lauga 2007; Smith *et al.* 2009), relating complex biological fluids to viscoelastic flow models can be challenging. We focus now on two applications, peristaltic pumping and swimming, which illustrate the effect of viscoelasticity upon these fundamental means of biological transport. The examples show complex dynamical behavior fundamentally different from that in a Newtonian fluid, and fall into the broad category of complex biofluids.

3.4.1 Pumping a viscoelastic fluid

We start by first considering peristaltic pumping, which is a fundamental transport mechanism for bulk fluids and materials, in both biological and industrial settings. Peristalsis occurs when contractile waves propagate down a fluid-containing tube and is schematically illustrated in Fig. 3.4a. It is the dominant means of material movement in digestive and reproductive tracts, and can involve complex biofluids such as cervical mucus. Fauci and Dillon (2006) have reviewed this area within the context of reproductive fluid dynamics.

As already stated, in many biological settings the pumped fluid is non-Newtonian. To study peristaltic pumping of such fluids, Teran *et al.* (2008) developed a numerical method based on the immersed boundary method of Peskin (2002) for the simulation of the Stokes-OB equations in a time-dependent geometry. For boundaries with moving peristaltic waves of deformation (see Fig. 3.4(a), Teran *et al.* showed that viscoelasticity produces fundamentally different results from the Newtonian case. Figure 3.4(b) shows the mean flow rate Θ as a function of the occlusion ratio χ (the ratio of the wave amplitude to the mean channel width; $\chi = 1$ means complete occlusion), for various Weissenberg numbers. The flow rate Θ was computed at long times in order to remove the effect of the particular initial data for the polymer extra stress. For small Weissenberg numbers, Θ increases monotonically with χ in a fashion similar to that for a Newtonian fluid, albeit with smaller values. For larger Weissenberg numbers, the flow rate reaches its maximum and then declines, well before complete occlusion occurs. In short, viscoelasticity can significantly limit pumping by peristalsis.

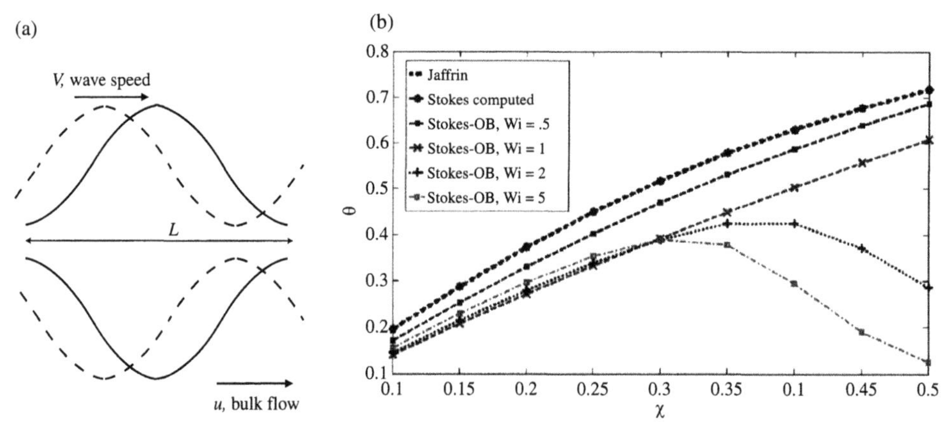

Fig. 3.4 (a) A peristaltic wave of wall deformation propagates with speed V down a channel in the direction of the bulk fluid flow. (b) The dimensionless mean flow rate Θ as a function of the wave amplitude ratio χ and Weissenberg number Wi. The upper curve compares the asymptotic result of Jaffrin and Shapiro (1971) with a Newtonian Stokes simulation (second curve from top). The other curves are for nonzero Wi. Reproduced with permission from Teran *et al.* (2008).

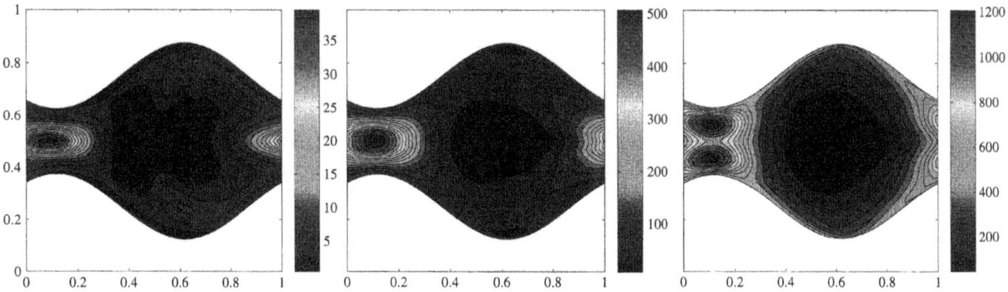

(a) Contour lines of σ_{11}

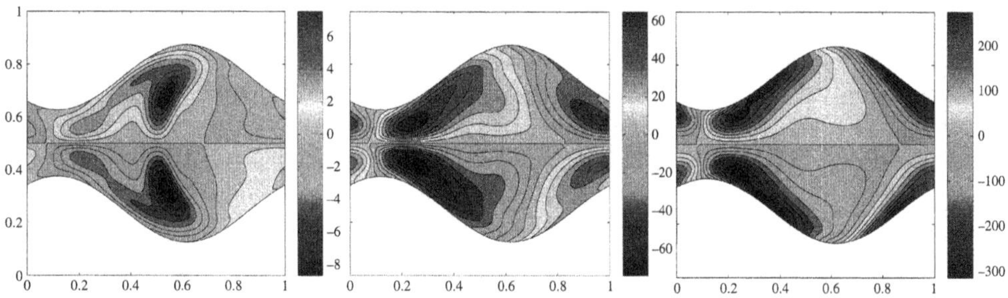

(a) Contour lines of σ_{12}

Fig. 3.5 Contours of the polymer stress components at $t_0 = 0.63$, $t_1 = t_0 + 2$, and $t_2 = t_0 + 9$, for $Wi = 5$ and $\chi = 0.5$. The peristaltic wave moves from right to left. Reproduced with permission from Teran *et al.* (2008).

As an illustrative example of the dramatic effects of viscoelasticity, Fig. 3.5, from Teran *et al.* (2008), shows the evolving polymer stress components σ_{11} and σ_{12} from simulations with $Wi = 5$ and $\chi = 0.5$. Unlike the case of a Newtonian Stokes flow, the material stresses now show strong time dependencies, and develop asymmetries that are associated with flow irreversibility. Over time, the viscoelastic polymer stresses also develop strong gradients in the neighborhood of the pump neck, and these structures are implicated in the strong viscoelastic refluxes that can inhibit pumping at even moderate occlusion ratios.

3.4.2 Swimming

The study of undulatory swimming at low Reynolds number in Newtonian fluids dates back to classical work by Taylor (1951), Lighthill (1960), and Purcell (1977). This continues to be a very active area, and recent example studies include those by Tam and Hosoi (2007) and Spagnolie and Lauga (2010), who both studied optimization of swimming strokes for speed and efficiency in a Newtonian fluid. In the Newtonian regime, there are a wealth of numerical and analytical tools, such as singularity and boundary integral methods that reduce representations of three-dimensional Stokes flows to two- or even one-dimensional problems; see Pozrikidis (1992). As with peristaltic pumping, the situation is more difficult when one studies the effect of viscoelasticity upon swimming.

3.4.2.1 Swimming sheet

For the Newtonian case, an analysis of swimming by a periodic bi-infinite sheet was provided by Taylor (1951), who performed a small-amplitude analysis and showed that with a sinusoidal deformation of amplitude A, the swimming speed scaled like $O(A^2)$. This problem was reexamined by Lauga (2007) for the viscoelastic case. He considered solutions that are steady in the traveling-wave frame, found the same small-amplitude scaling as Taylor, and showed that a swimmer in a Newtonian Stokes fluid was always faster than a Stokes-OB swimmer. A similar study was performed by Fu *et al.* (2007, 2009) for small-amplitude flagellar swimming.

To study the influence of viscoelasticity on the speed and efficiency of undulatory swimming in the time-dependent, large-amplitude case, Teran *et al.* (2010) developed a numerical model, based on the immersed boundary method of Peskin (2002), for the swimming of an undulating sheet immersed in a Stokes-OB fluid. The immersed sheet is an effectively inextensible surface along which a traveling bending wave moves. Wave motion is induced through an immersed boundary force derived from an elastic energy with a time-dependent preferred curvature (Fauci and Peskin 1988).

For small amplitudes and doubly periodic sheets, Teran *et al.* (2010) recovered the Taylor/Lauga scaling of swimming speed. Moreover, they found that as the amplitude increased, the Newtonian swimmer (i.e. an infinite sheet in a Newtonian liquid) was always faster than an Oldroyd-B swimmer, as predicted by Lauga (2007).

If the sheet is allowed to have a head and a tail, i.e. is a free swimmer, these results change drastically at large amplitudes. Figure 3.6(a) shows the horizontal displacement of the center of mass of the swimmer for varying De with $\beta = 1/2$, for a swimming stroke whose amplitude increases towards the tail, as is typical of sperm locomotion. Since the initial data for the polymer stress was the identity in all cases, all swimmers start off at the same speed. However, at long times, the swimming speeds rearrange themselves and reveal that the fastest swimmer has an $O(1)$ Deborah number. Figure 3.6(b) shows the fastest swimmer ($De = 1$), the slowest ($De = 5$, the largest value considered), and the Newtonian case, at the final time $t = 20$. Despite the substantially slower speed of the $De = 5$ swimmer, it is almost tied with the Newtonian free swimmer at this final time owing to its faster swimming speed at intermediate times.

Figure 3.7 shows graphically the distension of the polymer stress field at late times for the fastest swimmer ($De = 1$). In Fig. 3.7a, the free swimmer is very near the time of peak forward velocity during its stroke, when the backward moving wave has approached the tail, which is itself moving slightly upwards. Owing to the strong straining of the fluid by the motion of the tail, there is a large concentration of polymer stress concentration aft of the swimmer. Figure 3.7(b) is about one-quarter of a stroke later. The stress is now much more anisotropic, and the swimmer is actually slipping backwards. However, in comparison with the Newtonian case, the backwards slippage is of smaller magnitude, leading to an overall faster swimmer.

Note that these results for large-amplitude free swimmers are different from those of Lauga (2007) for small-amplitude periodic sheets. In particular, the study of Teran *et al.* (2010) predicts that both swimming speed and efficiency are maximized at a Deborah number of $O(1)$, when the stroke period is nearly matched to the relaxation

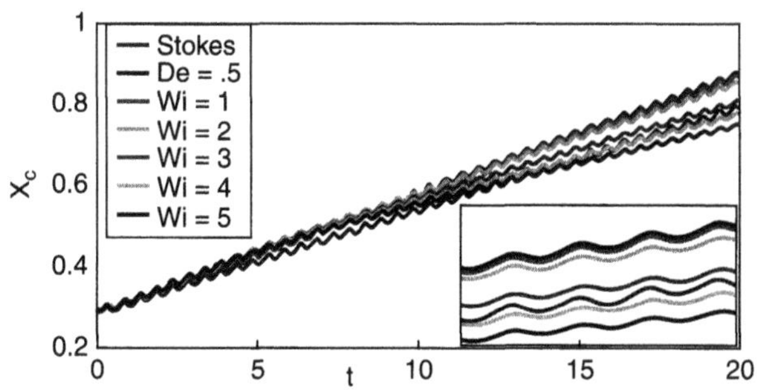

(a) Center of mass

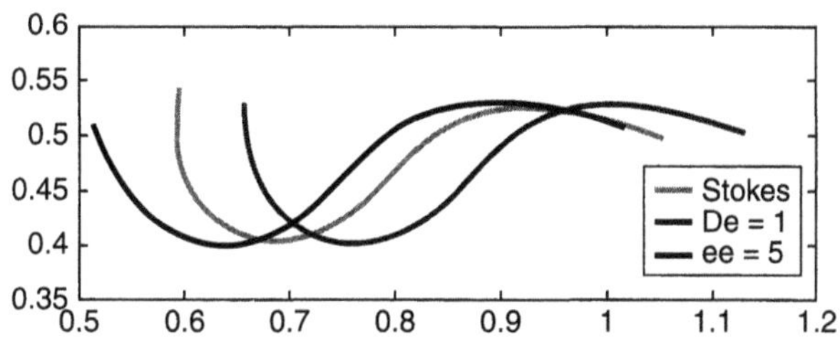

(b) Free-swimmer displacement

Fig. 3.6 (a) The location of the x-component of the center of mass of a free swimmer with $A = 0.05$ and various values of *Dc*. Insct: long-time behavior. (b) Shape and displacement of three free swimmers ($De = 0, 1, 5$) after 20 periods. Reproduced with permission from Teran *et al.* (2010).

time of the fluid. These results seem to be in qualitative agreement with recent experimental work of Smith *et al.* (2009), who found that human spermatozoa swim at a Deborah number of $O(1)$ in synthetic cervical mucus, and show greater displacement per stroke than in less viscoelastic media.

3.4.2.2 Active suspensions of swimming rods

The previous example considered the locomotion of single microswimmers in a viscoelastic fluid and showed the fundamental differences from locomotion in a Newtonian fluid. We now address instead the collective hydrodynamics of many active swimmers in a Newtonian fluid, i.e., an active suspension. Experiments by Kessler, Goldstein, and co-workers (Dombrowski *et al.* 2004; Tuval *et al.* 2005; Sokolov *et al.* 2007) showed that in the case of concentrated bacterial baths, large-scale vortices and jets emerged. These flow features were reproduced qualitatively by Graham and co-workers (Hernández-Ortiz *et al.* 2005; Underhill *et al.* 2008), using a force-dipole (dumbbell) model, and by Saintillan and Shelley (2007) using a more detailed model of the rod-like swimmer described in Example 4. Figure 3.8 shows a particle simulation from Saintillan and Shelley (2007) of 2500 Pusher particles (rods) in a box of size $10 \times 10 \times 3$ (in units of particle length). Initially, the rods are aligned, though with

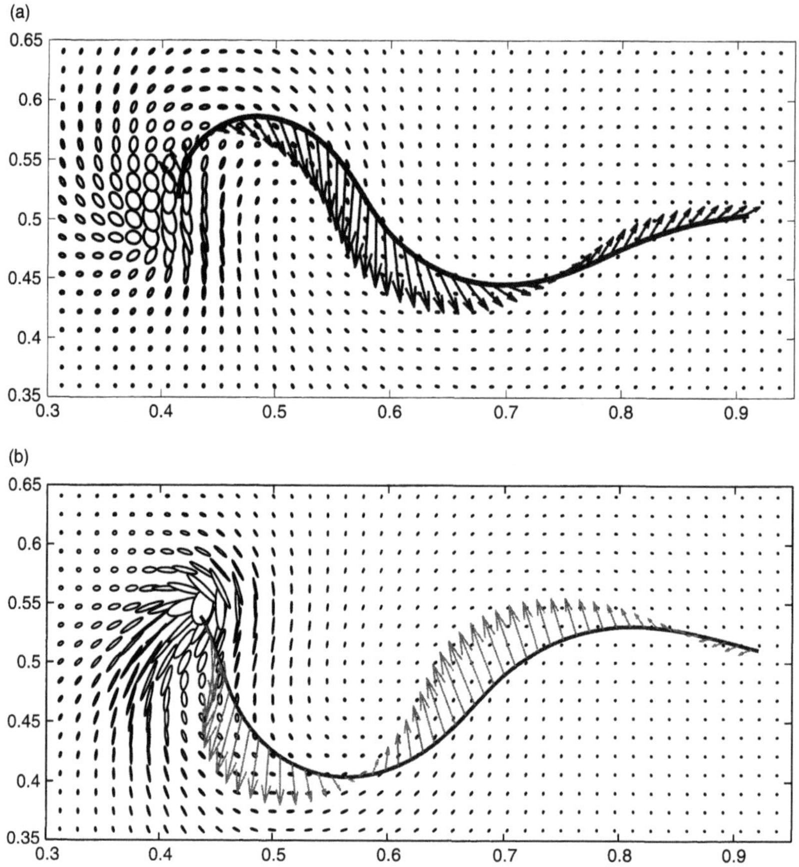

Fig. 3.7 Distension of polymer stress field by a free swimmer with $De = 1$ at late times. The ellipses represent $\boldsymbol{\sigma}$; the major axis is aligned with the principal eigenvector of $\boldsymbol{\sigma}$, with its length scaled by the associated eigenvalue, and the minor axis is associated with the second eigenvector. The vectors represent the fluid velocity on the swimmer. Reproduced with permission from Teran *et al.* (2010).

randomly chosen positions. As time increases, instabilities develop that move the rods away from global alignment, and eventually there emerge system-scale vortices and jets that are reminiscent of the experimental observations of Kessler, Goldstein, and co-workers (Dombrowski *et al.* 2004; Tuval *et al.* 2005; Sokolov *et al.* 2007). Although the initial global orientational order is destroyed, the subsequent dynamics preserves the local alignment of co-swimming particles over about a body length. Whereas the background flow is on the scale of the system, the individual swimmers show random walk statistics, where the random-walk seems to result from pair interactions, at least at lower volume concentrations. The onset of the large-scale instabilities appears to depend upon the effective volume concentration $\tilde{\nu} = \nu/8$ ($\nu = (Nl^3)/L^3$) of swimmers (see Fig. 2b of Saintillan and Shelley 2007).

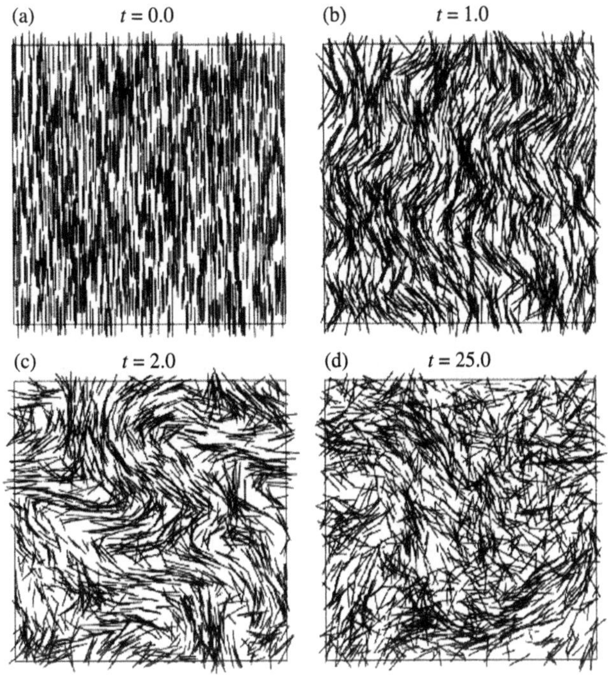

Fig. 3.8 Particle simulation of 2500 initially aligned Pushers in a box of size $10 \times 10 \times 3$ (in units of particle length) at an effective volume fraction $\tilde{\nu} = \nu/8 = 1$, showing the different stages of the dynamics (a)–(d). Reproduced with permission from Saintillan and Shelley (2007).

Saintillan and Shelley (2008*a*,*b*) subsequently developed a continuum model to describe large-scale macroscopic phenomena, based on the evolution of the probability distribution function (see Section 3.2), which we now discuss.

The configuration variables of a swimming rod are its center-of-mass position and its orientation. Therefore, the Smoluchowski equation (3.1) becomes

$$\frac{\partial \Psi}{\partial t} = -\nabla_x \cdot (\dot{\mathbf{x}}\Psi) - \nabla_p \cdot (\dot{\mathbf{p}}\Psi). \tag{3.33}$$

The particle fluxes are given by eqns (3.11) and (3.12). Recall that the normalization of Ψ is such that $\int dV_x \int dS_p \Psi = nL^3$:

$$\dot{\mathbf{x}} = U\mathbf{p} + \mathbf{u} - D_p \nabla_x \ln \Psi, \tag{3.34}$$

$$\dot{\mathbf{p}} = (\mathbf{I} - \mathbf{p}\mathbf{p}^{\mathrm{T}}) \cdot \nabla \mathbf{u}\, \mathbf{p} - d_p \nabla_p \ln \Psi. \tag{3.35}$$

The background flow satisfies the incompressible Stokes equations, forced by the extra stress generated by the swimming rods:

$$-\mu \Delta_x \mathbf{u} + \nabla_x q = \nabla_x \cdot \boldsymbol{\sigma}, \qquad \nabla_x \cdot \mathbf{u} = 0. \tag{3.36}$$

We consider two broad categories of swimmers: Pushers and Pullers. A Pusher, whose motion is actuated along the posterior of the body, served as motivation in Example 4. In contrast, the motion of a Puller is actuated along the anterior of the body, resulting in an oppositely signed extra-stress contribution (while moving in the same direction). Bacteria such as *B. subtilis* might be described as Pushers, while algae such as *Chlamydomonas* might be roughly described as Pullers. Figure 3.9 illustrates the flow lines for Pushers and Pullers and shows schematic pictures. We remark that there are other swimming microorganisms, such as the multicellular algae *Volvox* and densely covered ciliates, that may not fall into these categories. Following the discussion in Section 3.2, the extra stress for a Pusher is $\boldsymbol{\sigma} = -\sigma_0 \int dS_p(\mathbf{p}\mathbf{p}^{\mathrm{T}} - \mathbf{I}/3)\Psi$ (see also eqn (3.23)); for a Puller, $\boldsymbol{\sigma}$ is identical in magnitude but oppositely signed.

To nondimensionalize eqns (3.33)–(3.36), we choose the rescaling $x \to l_c\mathbf{x}$, $\mathbf{u} \to U\mathbf{u}$, $t \to (l_c/U)t$, and $\Psi \to n\Psi$. Here l_c is an intrinsic length scale to be determined. With the appropriate rescaled pressure, we obtain the following for the Smoluchowski equation and particle fluxes:

$$\frac{\partial \Psi}{\partial t} = -\nabla_x \cdot (\dot{\mathbf{x}}\Psi) - \nabla_p \cdot (\dot{\mathbf{p}}\Psi), \tag{3.37}$$

$$\dot{\mathbf{x}} = \mathbf{p} + \mathbf{u} - D\nabla_x \ln \Psi, \tag{3.38}$$

$$\dot{\mathbf{p}} = (\mathbf{I} - \mathbf{p}\mathbf{p}^{\mathrm{T}})\nabla_x \mathbf{u}\,\mathbf{p} - d\nabla_p \ln \Psi. \tag{3.39}$$

The forced Stokes equations become

$$-\Delta \mathbf{u} + \nabla q = \nabla \cdot \boldsymbol{\Sigma}^a, \qquad \nabla_x \cdot \mathbf{u} = 0, \tag{3.40}$$

$$\boldsymbol{\Sigma}^a = \alpha \int dV_x \int dS_p\, (\mathbf{p}\mathbf{p}^{\mathrm{T}} - \mathbf{I}/3)\Psi. \tag{3.41}$$

The nondimensional constants are $D = D_p/(l_c U_0)$, $d = d_p l_c/U_0$, $\tilde{L} = L/l_c$, and $\alpha = -(l_c\sigma_0 n)/(\mu U_0)$.

To determine l_c, we rewrite α using the expressions for σ_0 and U derived in Examples. 7 and 4, respectively, $\sigma_0 = \kappa_1 l^3 g$ and $U = \kappa_2 lg/\mu$. With these substitutions and $\nu = (Nl^3)/L^3$ for the volume concentration, we obtain $\alpha = -(l_c\nu/l)(\kappa_1/\kappa_2)$. Choosing $l_c = l/\nu$ turns α into a purely geometric constant, $\alpha = -\kappa_1/\kappa_2$.

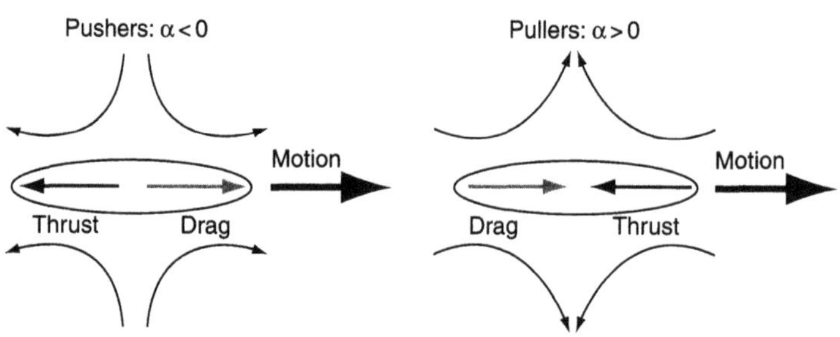

Fig. 3.9 Flow lines for Pushers ($\alpha < 0$) and Pullers ($\alpha > 0$).

In their particle simulations of rod-like swimmers, Saintillan and Shelley (2007) observed that at low volume concentrations (up to effective volume concentrations somewhat greater than $\tilde{\nu} = \nu/8 = 1$) $d_p \approx \nu \bar{d_p}$ and $D_p \approx \nu^{-1}\bar{D_p}$. If we assume these observed scalings of d_p and D_p with ν, the nondimensional diffusion coefficients become

$$d = \frac{l\bar{d_p}}{U_0} \quad \text{and} \quad D = \frac{\bar{D_p}}{lU_0}.$$

That is, d and D depend only upon the speed and length, of the swimmer, and the system size and the volume concentration of swimmers appear only in the normalized system size $\tilde{L}$.

We are interested in the nonlinear dynamics of this system, of partial differential equations, and its stability around simple steady states. Before getting into the specifics, we introduce the configurational entropy, which is a natural measure of the fluctuations in this system. The relative configurational entropy is defined as $\mathcal{S} = \int dV_x \int dS_p \, (\Psi/\Psi_0)\ln(\Psi/\Psi_0)$, where Ψ_0 is the constant value taken by Ψ when the system is uniform and isotropic (no concentration or orientation fluctuations). This quantity has the following properties:

1. *$\mathcal{S} \geq 0$, and $\mathcal{S} = 0$ if and only if $\Psi \equiv \Psi_0$. That is, a nonzero $\mathcal{S}$ measures the size of the fluctuations away from uniform isotropy, realized when $\Psi = \Psi_0$.*
2. *The entropy evolves via*

$$\Psi_0 \dot{\mathcal{S}} = -\frac{6}{\alpha}\int dV_x \, \mathbf{E} : \mathbf{E} - \int dS_p \int dV_x \Psi \left[D|\nabla_x \ln \Psi|^2 + d|\nabla_p \ln \Psi|^2\right],$$

where $\mathbf{E}$ is the symmetric rate-of-strain tensor. The first term on the right-hand side is proportional to the rate of viscous dissipation, and the second is strictly negative and reflects diffusional processes which serve to drive $\mathcal{S}$ towards its minimum (the uniform and isotropic state). The scalings of the system used here give $\Psi_0 = 4\pi$.

We omit the proofs, which are straightforward, of these statements.

The configurational entropy and its evolution establish the uniform isotropic state as a natural steady state to be examined for this system. Given their relevance to the study of biological flocking, the stability of aligned suspensions has also been examined, by Simha and Ramaswamy (2002) and by Saintillan and Shelley (2007, 2008*a*,*b*), among others. These authors find that aligned suspensions are generically unstable, particularly in the absence of diffusional processes, regardless of swimmer type (Pusher or Puller). Here we focus on the stability of the uniform isotropic state, which relates naturally to measures of fluctuations in the system, and whose analysis reveals strong differences depending on swimmer type and geometry.

We now turn our attention to the details of the stability of a uniform, isotropic suspension of active swimmers. First, some general conclusions can be drawn from Property 2 of the configurational entropy. For suspensions of Pullers, where $\alpha > 0$, we have $\dot{\mathcal{S}} < 0$, and fluctuations away from the uniform, isotropic state are expected to decay. On the other hand, for Pushers ($\alpha < 0$), the leading term is now positive, which allows the possibility of fluctuation growth and eventual balance with diffusion. Furthermore, if there is no diffusion, fluctuations should grow. These issues were analyzed in a detailed study of a system linearized near uniform isotropy (Hohenegger and Shelley 2010).

Let $\Psi = 1/(4\pi)\,(1+\varepsilon\psi)$, $\mathbf{v} = \varepsilon\mathbf{u}$, and $q \to \epsilon q$ with $\varepsilon \ll 1$. Keeping only linear-order terms and using the identity $\nabla_p \cdot (f_1\hat{\theta} + f_2\hat{\phi}) = (1/\sin\phi)(\partial_\theta f_1 + \partial_\phi(\sin\phi f_2))$, eqns (3.37)–(3.41) become

$$\psi_t + \mathbf{p}\cdot\nabla_x\psi - 3\mathbf{p}\mathbf{p}^{\mathrm{T}} : \nabla_x\mathbf{u} = D\nabla_x^2\psi + d\nabla_p^2\psi, \tag{3.42}$$

$$-\Delta_x\mathbf{u} + \nabla_x q = \nabla_x\cdot\boldsymbol{\sigma}, \qquad \nabla_x\cdot\mathbf{u} = 0, \tag{3.43}$$

where $\boldsymbol{\sigma} = (\alpha/4\pi)\int dS_p\,\psi\,(\mathbf{p}\mathbf{p} - \mathbf{I}/3)$.

The first step of the stability analysis is to transform the equations using a spatial Fourier transform in $\mathbf{x}$, $\tilde{f}_k = \int dV\,e^{-i\mathbf{k}\cdot\mathbf{x}} f(\mathbf{x})$, to decouple eqns (3.42) and (3.43) in the wave vector $\mathbf{k}$ ($k = |\mathbf{k}|$, $\mathbf{k} = k\hat{\mathbf{k}}$). From eqn (3.43), $\tilde{\mathbf{u}}_k = (i/k)(\mathbf{I} - \hat{\mathbf{k}}\hat{\mathbf{k}}^{\mathrm{T}})\cdot\tilde{\boldsymbol{\sigma}}_k\cdot\hat{\mathbf{k}}$, and eqn (3.42) then becomes

$$\partial_t\tilde{\psi}_k = -\,ik\hat{\mathbf{k}}\cdot\mathbf{p}\,\tilde{\psi}_k - Dk^2\tilde{\psi}_k + d\nabla_p^2\tilde{\psi}_k - 3\mathbf{p}\mathbf{p}^{\mathrm{T}} : (\mathbf{I} - \hat{\mathbf{k}}\hat{\mathbf{k}}^{\mathrm{T}})\cdot\tilde{\boldsymbol{\sigma}}_k\cdot\hat{\mathbf{k}}\hat{\mathbf{k}}. \tag{3.44}$$

The explicit dependence of eqn (3.44) on the direction of $\hat{\mathbf{k}}$ can be removed by a rotation, $\hat{\mathbf{k}} = \mathbf{R}\mathbf{z}$, which defines $\mathbf{q}$ through $\mathbf{p} = \mathbf{R}\mathbf{q}$. Let $\theta \in [0, 2\pi]$ be the azimuthal angle and let $\phi \in [0, \pi]$ be the polar angle on the unit $\mathbf{q}$-sphere. Equation (3.44) then becomes

$$\partial_t\tilde{\psi}_k = -\,ik\cos\phi\,\tilde{\psi}_k - Dk^2\tilde{\psi}_k + d\nabla_q^2\tilde{\psi}_k - 3\cos\phi\,\mathbf{q}\cdot(\mathbf{I} - \hat{\mathbf{z}}\hat{\mathbf{z}}^{\mathrm{T}})\cdot\mathbf{R}^{\mathrm{T}}\cdot\tilde{\boldsymbol{\sigma}}_k\cdot\mathbf{R}\cdot\mathbf{z}.$$

A further decoupling in θ is achieved by introducing a Fourier series in θ, $\tilde{\psi}_k = \sum_n A_{n,k}(\phi,t)e^{in\theta}$. After some algebra, we find

$$\begin{aligned}\partial_t A_{n,k} + A_{n,k}\left(ik\cos\phi + Dk^2\right) + d\left(\frac{n^2}{\sin^2\phi}A_{n,k} - \frac{1}{\sin\phi}\partial_\phi(\sin\phi\,\partial_\phi A_{n,k})\right)& \\ = -\frac{3\alpha}{4}\cos\phi\sin\phi\,F[A_{n,k}]\delta_{n,1},&\end{aligned} \tag{3.45}$$

where the scalar operator F is

$$F[h] = \int_0^\pi d\phi'\;h(\phi')\sin^2\phi'\cos\phi'$$

and $\delta_{n,1}$ is the Kronecker delta symbol.

The main feature of eqn (3.45) is that the entire stability is controlled by the first azimuthal mode on the sphere.

We now look for exponential solutions to eqn (3.45) in the special case when $D = d = 0$ and $\alpha < 0$ (Pushers). Hence we assume that $A_{1,k}(\phi,t) = \gamma_k(\phi)e^{\sigma t}$ and note that there is no a priori expectation that $A_{1,k}$ can be represented in this fashion, as eqn (3.45) depends explicitly on ϕ. Nonetheless, inserting the exponential ansatz into eqn (3.45) and applying F to both sides yields the eigenvalue relation

$$-\frac{3}{4}\alpha \int_0^{\pi} d\phi' \, \frac{\sin^3 \phi' \cos^2 \phi'}{\sigma + ik \cos \phi'} = 1. \tag{3.46}$$

This complex-valued integral can be evaluated by substitution and separation into real and imaginary parts. Some algebra produces the complex equation

$$-\alpha \left[4i\sigma k^3 + 6i\sigma^3 k - 3\sigma^2(\sigma^2 + k^2) \ln \frac{i\sigma - k}{i\sigma + k} \right] = 4ik^5, \tag{3.47}$$

where the complex logarithm is defined as $\ln(a + ib) = \ln \sqrt{a^2 + b^2} + i \arctan(b/a)$. Note that eqn (3.47) is valid only if $Re(\sigma) \neq 0$. Figure 3.10 shows a numerical solution of eqn (3.47), plotting the real and imaginary parts of σ for $\alpha = -1$. For small k, there are two branches of unstable eigenvalues with zero imaginary part, and the decrease in growth rate suggests a crossing of the k-axis to become negative at medium k. Saintillan and Shelley (2008*b*) gave an asymptotic solution for $k \ll 1$ for the upper branch in Fig. 3.10,

$$\sigma(k) = -\frac{\alpha}{5} + \left[\frac{15}{7\alpha} - D \right] k^2 + O(k^3). \tag{3.48}$$

For a suspension of Pushers ($\alpha < 0$), eqn (3.48) implies the existence of a long-wave instability with $\lim_{k \to 0} \sigma(k) = -\alpha/5 > 0$, corresponding to the upper branch in Fig. 3.10. In contrast, for Pullers, there is no long-wave instability.

One of the striking features of the analysis of the eigenvalue problem (Hohenegger and Shelley 2010) is that there is no crossing to negative k, contrary to the intuition gathered from Fig. 3.10. As a matter of fact, a singular behavior develops in the corresponding eigenvector as k approaches k_1, the zero-growth-rate value. The singular

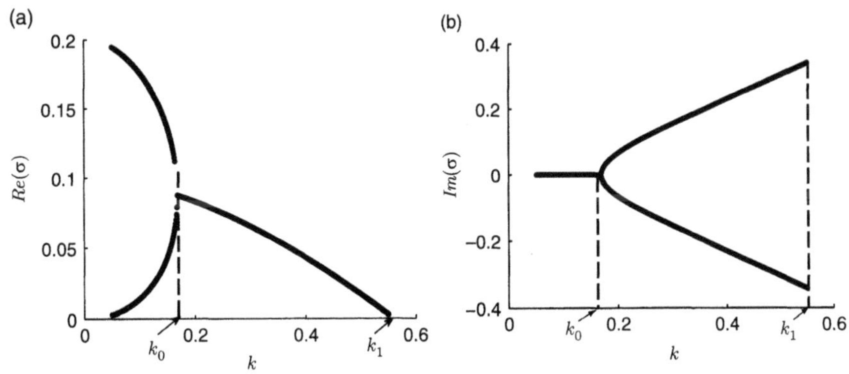

Fig. 3.10 Real and imaginary parts of the growth rate $\sigma(k)$ for $\alpha = -1$ (Pushers) with $D = d = 0$. Reproduced with permission from Hohenegger and Shelley (2010).

behavior results from a discontinuous pole singularity in the integral in eqn (3.46) as $Re(\sigma)$ approaches 0.

Finally, we remark that if $k = k_1$ is indeed a point of stability transition, then there exists a critical system size L or volume concentration ν above which the system becomes unstable. This follows from having scaled by the intrinsic length l_c, so that $k_1' = (k_1/2\pi)(\nu L/l)$. As the first allowable mode in the periodic box has $k' = 1$, the system is unstable if $k_1' \approx 0.089(\nu L/l) > 1$, and is stable otherwise.

As just highlighted, the eigenvalue analysis does not provide solutions for $k > k_1$. The dynamics in these cases was found numerically and was confirmed by a large k analysis (Hohenegger and Shelley 2010). Figure 3.11 illustrates these numerically determined solutions in the absence of diffusion. In each case, the initial condition is $A_{1,k}(\phi, 0) = \sin(\phi)$. In Fig. 3.11, the real and imaginary parts of $A_{1,k}$ for $k = 0.4$, $k = 0.8$, and $k = 10$ are plotted at $t = 50$ and $t = 100$ for Pushers ($\alpha = -1$). For $k = 0.4$, the real and imaginary parts (Fig. 3.11(a)) grow as predicted by the eigenvalue analysis. The values $k = 0.8$ and $k = 10$ are out of the range of exponential growth and both the real and the imaginary parts of the solution show oscillations. The number of oscillations increases with time, as illustrated with Figs. 3.11(b) and (c), and Fig. 3.11(d) shows that for large k, the envelope of the oscillations is determined by the initial condition (and is as determined analytically by Hohenegger and Shelley 2010).

We expect rotational diffusion (i.e. $d > 0$) to remove the singular behavior in the eigenvalue problem, and to yield exponentially decaying solutions past a crossing value. To investigate this, we assume again that $A_{1,k}(\phi, t) = \gamma_k(\phi)e^{\sigma t}$ in eqn (3.45) and renormalize the problem so that $F[\gamma_k] = 1$. Upon substitution into eqn (3.45), this leads to the system

$$\gamma_k(\sigma + ik\cos\phi + Dk^2) + d\left(\frac{1}{\sin^2\phi}\gamma_k - \frac{1}{\sin\phi}\partial_\phi\left(\sin\phi\,\partial_\phi\gamma_k\right)\right) = -\frac{3\alpha}{4}\cos\phi\sin\phi, \tag{3.49}$$

$$\int_0^\pi d\phi' \gamma_k(\phi', t)\sin^2\phi\cos\phi = 1. \tag{3.50}$$

Equations (3.49) and (3.50) were solved by discretizing the ϕ-derivatives and solving the resulting nonlinear system via Newton iteration starting from the known solution for $d = 0$.

In Fig. 3.12, the real part of $\sigma(k)$ obtained from eqns (3.49) and (3.50) with $d = 0$ and $d = 0.01$ is plotted. For this moderate degree of rotational diffusion, a downward shift of the branches is seen, leaving the long-wave instability intact but suppressing the lower branch. Not only does rotational diffusion reduce growth rates, but it also suppresses the eigenfunction singularity. For a suspension of Pullers, there are no positive growth rates for $d \geq 0$, as would be expected from the generic decay of the entropy.

The theoretical considerations above allow a comparison with the rod simulations of Saintillan and Shelley (2007). First, we remark that for saintillan and shelley's rod model, we estimate $\alpha \approx -0.9$, close to the value of -1 that we have used in the theoretical study described above. Using values for $\bar{d}_p$ and $\bar{D}_p$ gleaned from

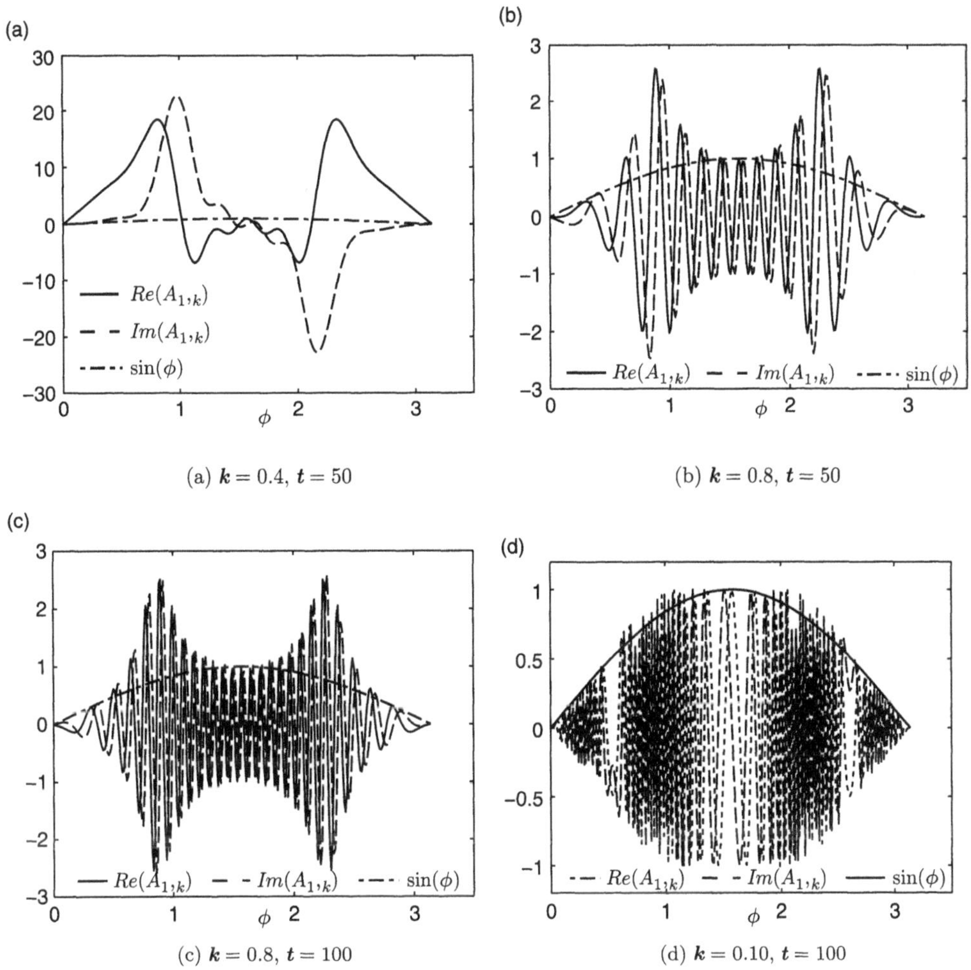

Fig. 3.11 Real and imaginary parts of the first Fourier modes $A_{1,k}(\phi, t)$ for $\alpha = -1$ (Pushers) and $D = d = 0$, from a single-mode sinusoidal initial condition. The increasing wavenumbers k show growth (top left), saturation in amplitude with an increasing number of oscillations in time (top right and bottom left), and convergence of the wave envelope to that of the initial condition. Reproduced with permission from Hohenegger and Shelley (2010).

Saintillan and Shelley (2007), we have solved the eigenvalue problem of eqns (3.49) and (3.50). This yields the crossing value $k_1 \approx 0.086$, and hence instability is found if $k_1' = (k_1/2\pi)(\nu L_p/l) > 1$. Taking $l = 1$ and $L_p = 10$, this analysis predicts the existence of a critical volume concentration $\tilde{\nu} = 0.9$ ($\tilde{\nu} = \nu/8$) for the emergence of a long-wave instability. This is consistent with the result of Saintillan and Shelley (2007) that organized dynamics emerges at volume concentrations in the neighborhood of $\tilde{\nu} = 0.5$.

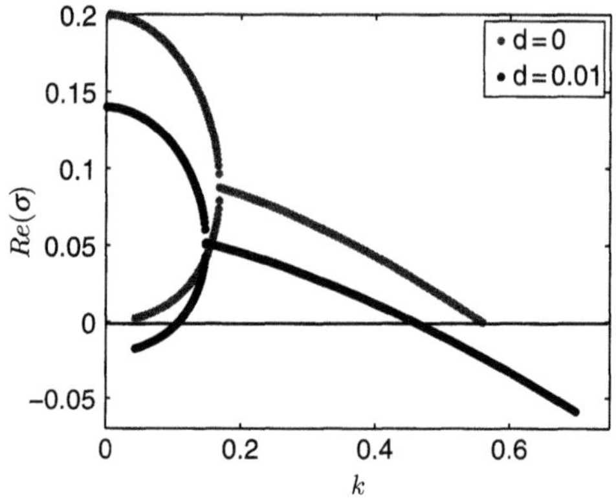

Fig. 3.12 Continuation of the real part of the growth rate $\sigma(k)$ for positive rotational diffusion $d = 0.01$, $\alpha = -1$ (Pushers), and $D = 0$. Reproduced with permission from Hohenegger and Shelley (2010).

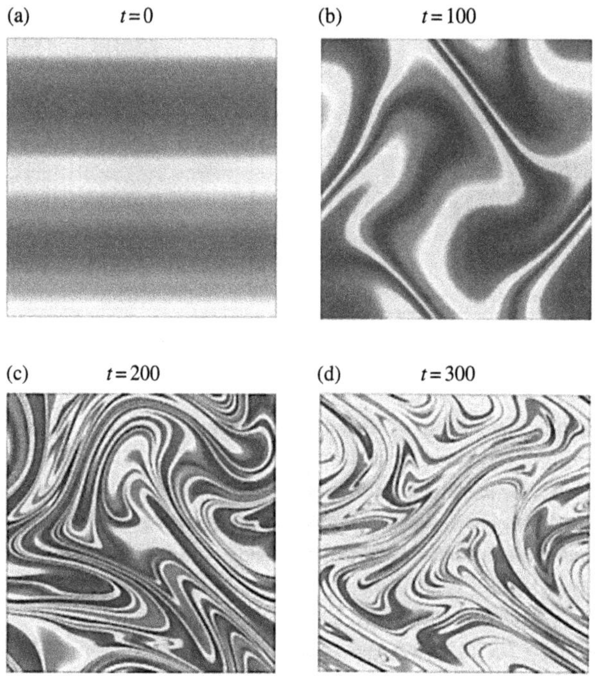

Fig. 3.13 Fluid mixing by an active suspension of Pushers ($\alpha = -1$). The gray levels show the configuration of a passive scalar field in the suspension at different times. Reproduced with permission from Saintillan and Shelley (2008*b*).

Using kinetic theory, Saintillan and Shelley (2008*a*,*b*) simulated this instability starting from uniform isotropy in the special case of two-dimensional flows. Figure 3.13 shows the evolution away from uniform isotropy through the dynamics of a scalar field being advected by the background velocity field. The fully developed nonlinear dynamics is quasi-periodic, has persistent concentration fluctuations (not predicted by linear theory), and yields efficient mixing of the scalar field through repeated folding and stretching of fluid elements. It was also found that the system reached a state of statistical equilibrium where the growth of configurational entropy saturates and where fluctuation growth is balanced, on average, by diffusional processes.

3.5 Conclusions

In our lectures and in these notes, we have reviewed the basics of non-Newtonian fluid mechanics and established formulae for the extra-stress contribution to the macroscopic stress tensor, in the particular cases of dumbbell- and- rod shaped particles. We discussed the derivation and properties of the Oldroyd-B model. We then focused on pumping and swimming in viscoelastic fluids as important biological examples. Both of these problems concern complex fluid–body interactions and required the development and use of sophisticated numerical methods (which we did not discuss in any detail). We then developed and analyzed the continuum theory of active particle suspensions put forward by Saintillan and Shelley (2008a,b), focusing on its structure near a state of uniform isotropy.

There are of course many other fascinating problems which could have been discussed if time had allowed. Among many such examples are the study of synthetic swimmers (e.g. Zerrouki *et al.* 2008; Keaveny and Shelley 2009), and microfluidic pumps produced by carpets of bacteria bound to channel walls (Kim and Breuer 2008).

Similarly, there are many directions in which the problems discussed here can evolve. We conclude by briefly mentioning some of these. Peristaltic pumping was shown to be hindered by viscoelasticity. Therefore a natural question to ask pertains to the optimal wave shape that minimizes the input power. Walker and Shelley (2010) developed a shape optimization algorithm for the Newtonian case, but the extension to a viscoelastic fluid is open. The continuum model of swimming rods gave rise to complex large-scale behavior, but the linearized analysis around uniformity and isotropy did not show growth in concentration fluctuations (on the contrary, these decayed). Furthermore, it is believed that in bacterial baths, other effects such as chemotaxis or oxygen-taxis (Sokolov *et al.* 2007), which were not included here, can play a significant role. Finally, motile microorganisms such as bacteria also thrive and move in other yet more complicated media, such as gel-like soft tissues and mucus.

Acknowledgments

The authors thank Saverio E. Spagnolie for useful comments. We are grateful for support from the Department of Energy and the National Science Foundation, as well as from the NYU-MRSEC Center.

References

Arratia P. E., Thomas C. C., Diorio J., and Gollub J. P. (2006). Elastic instabilities of polymer solutions in cross-channel flow. *Physical Review Letters*, **96**, 144502.

Batchelor G. K. (1970*a*). The stress system in a suspension of force-free particles. *Journal of Fluid Mechanics*, **41**, 545–570.

Batchelor G. K. (1970*b*). Slender-body theory for particles of arbitrary cross-section in Stokes flow. *Journal of Fluid Mechanics*, **44**, 419–440.

Batchelor G. K. (1977). The effect of Brownian motion on the bulk stress in a suspension of spherical particles. *Journal of Fluid Mechanics*, **83**, 97–117.

Bird R. B., Armstrong R. C., and Hassager O. (1987). *Dynamics of Polymeric Fluids: Kinetic Theory*. Wiley.

Doi M. and Edwards S. F. (1986). *The Theory of Polymer Dynamics*. Oxford University Press.

Dombrowski C., Cisneros L., Chatkaew S., Goldstein R. E., and Kessler J. O. (2004). Self-concentration and large-scale coherence in bacterial dynamics. *Physical Review Letters*, **93**, 098103.

Fauci L. J. and Peskin C. S. (1988). A computational model of aquatic animal locomotion. *Journal of Computational Physics*, **77**, 85–108.

Fauci L. J. and Dillon R. (2006). Biofluidmechanics of reproduction. *Annual Review of Fluid Mechanics*, **38**, 371–394.

Fu H. C., Powers T. R., and Wolgemuth C. W. (2007). Theory of swimming filaments in viscoelastic media. *Physical Review Letters*, **99**, 258101.

Fu H. C., Wolgemuth C. W., and Powers T. R. (2009). Swimming speeds of filaments in nonlinearly viscoelastic fluids. *Physics of Fluids*, **21**, 033102.

Groisman A. and Quake S. R. (2004). A microfluidic rectifier: Anisotropic flow resistance at low Reynolds numbers. *Physical Review Letters*, **92**, 094501.

Groisman A. and Steinberg V. (2001). Efficient mixing at low Reynolds numbers using polymer additives. *Nature*, **410**, 905–908.

Groisman A. and Steinberg V. (2004). Elastic turbulence in curvilinear flows of polymer solutions. *New Journal of Physics*, **6**, 29.

Hernández-Ortiz J. P., Stolz C. G., and Graham M. D. (2005). Transport and collective dynamics in suspensions of confined swimming particles. *Physical Review Letters*, **95**, 204501.

Hohenegger C. and Shelley M. J. (2010). Stability of active suspensions. *Physical Review E*, **81**, 046311.

Jaffrin M. and Shapiro A. (1971). Peristaltic pumping. *Annual Review of Fluid Mechanics*, **3**, 13.

Jeffery G. B. (1922). The motion of ellipsoidal particles immersed in a viscous fluid. *Proceedings of the Royal Society of London*, **102**, 161–179.

Johnson R. E. (1980). An improved slender-body theory for Stokes-flow. *Journal of Fluid Mechanics*, **99**, 411–431.

Keaveny E. E. and Shelley M. J. (2009). Hydrodynamic mobility of chiral colloidal aggregates. *Physical Review E*, **79**, 051405.

Keller J. B. and Rubinow S. I. (1976). Slender-body theory for slow viscous-flow. *Journal of Fluid Mechanics*, **75**, 705–714.

Kim M. J. and Breuer K. S. (2008). Self-organizing bacteria-powered microfluidic pump. *Small*, **4**, 111–118.

Larson R. G. (1995). *The Structure and Rheology of Complex Fluids*. Oxford University Press.

Lauga E. (2007). Propulsion in a viscoelastic fluid. *Physics of Fluids*, **19**, 081304.

Lighthill M. J. (1960). Note on the swimming of slender fish. *Journal of Fluid Mechanics*, **9**, 305–317.

Paxton W. F., Kistler K. C., Olmeda C. C., Sen A., St. Angelo S. K., Cao Y., Mallouk T. E., Lammert P. E., and Crespi V. H. (2004). Catalytic nanomotors: Autonomous movement of striped nanorods. *Journal of the American Chemical Society*, **126**, 13424.

Peskin C. (2002). The immersed boundary method. *Acta Numerica*, **11**, 479.

Pozrikidis C. (1992). *Boundary Integral and Singularity Methods for Linearized Viscous Flow*. Cambridge University Press.

Purcell E. M. (1977). Life at low Reynolds number. *American Journal of Physics*, **45**, 11.

Saintillan D. and Shelley M. J. (2007). Orientational order and instabilities in suspensions of self-locomoting rods. *Physical Review Letters*, **99**, 058102.

Saintillan D. and Shelley M. J. (2008*a*). Instabilities and pattern formation in active particle suspensions: Kinetic theory and continuum simulations. *Physical Review Letters*, **100**, 178103.

Saintillan D. and Shelley M. J. (2008*b*). Instabilities, pattern formation, and mixing in active suspensions. *Physics of Fluids*, **20**, 123304.

Simha R. A. and Ramaswamy S. (2002). Hydrodynamic fluctuations and instabilities in ordered suspensions of self-propelled particles. *Physical Review Letters*, **89**, 58101.

Smith D. J, Gaffney E. A., Gadêlha H., Kapur N., and Kirkman-Brown J. C. (2009). Bend propagation in the flagella of migrating human sperm, and its modulation by viscosity. *Cell Motility and the Cytoskeleton*, **66**, 220–236.

Sokolov A., Aranson I. S., Kessler J. O., and Goldstein R. E. (2007). Concentration dependence of the collective dynamics of swimming bacteria. *Experiments in Fluids*, **43**, 737.

Spagnolie S. E. and Lauga E. (2010). The optimal elastic flagellum. *Physics of Fluids*, **22**, 031901.

Tam D. and Hosoi A. E. (2007). Optimal stroke patterns for Purcell's three-link swimmer. *Physical Review Letters*, **98**, 68105.

Taylor G . I. (1951). Analysis of the swimming of microscopic organisms. *Proceedings of the Royal Society of London, Series A, Mathematical and Physical Sciences*, **209**, 447–461.

Teran J., Fauci L. J., and Shelley M. J. (2008). Peristaltic pumping and irreversibility of a Stokesian viscoelastic fluid. *Physics of Fluids*, **20**, 073101.

Teran J., Fauci L. J., and Shelley M. J. (2010). Viscoelastic fluid response can increase the speed and efficiency of a free swimmer. *Physical Review Letters*, **104**, 38101.

Thomases B. and Shelley M. J. (2007). Emergence of singular structures in Oldroyd-B fluids. *Physics of Fluids*, **19**, 103103.

Thomases B. and Shelley M. J. (2009). Transition to mixing and oscillations in a Stokesian viscoelastic flow. *Physical Review Letters*, **103**, 094501.

Tuval I., Cisneros L., Dombrowski C., Wolgemuth C. W., Kessler J. O., and Goldstein R. E. (2005). Bacterial swimming and oxygen transport near contact lines. *Proceedings of the National Academy of Sciences*, **102**, 2277.

Underhill P. T., Hernández-Ortiz J. P., and Graham M. D. (2008). Diffusion and spatial correlations in suspensions of swimming particles. *Physical Review Letters*, **100**, 248101.

Walker S. W. and Shelley M. J. (2010). Shape optimization of peristaltic pumping. *Journal of Computational Physics*, **229**, 1260–1291.

Young Y. N. and Shelley M. J. (2007). Stretch-coil transition and transport of fibers in cellular flows. *Physical Review Letters*, **99**, 58303.

Zerrouki D., Baudry J., Pine D., Chaikin P., and Bibette J. (2008). Chiral colloidal clusters. *Nature*, **455**, 380–382.

Part II

Biogels

4
Active fluids and gels

Frank Jülicher

Max Planck Institute for the Physics of Complex Systems, Dresden, Germany

4.1 Introduction

Cells and tissues are a form of soft condensed matter of extraordinary complexity. The most striking property of cellular systems is their inherent dynamics. Eukaryotic cells posess a cytoskeleton. The cell cytoskeleton is an active network of elastic filaments, which interact with motor molecules and many other associated molecules to organize themselves into gel-like materials or complex structures (Alberts *et al.* 2002). It provides cells with their material properties and is responsible for active and dynamic processes such as cell division and cell locomotion (Bray 2001). Cells are driven far from thermodynamic equilibrium by active molecular processes such as the action of motor proteins which transduce the chemical energy of the hydrolysis of adenosine triphosphate (ATP) to generate motion and mechanical work (Howard 2001).

The filaments of the cytoskeleton are rod-like elastic objects with a vectorial asymmetry. This polarity of the filaments defines the directionality of active processes such as the generation of forces by motor molecules. Complex dynamic behaviors such as dynamic patterns and contractile behaviors can emerge from the interactions of many motors and filaments in the cytoskeleton (Nédélec *et al.* 1997; Surrey *et al.* 1998; Kruse and Jülicher 2000; Nédélec *et al.* 2001; Kruse *et al.* 2001; Liverpool and Marchetti 2003; Ahmadi *et al.* 2005; Zumdieck *et al.* 2005; Aronson and Tsimring 2005; Salbreux *et al.* 2009).

On a coarse-grained scale, the cytoskeleton behaves as a complex fluid with visco-elastic properties (Bird and Armstrong 1987; Larson 1998). Because of the active processes on the molecular scale, this material is active and exhibits novel material properties such as spontaneous flows and active stresses. Such active behaviors do not exist in conventional gels in thermodynamic equilibrium. Furthermore, filament systems are anisotropic if the filaments tend to align locally on average to produce liquid crystalline order (De Gennes and Prost 1993). For example, if the filaments align along an axis, this can give rise to nematic order. Since the filaments are polar, they can align their polarity on average in the same direction and thereby generate polar order.

The large-scale properties of active fluids and gels can be characterized in a hydrodynamic limit by constitutive material equations (Martin *et al.* 1972). The behaviors in this hydrodynamic limit are very general and apply to active filament networks such as the cell cytoskeleton (Kruse *et al.* 2004, 2005, Voituriez *et al.* 2005, 2006, Joanny *et al.* 2007; Basu *et al.* 2008; Salbreux *et al.* 2009). Note that the hydrodynamic equations of an active fluid are generic and could also describe very different situations, ranging from driven granular materials (Aronson and Tsimring 2006) to suspensions of active particles (e.g. microswimmers) (Simha and Ramaswamy 2002) and flocks (Toner and Tu 1995). The active behaviors of tissues that consist of many cells and are dynamically remodeled also represent examples of active gels and fluids (Bittig *et al.* 2008, 2009, Basan *et al.* 2009).

In the following sections, we outline the general hydrodynamic theory of active polar fluids and gels (Kruse *et al.* 2004, 2005,Voituriez *et al.* 2005, 2006, Joanny *et al.* 2007; Basu *et al.* 2008). We start from a fluid in thermal equilibrium and use concepts from nonequilibrium thermodynamics to arrive at hydrodynamic equations (De Groot and Mazur 1984). We focus on the slow, or hydrodynamic, variables in the system and identify conjugate thermodynamic fluxes and forces. In linear response theory, the fluxes can be expressed using dissipative and reactive coupling coeficients. Together with conservation laws, the full hydrodynamic equations can be written. We discuss simple situations to illustrate the novel physics that emerges in such systems. Finally, we close with a brief discussion of biological examples for which such approaches are relevant.

4.1.1 Nonequilibrium thermodynamics

For simplicity, we consider isothermal systems which are in contact with a thermostat which keeps the system at a temperature T. We discuss the dynamics of systems close to thermodynamic equilibrium, following the general principles of irreversible thermodynamics (De Groot and Mazur 1984). In order to discuss the dynamics of slow modes, we divide the degrees of freedom of the system into two groups. There are many fast modes ψ_i, with $i = 1, \ldots, n$, which quickly relax and are in thermodynamic equilibrium on the timescales of interest. In addition, we consider the dynamics of a smaller number $N + \bar{N}$ of slowly relaxing modes ϕ_k, $k = 1, \ldots, N$, and $\bar{\phi}_l$, $l = 1, \ldots, \bar{N}$. Here we have introduced two sets of slow variables, with different signatures under time-reversal transformations. The variables ϕ_k transform under time inversion as $\phi_k(t) \to \phi_k(-t)$, i.e. their behavior is even under time reversal, with a positive signature $\epsilon = 1$. In contrast, the variables $\bar{\phi}_l$ transform as $\bar{\phi}_k(t) \to -\phi_k(-t)$, i.e. with $\epsilon = -1$ and thus an odd signature. Examples of the variables ϕ_k with time-reversal signature $+1$ are displacements, stresses, and chemical potentials. Examples of the variables $\bar{\phi}_k$ with signature -1 are momenta, velocities, and velocity gradients.

The behavior of the slow modes is then described using principles from thermodynamics together with the idea that the modes ψ_i are equilibrated. For fixed values of ϕ_k and $\bar{\phi}_l$, we can define the effective free energy

$$F(\phi_k, \bar{\phi}_l) = -k_{\mathrm{B}} T \ln \int d\psi_1 \ldots d\psi_n \, e^{-\beta H(\psi_i, \phi_k, \bar{\phi}_l)}. \tag{4.1}$$

As the slow variables change dynamically, the total change in entropy is $\Delta S = -\Delta F/T$. The rate of entropy production $\dot{S}$ therefore obeys

$$T\dot{S} = -\dot{F} = \sum_{k=1}^{N} f_k \frac{d\phi_k}{dt} + \sum_{l=1}^{\bar{N}} \bar{f}_l \frac{d\bar{\phi}_l}{dt} > 0, \tag{4.2}$$

where the dot denotes a time derivative. According to the second law of thermodynamics, $T\dot{S}$ must be positive definite. Here we have introduced the generalized thermodynamic forces $f_k = -\partial F/\partial \phi_k$ and $\bar{f}_l = -\partial F/\partial \bar{\phi}_l$. Note that $d\phi_k/dt$ is odd ($\epsilon = -1$) and $d\bar{\phi}_k/dt$ even ($\epsilon = 1$) under time reversal, which is the opposite to the behavior of ϕ_k and $\bar{\phi}_k$. Furthermore, since F is even under time reversal (it describes a reversible equilibrium), one can see that f_k is even ($\epsilon = 1$) and $\bar{f}_k$ is odd ($\epsilon = -1$). Clearly, $\dot{S}$ is odd under time reversal.

In thermodynamic equilibrium, F is minimized, S is maximized, $f_k = 0$, and $\dot{\phi}_k = 0$. In linear response theory, the fluxes are linear functions of the forces:

$$\frac{d\phi_k}{dt} = L_{kl}^{++} f_l + L_{kl}^{+-} \bar{f}_l + \eta_k, \tag{4.3}$$

$$\frac{d\bar{\phi}_k}{dt} = L_{kl}^{-+} f_l + L_{kl}^{--} \bar{f}_l + \bar{\eta}_k. \tag{4.4}$$

Here we use a summation convention and sum over paired indicies. Furthermore, we have introduced random terms η_k and $\bar{\eta}_k$ with $\langle \eta_k \rangle = 0$ and $\langle \bar{\eta}_k \rangle = 0$ that describe thermal fluctuations due to the equilibrated fast degrees of freedom. The coefficients L^{++} and L^{--} describe dissipative or irreversible processes which contribute to entropy production. They are linked to the variance of the thermal noise terms by generalized Einstein relations: $\langle \eta_k \eta_l \rangle = 2k_\mathrm{B}TL_{kl}^{++}$, $\langle \bar{\eta}_k \bar{\eta}_l \rangle = 2k_\mathrm{B}TL_{kl}^{--}$, and $\langle \eta_k \bar{\eta}_l \rangle = 0$. Because of these relations, the dissipative coefficients obey the Onsager symmetry relations

$$L_{kl}^{++} = L_{lk}^{++} \quad \text{and} \quad L_{kl}^{--} = L_{lk}^{--}. \tag{4.5}$$

The cross-coupling coefficients L_{kl}^{+-} and L_{kl}^{-+} are reactive, i.e. they do not contribute to entropy production. As a consequence, they are antisymmetric:

$$L_{kl}^{+-} = L_{lk}^{-+}. \tag{4.6}$$

We can introduce the dissipative and reactive parts of the thermodynamic fluxes as $\phi_k = \phi_k^\mathrm{d} + \phi_k^\mathrm{r}$ and $\bar{\phi}_k = \bar{\phi}_k^\mathrm{d} + \bar{\phi}_k^\mathrm{r}$, with $\phi_k^\mathrm{d} = L_{kl}^{++} f_l$, $\phi_k^\mathrm{r} = L_{kl}^{+-} \bar{f}_l$, and correspondingly for $\bar{\phi}_k^\mathrm{d}$ and $\bar{\phi}_k^\mathrm{r}$.

In the following, we will discuss the hydrodynamic equations of active gels based on the reactive and dissipative phenomenological coefficients introduced here.

4.1.2 Hydrodynamics of a simple fluid

In order to clarify the basic concepts, we briefly review the hydrodynamics of a simple, one-component fluid. We first recall the basic thermodynamic variables by considering a homogeneous fluid in equilibrium. In the rest frame of the system (i.e. the center of mass is at rest), the free energy is a function $F_0(N, V, T)$, where V denotes the volume, and N molecules with molecular mass m are present in the system. The mass density is $\rho = Nm/V$. The free-energy density $f(n, T) = F_0(N, V, T)/V$ depends on the number density $n = N/V$ only. The pressure is defined as $P = -\partial F_0/\partial V|_{N,T}$, and the chemical potential is given by $\mu = \partial F_0/\partial N|_{V,T}$.

The intensive quantities P and μ can be expressed using intensive quantities only: $P = -f + \mu n$, $\mu = \partial f/\partial n|_T$. Introducing the Gibbs free energy $G = F + PV = \mu N$, one can obtain the Gibbs–Duhem relation,

$$dP = n\, d\mu, \tag{4.7}$$

or, equivalently, $n\, \partial_\alpha \mu = \partial_\alpha P$.

To obtain the hydrodynamic equations, we consider an extended system in which local volume elements are equilibrated but, globally, the system is not in equilibrium.

Two conservation laws govern the large-scale dynamics of the system. Mass conservation reads

$$\partial_t \rho + \partial_\alpha(\rho v_\alpha) = 0. \tag{4.8}$$

This defines the center-of-mass velocity v_α of local volume elements. In the absence of external bulk forces, momentum conservation can be written as

$$\partial_t g_\alpha = \partial_\beta \sigma_{\alpha\beta}^{\text{tot}}. \tag{4.9}$$

Here, $g_\alpha = \rho v_\alpha$ is the momentum density and the stress tensor $\sigma_{\alpha\beta}^{\text{tot}}$ describes the (negative) momentum flux in the system.

We now discuss entropy production in an isothermal fluid to identify the relevant fluxes and forces. The total free energy, which also includes kinetic energy, is given by

$$F = \int d^3x \left(f\left(\frac{\rho}{m}\right) + \frac{1}{2}\rho \mathbf{v}^2 \right). \tag{4.10}$$

Considering an isothermal system with no fluxes and no work at the boundaries, we have

$$T\dot{S} = -\dot{F} = -\int d^3x \, \frac{\partial}{\partial t} \left(f\left(\frac{\rho}{m}\right) + \frac{1}{2}\rho \mathbf{v}^2 \right). \tag{4.11}$$

Using the conservation laws and the Gibbs–Duhem relation, we find, after performing partial integrations (and ignoring boundary terms which vanish according to our assumptions),

$$T\dot{S} = \int d^3x \left(\sigma_{\alpha\beta}^{\text{tot}} + \rho v_\alpha v_\beta + P\delta_{\alpha\beta} \right) \partial_\alpha v_\beta. \tag{4.12}$$

In order to separate the relevant variables clearly, we write $\partial_\alpha v_\beta = v_{\alpha\beta} + \omega_{\alpha\beta}$, where we introduce the symmetric and antisymmetric parts of the velocity gradients,

$$v_{\alpha\beta} = \frac{1}{2}(\partial_\alpha v_\beta + \partial_\beta v_\alpha), \tag{4.13}$$

$$\omega_{\alpha\beta} = \frac{1}{2}(\partial_\alpha v_\beta - \partial_\beta v_\alpha). \tag{4.14}$$

We define the symmetric and antisymmetric parts $\sigma_{\alpha\beta}$ and $\sigma_{\alpha\beta}^{\text{a}}$, respectively, of the deviatoric stress $\sigma_{\alpha\beta} + \sigma_{\alpha\beta}^{\text{a}} = \sigma_{\alpha\beta}^{\text{tot}} + \rho v_\alpha v_\beta + P\delta_{\alpha\beta}$:

$$\sigma_{\alpha\beta} = \sigma_{\alpha\beta}^{\text{tot}} + \rho v_\alpha v_\beta + P\delta_{\alpha\beta} - \sigma_{\alpha\beta}^{\text{a}}, \tag{4.15}$$

$$\sigma_{\alpha\beta}^{\text{a}} = \frac{1}{2}(\sigma_{\alpha\beta}^{\text{tot}} - \sigma_{\beta\alpha}^{\text{tot}}). \tag{4.16}$$

For an isotropic fluid, angular-momentum conservation implies that the antisymmetric part of the stress $\sigma_{\alpha\beta}^{\text{a}}$ vanishes. Writing $\sigma_{\alpha\beta} = \tilde{\sigma}_{\alpha\beta} + \bar{\sigma}\delta_{\alpha\beta}$, where $\tilde{\sigma}_{\alpha\beta}$ is the traceless part of the stress and $3\bar{\sigma}$ is the trace, we finally arrive at

$$T\dot{S} = \int d^3x \, (\tilde{\sigma}_{\alpha\beta}\tilde{v}_{\alpha\beta} + \bar{\sigma}v_{\gamma\gamma}). \tag{4.17}$$

Here, we have introduced the traceless part

$$\tilde{v}_{\alpha\beta} = v_{\alpha\beta} - \frac{1}{3}v_{\gamma\gamma}\delta_{\alpha\beta} \tag{4.18}$$

of $v_{\alpha\beta}$. We identify the conjugate fluxes and forces for the simple fluid as

$$\frac{d}{dt}(\bar{\phi}_1)_{\alpha\beta} = \tilde{\sigma}_{\alpha\beta} \leftrightarrow (\bar{f}_1)_{\alpha\beta} = \tilde{v}_{\alpha\beta}, \tag{4.19}$$

$$\frac{d}{dt}\bar{\phi}_2 = \bar{\sigma} \leftrightarrow \bar{f}_2 = v_{\gamma\gamma}. \tag{4.20}$$

In linear response theory, the symmetric deviatoric stress can now be expressed in terms of velocity gradients. Since $\tilde{v}_{\alpha\beta}$ and $v_{\gamma\gamma}$ are both odd under time reversal ($\epsilon = -1$) and $\sigma_{\alpha\beta}$ is even ($\epsilon = +1$), only dissipative couplings exist. Introducing the shear viscosity η and the bulk viscosity $\bar{\eta}$, we have

$$\sigma_{\alpha\beta} = 2\eta\left(v_{\alpha\beta} - \frac{1}{3}v_{\gamma\gamma}\delta_{\alpha\beta}\right) + \bar{\eta}v_{\gamma\gamma}\delta_{\alpha\beta}. \tag{4.21}$$

This constitutive relation (4.21) is a linear response relation of the form (4.4) with coefficients $L_{11}^{--} = \eta$ and $L_{12}^{--} = L_{21}^{--} = \bar{\eta}$. Together with the force balance relation

$$\partial_\beta(\sigma_{\alpha\beta} - \rho v_\alpha v_\beta - P\delta_{\alpha\beta}) = \frac{\partial}{\partial t}(\rho v_\alpha), \tag{4.22}$$

eqn(4.21) leads to the Navier–Stokes equations of a compressible fluid. In the incompressible limit, with $v_{\gamma\gamma} = 0$ and neglecting all inertial terms, we obtain the Stokes equation

$$\eta \, \Delta v_\alpha = \partial_\alpha P. \tag{4.23}$$

This equation describes the flow of a viscous fluid at low Reynolds numbers.

4.1.3 Viscoelastic gels: The Maxwell model

Complex fluids and, in particular, polymer systems are viscoelastic in general. A shear deformation will lead at short times to an elastic response because of transient cross-linking of molecular components. As a consequence, the elastic stress relaxes within a characteristic relaxation time. The stress relaxation is due to the dynamics of the cross-linkers. For example, physical cross-links have a finite lifetime. Entanglements of polymers are resolved in a diffusion process (reptation); the corresponding relaxation time depends on the length of the polymer molecules. In the cytoskeleton, stress relaxation can also be due to filament turnover or to the action of motors, which can disentangle filaments by directed motion along other filaments. Because of stress relaxation, such systems behave, when sheared at a constant shear rate, as a viscous fluid on long timescales (De Gennes 1979).

In general, many timescales of stress relaxation exist in complex fluids and there is no simple generic description of this process. However, if we focus on the longest relaxation time and ignore faster relaxation processes, we can describe this viscoelastic behavior by a single relaxation time using the Maxwell model. The basic idea is that the shear stress σ relaxes within a time τ as

$$\tau \frac{d\sigma}{dt} = -\sigma + 2\eta \dot{u}, \tag{4.24}$$

where $\dot{u}$ is a shear rate. At constant shear rate, $\sigma \simeq 2\eta\dot{u}$ is stationary, corresponding to viscous behavior. For short times, $\sigma \simeq 2\eta/\tau u$, and the response is elastic with an elastic modulus $2\eta/\tau$. In order to describe an incompressible bulk fluid, we write

$$\left(1 + \tau \frac{D}{Dt}\right) \tilde{\sigma}_{\alpha\beta} = 2\eta \tilde{v}_{\alpha\beta}, \tag{4.25}$$

describing the relaxation of shear stress. Since stress relaxation occurs in a material reference frame, we have to use a convected corotational time derivative to describe the system in a laboratory frame (Larson 1998):

$$\frac{D}{Dt}\sigma_{\alpha\beta} = \frac{\partial}{\partial t}\sigma_{\alpha\beta} + v_\gamma \partial_\gamma \sigma_{\alpha\beta} + \omega_{\alpha\gamma}\sigma_{\gamma\beta} + \omega_{\beta\gamma}\sigma_{\gamma\alpha}. \tag{4.26}$$

Since viscoelasticity combines dissipative and reactive behaviors, we decompose the stress into dissipative and reactive parts: $\sigma_{\alpha\beta} = \sigma^{\mathrm{d}}_{\alpha\beta} + \sigma^{\mathrm{r}}_{\alpha\beta}$. The constitutive material equations of the Maxwell model can then be written in the form

$$\left(1 - \tau^2 \frac{D^2}{Dt^2}\right) \tilde{\sigma}^{\mathrm{d}}_{\alpha\beta} = 2\eta \tilde{v}_{\alpha\beta}, \tag{4.27}$$

$$\tilde{\sigma}^{\mathrm{r}}_{\alpha\beta} = -\tau \frac{D}{Dt}\tilde{\sigma}^{\mathrm{d}}_{\alpha\beta}. \tag{4.28}$$

Note that this is equivalent to eqn. (4.25) because

$$\left(1 - \tau^2 \frac{D^2}{Dt^2}\right) \tilde{\sigma}^{\mathrm{d}}_{\alpha\beta} = \left(1 + \tau \frac{D}{Dt}\right)\left(1 - \tau \frac{D}{Dt}\right) \tilde{\sigma}^{\mathrm{d}}_{\alpha\beta} = \tilde{\sigma}_{\alpha\beta}. \tag{4.29}$$

Equation (4.25) describes the constitutive equation of a viscoelastic fluid or gel in the Maxwell model. It captures the fluid viscosity at times long compared with the viscoelastic relaxation time, and also the short-time elastic behavior. The constitutive relations for the Maxwell fluid are nonlinear in the laboratory reference frame because of the convective and corotational terms in the time derivative. This nonlinear dynamic of a viscoelastic fluid even in the low-Reynolds-number regime gives rise to complex fluid dynamic behaviors (Larson 1998).

4.2 Active polar gels

We now discuss the hydrodynamic equations of active polar gels using the concepts outlined in the introduction. We consider, for simplicity, a single-component complex fluid with local vectorial order described by a vector field $\mathbf{p}$. In the cytoskeleton, this order parameter corresponds to the average alignment of filaments in a volume element of the gel. It can be constructed by first assigning to each filament a unit tangent vector, pointing to the plus end. The average of all these tangent vectors in a volume element is given by $\mathbf{p}$.

We write equations for the center-of-mass velocity $\mathbf{v}$ and the total mass density ρ. The cytoskeleton is a multicomponent system. To describe its physics, we need in principle to introduce more density variables and currents. In the case of the cytoskeleton, these variables and currents might be, for example, the density and current of the actin gel, and the cytoplasm, and the concentrations of additional molecules such as motor proteins and ATP. The present case of a single component can be systematically generalized to multicomponent systems (Joanny *et al.* 2007).

Finally, we introduce a local active process by adding, as an additional pair of conjugate thermodynamic variables, the local rate of ATP consumption $r = -\dot{n}_{\mathrm{ATP}}$, and the chemical free energy transduced per ATP hydrolysis event $\Delta\mu = \mu_{\mathrm{ATP}} - \mu_{\mathrm{ADP}} - \mu_{\mathrm{P}}$. Here n_{ATP} is the number density of ATP molecules, and $\Delta\mu$ is the difference between the chemical potentials of the fuel μ_{ATP} and of the reaction products adenosine diphosphate (ADP) and inorganic phosphate μ_{ADP} and μ_{P}. The dissipation rate per unit volume due to ATP consumption is given by $r\ \Delta\mu$.

4.2.1 Entropy production

We begin our discussion with the free energy of a polar gel with a reservoir of ATP as an energy source,

$$F = \int d^3x \left(\frac{1}{2}\rho\mathbf{v}^2 + \left(f\left(\frac{\rho}{m}, p_\alpha, \partial_\alpha p_\beta\right)\right) + n_{\mathrm{ATP}}\ \Delta\mu\right). \tag{4.30}$$

In the absence of fluxes at the boundaries of the system, the entropy production is obtained as the rate of change of free energy,

$$-\dot{F} = -\int d^3x \left(\partial_t\left(\frac{1}{2}\rho\mathbf{v}^2\right) + \frac{\mu}{m}\frac{\partial\rho}{\partial t} - h_\alpha\frac{\partial p_\alpha}{\partial t} - r\ \Delta\mu\right). \tag{4.31}$$

Here we have introduced the field conjugate to the polarity vector,

$$h_\alpha(\mathbf{x}) = -\frac{\delta F}{\delta p_\alpha(\mathbf{x})}. \tag{4.32}$$

In the case including a polar order parameter $\mathbf{p}$, the Gibbs–Duhem relation can be written as

$$n\partial_\alpha\mu = -h_\gamma\partial_\alpha p_\gamma - \partial_\beta\sigma^{\mathrm{e}}_{\alpha\beta}, \tag{4.33}$$

where

$$\sigma^{\mathrm{e}}_{\alpha\beta} = (f - \mu n)\delta_{\alpha\beta} - \frac{\partial f}{\partial(\partial_\beta p_\gamma)}\partial_\alpha p_\gamma \tag{4.34}$$

is called the Ericson stress. The Ericson stress is the stress that exists in equilibrium. It is a generalization of the pressure P for an anisotropic liquid. The isotropic part of $\sigma^{\mathrm{e}}_{\alpha\beta}$ is given by the usual expression for the pressure $P = f - \mu n$, and gradients of the orientation field contribute additional anisotropic stresses.

For an anisotropic system with polar order, the total stress tensor can have an antisymmetric component. This follows from angular-momentum conservation, which implies that any torque $\mathbf{p} \times \mathbf{h}$ acting locally on the polarity $\mathbf{p}$ must be exerted by the torques in the fluid. This argument leads to

$$\sigma^{a}_{\alpha\beta} = \frac{1}{2}(\sigma^{\mathrm{tot}}_{\alpha\beta} - \sigma^{\mathrm{tot}}_{\beta\alpha}) = \frac{1}{2}(p_\alpha h_\beta - p_\beta h_\alpha). \tag{4.35}$$

The work done by the locally rotating fluid to turn the orientation field is thus

$$\sigma^{a}_{\alpha\beta}\omega_{\alpha\beta} = \omega_{\alpha\beta}h_\alpha p_\beta. \tag{4.36}$$

After using the conservation laws, the Gibbs–Duhem relation, and the antisymmetric part of the total stress, we obtain

$$T\dot{S} = \int d^3x \left[\sigma_{\alpha\beta}v_{\alpha\beta} + \left(\frac{\partial p_\alpha}{\partial t} + v_\gamma\partial_\gamma p_\alpha + \omega_{\alpha\beta}p_\beta\right)h_\alpha + r\,\Delta\mu\right]. \tag{4.37}$$

The symmetric part of the deviatoric stress is given by

$$\sigma_{\alpha\beta} = \sigma^{\mathrm{tot}}_{\alpha\beta} - \sigma^{\mathrm{a}}_{\alpha\beta} + \rho v_\alpha v_\beta - \sigma^{\mathrm{e,s}}_{\alpha\beta}, \tag{4.38}$$

where $\sigma^{\mathrm{e,s}}_{\alpha\beta}$ denotes the symmetric part of the Ericson stress $\sigma^{\mathrm{e}}_{\alpha\beta}$. We again introduce the traceless part $\tilde{\sigma}_{\alpha\beta}$ and the trace $\bar{\sigma} = (1/3)\sigma_{\gamma\gamma}$, writing $\sigma_{\alpha\beta} = \tilde{\sigma}_{\alpha\beta} + \bar{\sigma}\delta_{\alpha\beta}$.

The flux conjugate to the field h_α is the convected corotational time derivative of p_α,

$$\frac{Dp_\alpha}{Dt} = \frac{\partial p_\alpha}{\partial t} + v_\gamma\partial_\gamma p_\alpha + \omega_{\alpha\beta}p_\beta. \tag{4.39}$$

We can now identify four pairs of conjugate thermodynamic variables:

$$\tilde{\sigma}_{\alpha\beta} \leftrightarrow \tilde{v}_{\alpha\beta}, \qquad \epsilon = -1, \tag{4.40}$$

$$\bar{\sigma} \leftrightarrow v_{\gamma\gamma}, \qquad \epsilon = -1, \tag{4.41}$$

$$\frac{Dp_\alpha}{Dt} \leftrightarrow h_\alpha, \qquad \epsilon = 1, \tag{4.42}$$

$$r \leftrightarrow \Delta\mu, \qquad \epsilon = 1. \tag{4.43}$$

Based on these pairs of conjugate fluxes and forces, the constitutive material relations can now be derived systematically.

4.2.2 Reactive and dissipative couplings

We now write the fluxes as sums of dissipative and reactive components:

$$\tilde{\sigma}_{\alpha\beta} = \tilde{\sigma}^{\mathrm{d}}_{\alpha\beta} + \tilde{\sigma}^{\mathrm{r}}_{\alpha\beta}, \tag{4.44}$$

$$\bar{\sigma} = \bar{\sigma}^{\mathrm{d}} + \bar{\sigma}^{\mathrm{r}}, \tag{4.45}$$

$$\frac{Dp_\alpha}{Dt} = \frac{Dp^{\mathrm{d}}_\alpha}{Dt} + \frac{Dp^{\mathrm{r}}_\alpha}{Dt}, \tag{4.46}$$

$$r = r^{\mathrm{d}} + r^{\mathrm{r}}. \tag{4.47}$$

The dissipative couplings that are allowed by symmetry read

$$\tilde{\sigma}^{\mathrm{d}}_{\alpha\beta} = 2\eta\tilde{v}_{\alpha\beta}, \tag{4.48}$$

$$\bar{\sigma}^{\mathrm{d}} = \bar{\eta}v_{\gamma\gamma}, \tag{4.49}$$

$$\begin{aligned}\frac{Dp^{\mathrm{d}}_\alpha}{Dt} &= \frac{1}{\gamma_1}h_\alpha + \lambda_1\,\Delta\mu\, p_\alpha + \lambda_2\,\Delta\mu\, p_\gamma(\partial_\gamma p_\alpha)\\ &\quad + \lambda_3\,\Delta\mu\,(\partial_\gamma p_\gamma)p_\alpha + \lambda_4\,\Delta\mu\,\partial_\alpha(p_\gamma p_\gamma),\end{aligned} \tag{4.50}$$

$$\begin{aligned}r^{\mathrm{d}} &= \Lambda\,\Delta\mu + \lambda_1 p_\alpha h_\alpha + \lambda_2 p_\gamma(\partial_\gamma p_\alpha)h_\alpha\\ &\quad + \lambda_3(\partial_\gamma p_\gamma)p_\alpha h_\alpha + \lambda_4 h_\alpha\partial_\alpha(p_\gamma p_\gamma).\end{aligned} \tag{4.51}$$

Here, the dissipative coefficients obey Onsager symmetry relations. We have, for simplicity, neglected anisotropies in the viscosity due to the polar anisotropies in the active polar gel, and we have ignored the viscoelastic relaxation time. The reactive couplings are described by phenomenological coefficients which enter antisymmetrically. The reactive couplings read

$$\tilde{\sigma}^{\mathrm{r}}_{\alpha\beta} = \frac{\nu_1}{2}\left(p_\alpha h_\beta + p_\beta h_\alpha - \frac{2}{3}p_\gamma h_\gamma\delta_{\alpha\beta}\right) - \zeta\,\Delta\mu\,\tilde{q}_{\alpha\beta}, \tag{4.52}$$

$$\bar{\sigma}^{\mathrm{r}} = \bar{\nu}_1 h_\gamma p_\gamma - \bar{\zeta}\Delta\mu - \zeta\,'\Delta\mu\, p_\gamma p_\gamma, \tag{4.53}$$

$$\frac{Dp^{\mathrm{d}}_\alpha}{Dt} = -\nu_1 p_\beta\tilde{v}_{\alpha\beta} - \bar{\nu}_1 p_\alpha v_{\gamma\gamma}, \tag{4.54}$$

$$r^{\mathrm{r}} = \zeta\tilde{q}_{\alpha\beta}\tilde{v}_{\alpha\beta} + \zeta' p_\gamma p_\gamma v_{\delta\delta} + \bar{\zeta}v_{\gamma\gamma}. \tag{4.55}$$

Here, we have introduced the traceless symmetric tensor

$$\tilde{q}_{\alpha\beta} = p_\alpha p_\beta - \frac{1}{3}p_\gamma p_\gamma\delta_{\alpha\beta}, \tag{4.56}$$

which describes the anisotropies of the local filament orientation. This is a nematic tensor defining an axis of filament alignment but is insensitive to filament direction.

4.2.3 Constitutive equations

By combining the reactive and dissipative parts of the thermodynamic fluxes, we obtain the constitutive equations of the active gel which describe the active material properties. We can also systematically include viscoelastic behavior of the gel described by the Maxwell model. The full constitutive equations then read:

$$2\eta\tilde{v}_{\alpha\beta} = \left(1+\tau\frac{D}{Dt}\right)\left\{\tilde{\sigma}_{\alpha\beta} + \zeta\ \Delta\mu\ \tilde{q}_{\alpha\beta} - \frac{\nu_1}{2}(p_\alpha h_\beta + p_\beta h_\alpha - \frac{2}{3}\delta_{\alpha\beta})\right\}, \tag{4.57}$$

$$\bar{\eta}v_{\gamma\gamma} = \left(1+\tau\frac{D}{Dt}\right)\left\{\bar{\sigma} + \zeta'\ \Delta\mu\ p_\gamma p_\gamma + \bar{\zeta}\ \Delta\mu - \bar{\nu}_1 h_\gamma p_\gamma\right\}, \tag{4.58}$$

$$\frac{Dp_\alpha}{Dt} = \left(1+\tau\frac{D}{Dt}\right)\frac{1}{\gamma_1}h_\alpha + \lambda_1 p_\alpha\ \Delta\mu + \lambda_2 p_\gamma\partial_\gamma p_\alpha + \dots$$
$$-\nu_1 p_\beta\tilde{v}_{\alpha\beta} - \bar{\nu}_1 p_\alpha v_{\gamma\gamma}, \tag{4.59}$$

$$r = \Lambda\ \Delta\mu + \lambda_1 p_\alpha h_\alpha + \lambda_2 p_\gamma(\partial_\gamma p_\alpha)h_\alpha + \dots$$
$$+\zeta\tilde{q}_{\alpha\beta}\tilde{v}_{\alpha\beta} + \zeta' p_\gamma p_\gamma v_{\delta\delta} + \bar{\zeta}v_{\gamma\gamma}, \tag{4.60}$$

where the terms proportional to the coefficients λ_3 and λ_4 have been indicated by dots to keep the expressions simpler.

These constitutive equations, together with the conservation laws for mass and momentum, define the dynamics of the system. Momentum conservation or force, balance reads

$$\partial_\beta\left(\tilde{\sigma}_{\alpha\beta} + \frac{1}{3}\bar{\sigma}\delta_{\alpha\beta} + \sigma^a_{\alpha\beta} + \sigma^{\mathrm{e,s}}_{\alpha\beta}\right) = 0, \tag{4.61}$$

where the antisymmetric part of the stress is given by

$$\sigma^a_{\alpha\beta} = \frac{1}{2}(p_\alpha h_\beta - p_\beta h_\alpha) \tag{4.62}$$

and $\sigma^{\mathrm{e,s}}_{\alpha\beta}$ is the symmetric part of the Ericson stress given in eqn (4.34). This theory includes as special cases the hydrodynamics of polar and nematic liquid crystals (for $\Delta\mu = 0$), and the Maxwell model of viscoelastic gels.

Novel physics arises from the active terms proportional to $\Delta\mu$. The most prominent features are the isotropic and anisotropic active stresses $\bar{\zeta}\ \Delta\mu$ and $\zeta\ \Delta\mu\ \tilde{q}_{\alpha\beta}$. In an incompressible system, the isotropic active stress does not generate new effects. In general, active anisotropic stresses are responsible for unconventional properties and dynamics. Such anisotropic stresses induce spontaneous shear of the active material, which in turn couples to reorient the polar order $\mathbf{p}$ and thus the axis of stress anisotropy. This interplay of spontaneous shear and reorientation of order is the basis of some interesting dynamics and instabilities.

Active terms also occur in the dynamics of polarity reorientation. However, the dominant term $\lambda_1 p_\alpha\ \Delta\mu$ essentially renormalizes a similar term linear in p_α that generally occurs in the expression for h_α stemming from the free energy. Such a term modifies the onset of the isotropic-to-polar or isotropic-to-nematic transition, as it affects the magnitude of $\mathbf{p}$. The term $\lambda_2 p_\gamma \partial_\gamma p_\alpha$ does not exist in the field h_α and is a truly nonequilibrium phenomenon (Toner and Tu 1995). Note, however, that this term is of higher order in a gradient expansion and thus weak in the hydrodynamic limit of large wavelengths.

4.3 Examples of active gels

The constitutive equations for active gels can be used to discuss the dynamics of active systems in various geometries. Cell biological processes such as cell locomotion and the dynamics of the cell cortex result from the active dynamics of the actin cytoskeleton, which, on large scales, can be described by the properties of active viscoelastic gels.

4.3.1 Thin layer of an active gel on a substrate

As simple and general example to illustrate the consequences of active anisotropic stresses, we consider the simple case of a layer of a polar gel of thickness H on a solid surface in the viscous regime ($\tau = 0$) (Voituriez *et al.* 2005). We consider, for simplicity, the case of an incompressible gel ($\partial_\gamma v_\gamma = 0$), with a polarity vector $\mathbf{p} = (p_x, p_z)$, where $p_x = \sin\theta$ and $p_z = \cos\theta$. The system is considered homogeneous in the y-direction so that we can use an effective two-dimensional description in the $x - z$ plane. The angle θ describes the direction of polarity, with $\mathbf{p}^2 = 1$. In this situation, the hydrostatic pressure P and the parallel component $h_\parallel = h_\gamma p_\gamma$ of the field h_α play the roles of Lagrange multipliers to enforce the constraints of incompressibility, $\partial_\gamma v_\gamma = 0$, and of constant magnitude of polarization, $\mathbf{p}^2 = 1$. The constitutive equations then read

$$2\eta \tilde{v}_{\alpha\beta} = \tilde{\sigma}_{\alpha\beta} + \zeta\, \Delta\mu\, \tilde{q}_{\alpha\beta} - \frac{\nu_1}{2}\left(p_\alpha h_\beta + p_\beta h_\alpha - \frac{2}{3} p_\gamma h_\gamma \delta_{\alpha\beta}\right), \tag{4.63}$$

$$\frac{\partial p_\alpha}{\partial t} = -v_\gamma \partial_\gamma p_\alpha - \omega_{\alpha\beta} p_\beta - \nu_1 \tilde{v}_{\alpha\beta} p_\beta + \frac{1}{\gamma_1} h_\alpha, \tag{4.64}$$

where we have ignored higher-order gradient terms in p_α. We consider solutions that are translationally invariant in the x–y plane of the substrate: $v_x = v(z)$, $v_y = 0$, $v_z = 0$. We introduce $u(z) = (1/2)dv/dz$ and have

$$\tilde{v}_{\alpha\beta} = \begin{pmatrix} 0 & u \\ u & 0 \end{pmatrix}, \quad \omega_{\alpha\beta} = \begin{pmatrix} 0 & -u \\ u & 0 \end{pmatrix}. \tag{4.65}$$

For $\mathbf{p}^2 = 1$, the polarity dynamics then implies the condition on the parallel field $h_\parallel$

$$u\nu_1 \sin 2\theta = \frac{h_\parallel}{\gamma_1}. \tag{4.66}$$

The dynamics of reorientation of the polarity angle obeys

$$\partial_t \theta = \frac{h_\perp}{\gamma_1} - u(\nu_1 \cos 2\theta - 1). \tag{4.67}$$

The equation relating stress and strain rate can be written as

$$2\eta u = \tilde{\sigma}_{xz} + \frac{\zeta\,\Delta\mu}{2}\sin 2\theta - \frac{\nu_1}{2}(h_{\parallel}\sin 2\theta + h_{\perp}\cos 2\theta). \tag{4.68}$$

where $\tilde{\sigma}_{xz} = (\sigma_{xz} + \sigma_{zx})/2$. With the antisymmetric stress

$$\sigma^{\mathrm{a}}_{\alpha\beta} = \frac{1}{2}(p_\alpha h_\beta - p_\beta h_\alpha) = \begin{pmatrix} 0 & -h_\perp/2 \\ h_\perp/2 & 0 \end{pmatrix}, \tag{4.69}$$

we have $\sigma_{xz} - \sigma_{zx} = -h_\perp$. The x-component of the force balance equation reads

$$\partial_x(\sigma_{xx} - P) + \partial_z \sigma_{xz} = 0 \tag{4.70}$$

and can be used to determine the stress profile. Since the system is homogeneous in the $x-$direction, all $x-$derivatives vanish and $\sigma_{xz} = \sigma^{\mathrm{ext}}_{xz}$ does not depend on z; it is set by the externally applied shear stress $\sigma^{\mathrm{ext}}_{xz}$. We thus find $\tilde{\sigma}_{xz} = \sigma^{\mathrm{ext}}_{xz} + h_\perp/2$. By eliminating $h_\parallel$, we find the flow field

$$u = \frac{\zeta\,\Delta\mu \sin 2\theta + h_\perp(1 - \nu_1\cos 2\theta - 1) + 2\sigma^{\mathrm{ext}}_{xz}}{4\eta + \nu_1^2\gamma_1\sin^2 2\theta}. \tag{4.71}$$

This equation reveals that active stresses can generate spontaneous flows in the absence of external stresses ($\sigma^{\mathrm{ext}}_{xz} = 0$). Furthermore, in the passive case ($\Delta\mu = 0$), the coefficient ν_1 renormalizes the fluid visocity. Together with the dynamics of polarity reorientation (4.67), this determines the flow profile of the gel layer.

The perpendicular field $h_\perp$ can be determined from the free energy of the orientation field. In the simple case where only one elastic coefficient K describes deformations of the orientation field, we have

$$h_\perp = \frac{K}{2}\partial_z^2\theta. \tag{4.72}$$

This corresponds to a Franck free energy with equal elastic moduli.

We discuss first the simple situation where the angle θ is constant in space and thus $h_\perp = 0$. In the absence of external stresses, $\sigma^{\mathrm{ext}}_{xz} = 0$, but with $\zeta\,\Delta\mu \neq 0$, steady states exist for $\theta = 0$ and $\theta = \pi/2$ for which no motion occurs, i.e. $u = 0$. The orientation equation

$$\partial_t\theta = u(1 - \nu_1\cos 2\theta) \tag{4.73}$$

has, for $|\nu_1| > 1$, additional steady states for angles $\theta = \theta_0$ with

$$\cos 2\theta_0 = \frac{1}{\nu_1}. \tag{4.74}$$

In these states, the active stress generates the shear rate

$$u = \frac{\zeta\,\Delta\mu\sin 2\theta_0}{4\eta + \nu_1^2\gamma_1\sin^2 2\theta_0}. \tag{4.75}$$

These solutions correspond to a spontaneous shear flow generated by the active stress $\zeta\ \Delta\mu$, which in turn aligns the polarity orientation to a stationary angle θ_0. Two distinct angles θ_0 satisfy the condition (4.74). One of these angles is a stable state, and the other is unstable. The stability of the steady states depends on the sign of $\zeta\ \Delta\mu$, (see eqn 4.73). The stationary angle of the polarity orientation under shear is determined by the reactive phenomenological coefficient ν_1. This coefficient is a property that is known for passive nematic liquid crystals. There, the nematic order realigns under shear and has stationary angles θ_0 determined by the coefficient ν_1. This passive case is recovered in our analysis if no active stress is present, $\zeta\ \Delta\mu = 0$, and an external stress $\sigma_{xz}^{\rm ext}$ is applied. In this case a passive shear flow $u^{\rm pass}$ exists with $2\eta^{\rm eff}u^{\rm pass} = \sigma_{xz}^{\rm ext}$, with effective viscosity

$$\eta^{\rm eff} = \eta + \frac{1}{4}\gamma_1(\nu_1^2 - 1). \tag{4.76}$$

The polarity orientation is again aligned with the flow at an angle θ_0 determined by the condition (4.74). The polarity dynamics leads to a renormalization of the viscosity as given in eqn (4.76).

4.3.2 Dynamic instabilities

The simple solutions discussed in the previous section are independent of the distance z from the substrate and thus not compatible with most boundary conditions. Also, as we discuss below, such solutions can become unstable with respect to finite-wavelength modes. Therefore it is important to study the dynamics of active gel layers with finite thickness and specific boundary conditions on the flow field and the polarity field. Such solutions, in general, depend on the height z from the substrate. The steady-state condition $\partial_t\theta(z) = 0$ then implies (Voituriez *et al.* 2005)

$$\partial_z^2\theta = \Phi(\theta(z)), \quad \text{where} \tag{4.77}$$

$$\Phi(\theta) \equiv \frac{\zeta\ \Delta\mu \sin 2\theta(\nu_1\cos 2\theta - 1)}{K(4\eta/\gamma_1 + \nu_1^2 - 2\nu_1\cos 2\theta + 1)}. \tag{4.78}$$

The function $\Phi(\theta)$ has zeros for $\theta = 0, \pi/2$, and θ_0, which correspond to the solutions with θ constant discussed above. As a simple example, we consider the steady state $\theta = \pi/2$ and assume that at $z = 0$ and $z = H$ the angle is imposed at the boundaries: $\theta(0) = \theta(H) = \pi/2$. The linear stability of this solution can be tested by studying small perturbations $\delta\theta(z)$ such that the polarity orientation angle is of the form $\theta(z) = \pi/2 + \delta\theta(z)$. Linearizing eqn (4.77) implies $\partial_z^2\ \delta\theta = -\alpha\ \delta\theta$, where

$$\alpha = -\left.\frac{d\Phi}{dz}\right|_{\theta=\pi/2} = \frac{-\zeta\ \Delta\mu\ (\nu_1 + 1)}{2K(4\eta/\gamma_1 + (\nu_1+1)^2)}. \tag{4.79}$$

The solutions of this equation are of the form

$$\delta\theta = A\sin(\alpha^{1/2}z) + B\cos(\alpha^{1/2}z), \tag{4.80}$$

where A and B are integration constants. An instability of the homogeneous state occurs if a finite-wavelength solution appears that satisfies the boundary conditions $\delta\theta(0) = \delta\theta(H) = 0$. This is the case if $\alpha^{1/2} = \pi/H$:

$$\delta\theta = A \sin(z\pi/H). \tag{4.81}$$

These arguments imply that for a given value of the active stress $\zeta\Delta\mu < 0$, the steady state $\theta = \pi/2$ becomes unstable when H exceeds a critical value $H_c = \alpha^{-1/2}$. This value diverges in the limit where the active stress vanishes. Equivalently, a system of given height becomes unstable if the active stress is negative (contractile) and reaches a critical value $|\zeta\ \Delta\mu| = \zeta\ \Delta\mu_c$, where

$$(\zeta\ \Delta\mu)_c = \frac{\pi^2 K(4\eta/\gamma_1 + (\nu_1+1)^2)}{2H^2(\nu_1+1)}. \tag{4.82}$$

For positive $\zeta\ \Delta\mu$, the steady state $\theta = \pi/2$ is stable. If we consider instead the stability of the other steady state $\theta = 0$, we find that it is stable for negative active stresses but becomes unstable as $\zeta\ \Delta\mu$ exceeds a positive critical value.

At the instability, the system reaches a new steady state with an inhomogeneous angular profile $\theta(z)$. For large H, this steady state consists of a bulk region where $\theta(z) \simeq \theta_0$ is close to the polarity orientation angle described above, which is generated by shear flows. In this case, the shear flow is self-generated by the active stresses at the instability and subsequently orients the polarity field. Near the boundaries, $\theta(z)$ approaches the boundary values within a boundary layer of thickness $H_B \simeq \alpha^{-1/2}$.

4.3.3 Lamellipodium motion

Many cells can move on solid substrates. A model example is keratocytes, which exhibit continuous gliding motions with an approximately constant shape. This motion results from a very complex interplay of the dynamics of the actin cytoskeleton with adhesion and the localized polymerization and depolymerization of the actin gel in the front and back of the polarized cell.

Key to cell locomotion is the formation and extension of a lamellipodium which is a thin actin sheet enclosed by the cell membrane and which adheres to the substrate via specific adhesion molecules. Actin polymerization and cross-linking occur locally near the cell membrane in the front of the lamellipodium. This process is regulated by WASP proteins localized in the front part of the cell membrane. WASP activates Arp2/3, which serves as a nucleator of branches. The branching actin networks then entangle and form an active gel driven by actin treadmilling and the action of myosin.

In order to capture the key large-scale features and force balances in the lamellipodium, we describe the lamellipodium as a layer of an active viscous fluid and assume that we can neglect the cytoplasmic solvent in our description. The polarity of the cell is assumed to be provided by other mechanisms and the average filament polarity is determined by the polymerization in the front, setting a constant polarity vector $\mathbf{p} = \mathbf{e}_x$, which is the unit vector in the x-direction. For simplicity, we consider an incompressible gel in the viscous limit and consider the case where the system is independent of the y-direction and the distance from the substrate is measured in the z-direction (Kruse *et al.* 2006).

The material properties of the active gel are then described by

$$2\eta\tilde{v}_{\alpha\beta} = \tilde{\sigma}_{\alpha\beta} + \zeta\,\Delta\mu\,\tilde{q}_{\alpha\beta}, \tag{4.83}$$

where $\alpha = x, y$. With $p_\alpha = (1, 0)$, we have

$$\tilde{q}_{\alpha\beta} = \begin{pmatrix} 1 & 0 \\ 0 & 0 \end{pmatrix} - \begin{pmatrix} 1/2 & 0 \\ 0 & 1/2 \end{pmatrix} = \begin{pmatrix} 1/2 & 0 \\ 0 & -1/2 \end{pmatrix}. \tag{4.84}$$

The condition of incompressibility implies $v_{xx} + v_{zz} = 0$.

By averaging over a thin film, we find an equation for the velocity gradients

$$4\eta\frac{\partial v}{\partial x} = \sigma - P + \zeta\,\Delta\mu, \tag{4.85}$$

where v and σ denote the velocity component v_x and the stress σ_{xx} averaged over the film thickness h, respectively. Furthermore, we have used $\sigma_{zz} \simeq P$, where P denotes the external pressure. Force balance on a solid substrate implies

$$\frac{d}{dx}(h(\sigma - P)) = \xi v. \tag{4.86}$$

Finally, mass conservation implies

$$\frac{d}{dx}((u + v)h) = k_p\rho_{\mathrm{wa}}(x). \tag{4.87}$$

Here, u is the velocity at which the lamellipodium advances on the substrate, $h(x,t) = h(x - ut)$, and $\rho_{\mathrm{wa}} = \rho_0 e^{-x/\lambda}$ is the density of the branching factors on the cell membrane that locally control gel assembly. The assembly rate is characterized by the parameter k_{p}.

By introducing the total stress $F = h(\sigma - P)$ along the lamellipodium axis, we obtain three first-order differential equations for the shape and flow profiles (Kruse *et al.* 2006):

$$\frac{dF}{dx} = \xi v, \tag{4.88}$$

$$4\eta\frac{dv}{dx} = \frac{F}{h} + \zeta\,\Delta\mu, \tag{4.89}$$

$$4\eta(u + v)\frac{dh}{dx} = -F - h\zeta\,\Delta\mu + 4\eta k_{\mathrm{p}}\rho_{\mathrm{wa}}. \tag{4.90}$$

For a lamellipodium with small height variations around a fixed height $h \simeq \bar{h}$, eqns (4.88) and (4.89) become

$$4\eta\bar{h}\frac{d^2 v}{dx^2} = \xi v, \tag{4.91}$$

which implies that the velocity field decays exponentially as

$$v = v_0 e^{-x/d} \tag{4.92}$$

on a characteristic length scale $d = (4\eta h_0/\xi)^{1/2}$. With eqn (4.87), we have

$$h(x) = \frac{k_\mathrm{p}}{u + v(x)} \int_0^x dx' \rho_\mathrm{wa}(x'). \tag{4.93}$$

In the region $x \gg d$, where the height is close to h_0 and $v = 0$, we find that $\bar{h} \simeq k_\mathrm{p} \lambda \rho_0 / u$.

Equations (4.88) and (4.89) represent coupled first-order equations for the force F, the gel flow v and the height profile. They can be solved by imposing boundary conditions on the force at $x = 0$ (the lamellipodium front) and $x = L$ (the connection of the lamellipodium to the cell body), and $h(x = 0) = 0$. In order to take into account the combined effects of adhesion and external forces, we impose $F(0) = -(W + f_\mathrm{ext}^0$ at the front and $F(L) = W' + f_\mathrm{ext}^L$ at the rear. External forces f_ext act on the lamellipodium at its front and rear. The force at the rear includes that exerted by the cell body on the lamellipodium. We have ignored the effect of the tension and bending rigidity of the membrane at the free surface of the cell, but have included the effects of the adhesion energies per unit area of the lamellipodium (W) and of the cell body (W') on the substrate.

With these conditions, for a given length L there exists one steady-state solution for any positive value of the velocity u. The value of the velocity is selected by the fact that at the rear, the cell body imposes a height $h(L) = h_0$. Depolymerization in the rear is essential for this solution to exist. The depolymerization velocity v_d in the steady state is related to the total velocity at the back by the conservation of the gel flux, $v_\mathrm{d} = u + v(L)$. This conservation allows us to obtain an approximate equation for the velocity of motion of the lamellipodium,

$$u \simeq v_\mathrm{d} - \left(\frac{h_0}{4\eta\xi}\right)^{1/2} \left(\zeta\ \Delta\mu + \frac{f_\mathrm{ext}^L}{h_0}\right). \tag{4.94}$$

The cell exerts stresses on the substrate by friction. Although the total force exerted on the substrate vanishes, there exists a force dipole

$$Q = \int_0^L \xi x v\, dx \simeq L\bar{h}\zeta\ \Delta\mu < 0. \tag{4.95}$$

This implies that the traction forces which can be measured experimentally are due to active stresses in the actin gel.

Taking the example of a crawling keratocyte, we estimate $L \simeq 10\,\mu\mathrm{m}$ and $\bar{h} \simeq 1\,\mu\mathrm{m}$. The velocity of advance is $u \simeq 10\,\mu\mathrm{m/min}$, and a retrograde flow exists at the front with $v(0) \simeq 1\mu\mathrm{m/min}$ (Kucik *et al.* 1991; Vallotton *et al.* 2005). The elastic modulus E of an actin gel in a cell is of the order of 10^4 Pa (Gerbal *et al.* 2000; MacKintosh *et al.* 1995; Gardel *et al.* 2004), and the typical viscoelastic relaxation time τ is of the order of 10 s (Wottawah *et al.* 2005). From existing measurements of both the surface stress $\sigma_{xz} \simeq 4 \cdot 10^2$ N/m^2 (Oliver *et al.* 1999) and the retrograde flow velocity $v(0)$, we can deduce the friction coefficient of a lamellipodium, $\xi \simeq \sigma_{xz}/v(0) \simeq 10$ Pa s/m. We estimate a length $d \simeq 6$ μm. The anisotropic active stress is obtained from the retrograde flow velocity as $-\zeta \; \Delta\mu \simeq 4\eta v(0)/d \simeq 10^3$ Pa. Note that in this case the force transmitted by the lamellipodium does not vanish, since it balances the force f^L_{ext} exerted by the cell body on the lamellipodium, where $f^L_{\text{ext}} \simeq \xi v(0) d \simeq -\zeta \; \Delta\mu \; \bar{h} \simeq 1$ nN/μm. For a lamellipodium with an effective lateral size $L_y \simeq 30$ μm, the total force is therefore 30 nN; this order of magnitude is consistent with the force of 10 nN obtained by Oliver *et al.* (1999) by integration of the stress. The force dipole of the cell is $QL_y \simeq f^L_{\text{ext}} L_c L_y \simeq 6 \cdot 10^{-13}$ J, where $L_c \simeq 20$ μm is the total length of the cell.

4.4 Discussion

We have outlined some general concepts for understanding the active behaviors of viscous fluids and viscoelastic gels which are inherently active. This theory presents a paradigm for biological matter which is driven by active processes on the molecular scale. The cell cytoskeleton and, in particular, actin networks in the cell are specific examples of such active materials. The theory outlined here concerns the generic properties of such systems in the hydrodynamic limit, i.e. the dynamics of slow modes on large scales. If one studies such systems at shorter times and on smaller scales, additional features become relevant that depend on the specific organization of the system under investigation. In addition, fluctuations become important (Basu *et al.* 2008), which we have neglected here.

The generic hydrodynamic theory presented here applies also on larger scales to the large-scale properties of multicellular systems. During the development of an organism from a fertilized egg, large numbers of cells are created by cell division. These cells collectively organize into tissues, where they form patterns and complex morphologies (Wolpert 2006). The collective behaviors of many cells again generate an active material. Cell division, cell death, and active cellular force generation generate active stresses in the cell collection, which drive collective rearrangements that can be described as flow fields (Bittig *et al.* 2008, 2009; Basan *et al.* 2009). All of the general features of the hydrodynamic approach developed here apply again. We suggest that the generation of movements and flows in biological matter can be discussed and interpreted in the generic framework of active fluids and gels.

Acknowledgments

I thank Abhik Basu, Justin Bois, Martin Depken, Sebastian Fürthauer, Stephan Grill, Karsten Kruse, Jean-Francois Joanny, Jacques Prost, Jonas Ranft, and Ken Sekimoto for stimulating collaborations on active gels.

References

Alberts, B. *et al.* (2002). *Molecular Biology of the Cell* (4th edn). Garland, New York.

Ahmadi, A., Liverpool, T. B., and Marchetti, M. C. (2005). Nematic and polar order in active filament solutions. *Phys. Rev. E*, **72**, 060901.

Aronson, I. S. and Tsimring, L. S. (2005). Pattern formation of microtubules and motors: Inelastic interaction of polar rods. *Phys. Rev. E*, **71**, 050901(R).

Aronson, I. S. and Tsimring, L. S. (2006). Patterns and collective behavior in granular media: Theoretical concepts. *Rev. Mod. Phys.*, **78**, 641.

Basan, M., Risler, T., Joanny, J.-F., Sastre-Garau, X., and Prost, J. (2009). Homeostatic competition drives tumor growth and metastasis nucleation. *HFSP J.*, **3**, 265.

Basu, A., Joanny, J. F., Jülicher, F. and Prost, J. (2008). Thermal and non-thermal fluctuations in active polar gels. *Eur. Phys. J. E*, **9**, 10364.

Bird, R. B. and Armstrong, R. C. (1987). *Curtiss, Dynamics of Polymeric Liquids* (2nd edn). Wiley, New York.

Bittig, T., Wartlick, O., Kicheva, K., Golzalez-Gaitan, M., and Jülicher, F. (2008). Dynamics of anisotropic tissue growth. *New J. Phys.*, **10**, 063001.

Bittig, T., Wartlick, O., Golzalez-Gaitan, M., and Jülicher, F. (2009). Quantification of growth asymmetries in developing epithelia. *Eur. Phys. J. E*, **30**, 93.

Bray, D. (2001). *Cell Movements* (2nd edn). Garland, New York.

De Gennes, P. G. (1979). *Scaling Concepts in Polymer Physics.* Cornell University Press, Ithaca, NY.

De Gennes, P. G. and Prost, J. (1993). *The Physics of Liquid Crystals.* Clarendon Press, Oxford.

De Groot, S. R. and Mazur, P. (1984). *Non-Equilibrium Thermodynamics.* Dover, New York.

Gardel, M. L. *et al.* (2004). Elastic behavior of cross-linked and bundled actin networks. *Science*, **304**, 1301.

Gerbal, F., Chaikin, P., Rabin, Y., and Prost, J. (2000). An elastic analysis of *Listeria monocytogenes* propulsion. *Biophys. J.*, **79**, 2259.

Howard, J. (2001). *Mechanics of Motor Proteins and the Cytoskeleton.* Sinauer Associates Sunderland, MA.

Joanny, J. F., Jülicher, F., Kruse, K., and Prost, J. (2007). Hydrodynamic theory for multi-component active polar gel. *New J. Phys.*, **9**, 422.

Kruse, K. and Jülicher, F. (2000). Actively contracting bundles of polar filaments. *Phys. Rev. Lett.*, **85**, 1778.

Kruse, K., Camalet, S., and Jülicher, F. (2001). Self-propagating patterns in active filament bundles. *Phys. Rev. Lett.*, **87**, 138101.

Kruse, K., Joanny, J. F., Jülicher, F., Prost, J., and Sekimoto, K. (2004). Asters, vortices, and rotating spirals in active gels of polar filaments. *Phys. Rev. Lett.*, **92**, 078101.

Kruse, K., Joanny, J. F., Jülicher, F., Prost, J. and Sekimoto, K. (2005). Generic theory of active polar gels: A paradigm for cytoskeletal dynamics. *Eur. Phys. J. E*, **16**, 5.

Kruse, K., Joanny, J. F., Jülicher, F., and Prost, J. (2006). Contractility and retrograde flow in lamellipodium motion. *Phys. Biol.*, **3**, 130.

Kucik, D. F., Kuo, S. C., Elson, E. L. and Sheetz, M. P. (1991). Preferential attachment of membrane glycoproteins to the cytoskeleton at the leading edge of lamella. *J. Cell Biol.,* **114**, 1029.

Larson, R. (1998). *Constitutive Equations for Polymer Melts and Solutions.* Butterworth-Heinemann.

Liverpool, T. B. and Marchetti, M. C. (2003). Instabilities of isotropic solutions of active polar filaments. *Phys. Rev. Lett.*, **90**, 138102.

MacKintosh, F. C., Käs, J., and Janmey, P. A. (1995). Elasticity of semiflexible biopolymer networks. *Phys. Rev. Lett.*, **75**, 4425.

Martin, P. C., Parodi, O., and Pershan, P. (1972). Unified hydrodynamic theory for crystals, liquid crystals, and normal fluids. *Phys. Rev. A*, **6**, 2401.

Nédélec, F. J., Surrey, T., Maggs, A. C., and Leibler, S. (1997). Self-organization of microtubules and motors. *Nature*, **389**, 305.

Nédélec, F. J., Surrey, T., and Maggs, A. C. (2001). Dynamic concentration of motors in microtubule arrays. *Phys. Rev. Lett.*, **86**, 3192.

Oliver, T., Dembo, M., and Jacobson, K. (1999). Separation of propulsive and adhesive traction stresses in locomoting keratocytes. *J. Cell Biol.*, **145**, 589.

Salbreux, G., Joanny, J. F., and Prost, J. (2009). Hydrodynamics of cellular cortical flows and the formation of contractile rings. *Phys. Rev. Lett.*, **103**, 058102.

Simha, R. A. and Ramaswamy, S. (2002). Hydrodynamic fluctuations and instabilities in ordered suspensions of self-propelled particles. *Phys. Rev. Lett.*, **89**, 058101.

Surrey, T., Elowitz, M. B., Wolf, P.-E., Yang, F., Nédélec, F. J., Shokat, K., and Leibler, S. (1998). Chromophore-assisted light inactivation and self-organization of microtubules and motors. *Proc. Natl. Acad. Sci.*, **95**, 4293.

Toner, J. and Tu, Y. (1995). Long-range order in a two-dimensional dynamical XY model: How birds fly together. *Phys. Rev. Lett.*, **75**, 4326.

Vallotton, P. *et al.* (2005). Tracking retrograde flow in keratocytes: News from the front. *Mol. Biol. Cell*, **16**, 1223.

Voituriez, R., Joanny, J. F., and Prost, J. (2005). Spontaneous flow transition in active polar gels. *Eur. Phys. Lett.*, **70**, 404.

Voituriez, J., Joanny, J. F., and Prost, J. (2006). Generic phase diagram of active polar films. *Phys. Rev. Lett.*, **96**, 028102.

Wolpert, L. (2006). *Principles of Development.* 3rd edition. Oxford University Press.

Wottawah, F. *et al.* (2005). Optical rheology of biological cells. *Phys. Rev. Lett.*, **94**, 098103-1.

Zumdieck, A., Lagomarsino, M. C., Tanase, C., Kruse, K., Mulder, B., Dogterom, M., and Jülicher, F. (2005). Continuum description of the cytoskeleton: Ring formation in the cell cortex. *Phys. Rev. Lett.*, **95**, 258103.

5
Elasticity and dynamics of cytoskeletal filaments and networks of them

Fred C. MacKintosh

Vrije Universiteit, The Netherlands

5.1 Cytoskeletal networks

5.1.1 Introduction

Biological cells both experience external stresses and exert forces on their surroundings, for instance in cell locomotion. There are many important and outstanding problems in cell biology concerning the origins and regulation of the mechanical properties of cells (Alberts *et al.* 1994; Boal 2002). Such mechanical factors can determine how a cell maintains and modifies its shape, how it moves, and even what type of specialized cell it will become. Mechanical stimuli can also result in changes in gene expression in the cell.

As materials or mechanical systems, cells exhibit rich composite structures on the nanometer to micrometer scale, involving rather soft membranes and rather rigid filamentous proteins or biopolymers. Most plant and animal cells possess a complex network structure of biopolymers and associated proteins and enzymes for bundling, cross-linking, and active force generation. This *cytoskeleton* is often the principal determinant of cell elasticity and mechanical stability. Over the last few years, the single-molecule properties of many of the important constituents of the cytoskeleton have been studied in detail by biophysical techniques such as high-resolution microscopy, scanning force microscopy, and optical tweezers. At the same time, numerous *in vitro* experiments aimed at understanding some of the unique mechanical and dynamic properties of solutions and networks of cytoskeletal filaments have been performed. In parallel with these experiments, theoretical models have emerged that have served both to explain many of the essential material properties of these networks, and to motivate quantitative experiments to determine, for example, the concentration dependence of shear moduli and the effects of cross-links. In this chapter, we focus on the state of the theoretical modeling of both cytoskeletal solutions and networks, and on the properties of synthetic semiflexible polymer systems with similar underlying properties.

One of the principal components of the cytoskeleton, and in fact one of the most prevalent proteins in many cells, is actin, which exists in both monomeric/globular (G-actin) and polymeric/filamentary (F-actin) forms. Especially in animal cells, this often forms a network of entangled, cross-linked filaments known as the actin cortex, which is frequently found near the periphery of cells, and appears to be strongly associated with the outer membrane. *In vivo*, this network is far from passive, with both active motion and force generation during cell locomotion, and with a strong coupling to membrane proteins that appears to play a key role in the ability of cells to sense and respond to external stresses. Individual actin filaments are rather rigid to bending, exhibiting only small lateral bending fluctuations on the micrometer scale in thermal equilibrium. Other key cytoskeletal filaments include the much more rigid *microtubules*, and a much more flexible family of *intermediate filaments*, which typically play more specialized roles in specific cell types.

Fundamental to any understanding of the complex cytoskeleton is a quantitative model for the structure, interactions, and response of networks such as the actin cortex. Unlike conventional polymer networks and gels, however, these networks have been clearly shown to possess properties that cannot be modeled by existing polymer theories. These properties include anomalously large shear moduli, strong signatures of a nonlinear response (in which, for example, the shear modulus can increase by a full factor of 10 or more under modest strains of only 10% or so) (Janmey *et al.* 1994), and unique dynamics. In a very close and active collaboration between theory and experiment over the past decade or so, a standard model of sorts for the material properties of such networks has emerged. Central to these models has been the semiflexible nature of the constituent filaments, which is both a fundamental property of almost any filamentous protein and a clear departure from conventional polymer physics, which has focused on flexible and rod-like limits. In contrast, biopolymers such as F-actin and microtubules are nearly rigid on the scale of a micrometer, while at the same time showing significant thermal fluctuations on the cellular scale of many microns.

We begin with an introduction to the models of the response and dynamics of single filaments. Since cytoskeletal filaments are often the most important structural components in cells, a quantitative understanding of their mechanical response to bending, stretching, and compression is crucial for any model of the mechanics of networks of these filaments. We shall see how the fundamental properties of the individual filaments can explain many of the properties of solutions and networks of biopolymers. Finally, we will examine some of the nonequilibrium aspects of cells and reconstituted cytoskeletal networks.

5.1.2 Single-filament properties

Biopolymers such as those that make up the cytoskeleton consist of aggregates of large globular proteins that are bound together rather weakly, as compared with most synthetic, covalently bonded polymers. Nevertheless, they can in many cases form filaments of surprising stability and strength. Being rather bulky, however, with diameters as large as 10–25 nm, they are far more rigid than most common synthetic polymers. In fact, the most fundamental aspect in determining their long-wavelength behavior can be said to be their bending rigidity. In many cases, however, the molecular weight or contour length of these filaments is still large enough that they may begin to show significant thermal bending fluctuations. Thus they are said to be *semiflexible*, or *worm-like*.

The length at which significant bending fluctuations occur is a convenient characterization of the mechanical stiffness of such polymers. This thermal bending length, or *persistence length,* ℓ_{p} is defined more precisely in terms of the angular correlations of the local tangent along the polymer backbone, which decay exponentially with a characteristic length ℓ_{p}. The persistence lengths of a few important biopolymers are given in Table 5.1, along with their approximate diameter and length.

5.1.2.1 Worm-like Chain

When modeling such a polymer that is rigid on the molecular scale of a few nanometers, it is natural to think of the chain as a continuous line with a finite resistance to bending. This is the essence of the so-called worm-like-chain model. This can be described by an energy or Hamiltonian of the form,

Table 5.1 Persistence lengths and other parameters of various biopolymers (Howard 2001; Gittes *et al.* 1993).

Type	Approximate diameter	Persistence length	Contour length
DNA	2 nm	50 nm	$\lesssim$ 1 m
F-actin	7 nm	17 μm	$\lesssim$ 50 μm
Microtubule	25 nm	$\sim$ 1–5 mm	10s of μm

$$H_{\text{bend}} = \frac{\kappa}{2} \int ds \left| \frac{\partial \vec{t}}{\partial s} \right|^2, \tag{5.1}$$

where κ is the *bending modulus* and $\vec{t}$ is a (unit) tangent vector along the chain. Here, the chain position $\vec{r}(s)$ is described by an arc-length coordinate s along the chain backbone. Hence, the tangent vector is given by

$$\vec{t} = \frac{\partial \vec{r}}{\partial s}.$$

These quantities are illustrated in Fig. 5.1.

The bending modulus κ has units of energy times length. A natural energy scale for a rod subject to Brownian fluctuations is kT, where T is the temperature and k is Boltzmann's constant. Thus, $\ell_{\text{p}} = \kappa/(kT)$ is a length. This is a fundamental length, known as the *persistence length*, which has a very simple physical interpretation. It is the typical length scale over which the polymer forgets its orientation, i.e., the maximum length over which the polymer will appear straight in the presence of the constant Brownian forces it experiences in a medium at finite temperature.

More precisely, for a homogeneous rod of diameter $2a$ consisting of a homogeneous material, the bending modulus should be proportional to the Young's modulus E, which has units of energy per volume. Thus, on dimensional grounds, we expect that $\kappa \sim Ea^4$. In fact (Landau and Lifshitz 1986),

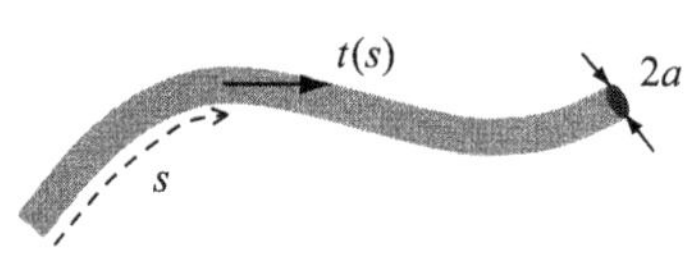

Fig. 5.1 A filamentous protein can be regarded as an elastic rod of radius a. Provided the length of the rod is very long compared with the monomeric dimension a and that the rigidity is high (specifically, the persistence length $\ell_{\text{p}} \gg a$), this can be treated as an abstract line or curve, characterized by the length s along its backbone. A unit vector $\vec{t}$ tangent to the filament defines the local orientation of the filament. Curvature is present when this orientation varies with s. For bending in a plane, it is sufficient to consider the angle $\theta(s)$ that the filament makes with respect to some fixed axis. The curvature is then $\partial\theta/\partial s$.

$$\kappa = \frac{\pi}{4} E a^4.$$

This is often expressed as $\kappa = EI$, where I is the moment of inertia of the cross section. For a rod-like object consisting of a homogeneous elastic material with modulus E, the bending modulus can only depend linearly on E, with a purely geometric prefactor depending only on the cross section. The factor $\pi a^4/4$ happens to be the right one for a solid rod of radius a. For a hollow tube, the prefactor would be different, but still of order a^4, where a is the (outer) radius.

In general, for bending in 3D, there are two independent directions for deflections of the rod or polymer transverse to its local axis. It is often instructive to consider a simpler case of a single transverse degree of freedom, i.e., motion confined to a plane. Here, the integrand in eqn (5.1) becomes $(\partial\theta/\partial s)$, where θ is simply the local angle that the chain axis makes relative to any fixed axis. A discrete approximation to the corresponding energy is

$$H \simeq \frac{\kappa}{2} \sum_i \frac{(\Delta\theta_i)^2}{\Delta s,} \tag{5.2}$$

where $\Delta\theta_i = \theta_i - \theta_{i-1}$.

Thus, from the equipartition theorem,

$$(\Delta\theta_i)^2 = \frac{kT\ \Delta s}{\kappa}. \tag{5.3}$$

This can be used to determine the correlations of orientations from one point along the chain to another. Specifically, we note that

$$\begin{aligned} \cos(\theta_m - \theta_n) &= \cos(\Delta\theta_m + \theta_{m-1} - \theta_n) \\ &= \cos(\Delta\theta_m)\cos(\theta_{m-1} - \theta_n) - \sin(\Delta\theta_m)\sin(\theta_{m-1} - \theta_n) \end{aligned} \tag{5.4}$$

and, thus,

$$\begin{aligned} \langle\cos(\theta_m - \theta_n)\rangle &= \langle\cos(\Delta\theta_m)\rangle\langle\cos(\theta_{m-1} - \theta_n)\rangle \\ &\dots \\ &= \langle\cos(\Delta\theta_m)\rangle^{m-n-1}. \end{aligned} \tag{5.5}$$

Here, we have used the independence of the various $\Delta\theta_i$, and the fact that $\langle\sin(\Delta\theta_m)\rangle = 0$.

From this, we can find the correlation function

$$\langle\vec{t}(s)\cdot\vec{t}(s')\rangle \simeq \langle\cos(\Delta\theta)\rangle^{|s-s'|/\Delta s} \simeq e^{-kT|s-s'|/2\kappa}. \tag{5.6}$$

Here, we have used the fact that $\langle\cos(\Delta\theta)\rangle \simeq 1 - \frac{1}{2}\langle\Delta\theta^2\rangle$ for small Δs and $\Delta\theta$. Of course, this is all for motion confined to a plane. Taking into account the two independent transverse directions for thermal fluctuations, one finds in 3D that

$$\langle\vec{t}(s)\cdot\vec{t}(s')\rangle = e^{-|s-s'|/\ell_p}, \tag{5.7}$$

where ℓ_p is the persistence length defined above. This persistence length provides a *geometric* measure of the *mechanical* stiffness of the rod, provided that it is in equilibrium at temperature T.

This provides, in principle, a way to measure the persistence length and thus the bending modulus of filaments, by imaging the angular correlations along a filament. As discussed below, however, one must also be careful to account for the dynamics of filaments.

5.1.2.2 Force–extension relation

At the heart of any model for network properties must be at least two things: a detailed model for the response at a single-molecule level, and a characterization of the way in which the single polymers are connected or otherwise interact with each other. The latter, being inherently a collective property, we reserve for later discussion.

A single filament can respond to forces in at least two ways. It can respond to both transverse and longitudinal forces by either bending or stretching/compressing. On length scales shorter than the persistence length, bending can be described in mechanical terms, as above for elastic rods. By contrast, stretching and compression can involve both a purely elastic or mechanical response (again, as in the stretching, compression, or even buckling of macroscopic elastic rods) and entropic response. The latter comes from the thermal fluctuations of the filament. Perhaps surprisingly, as we shall see below, the longitudinal response can be dominated by entropy even on length scales small compared with the persistence length. Thus, it is incorrect to think of a filament as truly rod-like, even on length scales short compared with ℓ_p.

The longitudinal single-filament response is often described in terms of a so–called force-extension relationship. Here, the force required to extend the filament is measured or calculated in terms of the degree of extension along a line. At any finite temperature, there is a resistance to such extension due to the presence of thermal fluctuations that make the polymer deviate from a straight conformation. This has been the basis of mechanical studies of long DNA, for example (Bustamante *et al.* 1994). In the limit of large persistence length, this can be calculated as follows (MacKintosh *et al.* 1995).

We consider a filament segment of length ℓ that is short compared with the persistence length ℓ_p. It is then nearly straight, with small transverse fluctuations. We let the x-axis define the average orientation of the chain segment, and let u and v represent the two independent transverse degrees of freedom. These can then be thought of as functions of x and time t in general. For simplicity, we shall mostly consider just one of these coordinates, $u(x, t)$. The bending energy is then

$$H_{\text{bend}} = \frac{\kappa}{2}\int dx \left(\frac{\partial^2 u}{\partial x^2}\right)^2 = \frac{\ell}{4}\sum_q \kappa q^4 u_q^2, \tag{5.8}$$

where we have represented $u(x)$ by a Fourier series

$$u(x, t) = \sum_q u_q \sin(qx). \tag{5.9}$$

Fig. 5.2 If one end of a filament is fixed, both in position and in orientation, while the other is free, the filament tends to wander in a way that can be characterized by $u(x)$, a transverse displacement field. For a fixed total arc length of the filament, thermal fluctuations result in a contraction of the end-to-end distance, which is denoted by $\Delta\ell$. In fact, this contraction is actually distributed about a thermal average value $\langle\Delta\ell\rangle$. The mean square (longitudinal) fluctuations about this average are denoted by $\langle\delta\ell^2\rangle$, and the mean square lateral fluctuations (i.e., with respect to the dashed line) are denoted by $\langle u^2\rangle$.

As illustrated in Fig. 5.2, the local orientation of the filament is given by the slope $\partial u/\partial x$, and the local curvature is given by the second derivative $\partial^2 u/\partial x^2$. Such a description is appropriate for the case of a nearly straight filament with fixed boundary conditions $u = 0$ at the ends, $x = 0, \ell$. For this case, the wave vectors q are equal to $n\pi/\ell$, where $n = 1, 2, 3, \ldots$.

We assume that the chain has no compliance in its contour length, i.e., that the total arc length

$$\int dx \sqrt{1 + \left|\frac{\partial u}{\partial x}\right|^2} \tag{5.10}$$

is fixed. Then, the contraction of the chain relative its full contour length in the presence of thermal fluctuations in u is

$$\Delta\ell = \int dx \left(\sqrt{1 + \left|\frac{\partial u}{\partial x}\right|^2} - 1\right) \simeq \frac{1}{2}\int dx \left|\frac{\partial u}{\partial x}\right|^2. \tag{5.11}$$

The integration here is actually over the projected length of the chain. But, to leading (quadratic) order in the transverse displacements, we make no distinction between projected and contour lengths here, nor in the case of $H_{\rm bend}$ above.

Thus, the contraction is

$$\Delta\ell = \frac{\ell}{4}\sum_q q^2 u_q^2. \tag{5.12}$$

Conjugate to this variable is the tension τ in the chain. Thus, we consider the effective energy

$$H = \frac{1}{2}\int dx \left[\kappa\left(\frac{\partial^2 u}{\partial x^2}\right)^2 + \tau\left(\frac{\partial u}{\partial x}\right)^2\right] = \frac{\ell}{4}\sum_q \left(\kappa q^4 + \tau q^2\right) u_q^2. \tag{5.13}$$

Under a constant tension τ, therefore, the equilibrium amplitudes u_q must satisfy

$$\langle |u_q|^2 \rangle = \frac{2kT}{\ell\,(\kappa q^4 + \tau q^2)}. \tag{5.14}$$

and the contraction satisfies

$$\langle \Delta\ell \rangle = kT \sum_q \frac{1}{(\kappa q^2 + \tau)}. \tag{5.15}$$

There are, of course, two transverse degrees of freedom, and so this last answer incorporates a factor of two appropriate to a chain fluctuating in 3D.

Semiflexible filaments exhibit a strong suppression of bending fluctuations for modes of wavelength less than the persistence length ℓ_p. More precisely, as we see from eqn (5.14), the mean square amplitudes of shorter-wavelength modes are increasingly suppressed as the fourth power of the wavelength. This has important consequences for many of the thermal properties of such filaments. In particular, it means that the longest unconstrained wavelengths tend to be dominant in most cases. This allows us, for instance, to anticipate the scaling form of the end-to-end contraction $\Delta\ell$ between points separated by an arc length ℓ in the absence of an applied tension. We note that it is a length, and it must vary inversely with the stiffness κ and must increase with temperature. Thus, since the dominant mode of the fluctuations is that of the maximum wavelength, ℓ, we expect the contraction to be of the form $\langle \Delta\ell \rangle_0 \sim \ell^2/\ell_p$. More precisely, for $\tau = 0$,

$$\langle \Delta\ell \rangle_0 = \frac{kT\ell^2}{\kappa\pi^2} \sum_{n=1}^{\infty} \frac{1}{n^2} = \frac{\ell^2}{6\ell_p}. \tag{5.16}$$

Similar scaling arguments to those above lead us to expect that the typical transverse amplitude of a segment of length ℓ is approximately given by

$$\langle u^2 \rangle \sim \frac{\ell^3}{\ell_p} \tag{5.17}$$

in the absence of applied tension. The precise coefficient for the mean square amplitude of the midpoint between ends separated by ℓ (with vanishing deflection at the ends) is 1/24.

For a finite tension τ, however, there is an extension of the chain (toward full extension) by an amount

$$\delta\ell = \langle \Delta\ell \rangle_0 - \langle \Delta\ell \rangle_\tau = \frac{kT\ell^2}{\kappa\pi^2} \sum_n \frac{\phi}{n^2\,(n^2 + \phi)}, \tag{5.18}$$

where $\phi = \tau\ell^2/(\kappa\pi^2)$ is a dimensionless force. The characteristic force $\kappa\pi^2/\ell^2$ that enters here is the critical force in the classical Euler buckling problem (discussed below). Thus, the force–extension curve can be found by inverting this relationship. In the linear regime, this becomes

$$\delta\ell = \frac{\ell^2}{\ell_p \pi^2}\phi \sum_n \frac{1}{n^4} = \frac{\ell^4}{90\ell_p\kappa}\tau, \tag{5.19}$$

i.e., the effective spring constant for longitudinal extension of the chain segment is $90\kappa\ell_p/\ell^4$. The scaling form of this could also have been anticipated, based on very simple physical arguments similar to those above. In particular, given the expected dominance of the longest-wavelength mode (i.e., ℓ), we expect that the end-to-end contraction will scale as $\delta\ell \sim \int (\partial u/\partial x)^2 \sim u^2/\ell$. Thus, $\langle\delta\ell^2\rangle \sim \ell^{-2}\langle u^4\rangle \sim \ell^{-2}\langle u^2\rangle^2 \sim \ell^4/\ell_p^2$, which is consistent with the effective (linear) spring constant derived above. The full nonlinear force–extension curve can be calculated numerically by inversion of the expression above. This is shown in Fig. 5.3. Here, one can see both the linear regime for small forces, with the effective spring constant given above, and a divergent force near full extension. In fact, the force diverges in a characteristic way, as the inverse square of the distance from full extension: $\tau \sim |\delta\ell - \Delta\ell|^{-2}$ (Fixman and Kovac 1973).

Before concluding our discussion of the longitudinal response of semiflexible polymers, it is worth asking about another obvious contribution to their response. This we can think of as the *zero-temperature* enthalpic or *purely mechanical* response. After all, we are treating semiflexible polymers as small bendable rods. To the extent that they behave of rigid rods, we might expect them to respond to longitudinal stresses by, for example, increasing/decreasing their arc length. Based on the arguments above, it seems that the persistence length ℓ_p determines the length below which filaments behave like rods, and above which they behave like flexible polymers with significant thermal fluctuations. Perhaps surprisingly, however, even for short segments of semiflexible polymers of length much less than the persistence length, their longitudinal response can be dominated by the entropic force–extension relation described above, i.e., in which the response is due to transverse thermal fluctuations.

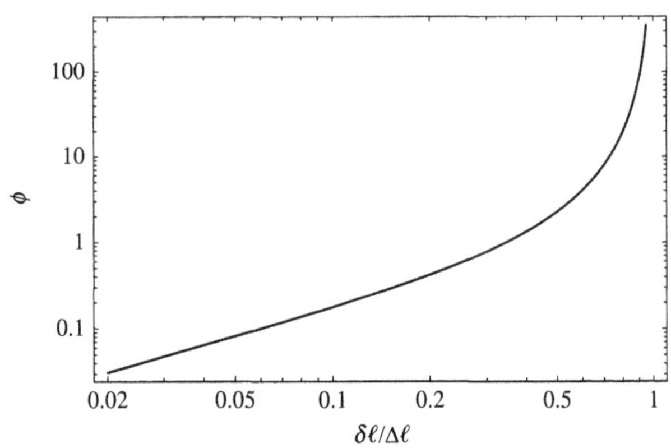

Fig. 5.3 The dimensionless force ϕ as a function of extension $\delta\ell$, relative to maximum extension $\Delta\ell$. For small extension, the response is linear.

In order to examine this, we consider a simple model of a semiflexible polymer, in which it is treated as a homogeneous elastic rod of radius a. We have already seen that the bending modulus κ is of order Ea^4. Likewise, the (linear) stretching/compression of such an elastic rod is described by the Hamiltonian

$$H_{\text{stretch}} = \frac{1}{2}\mu \int \mathrm{d}s \left(\frac{\mathrm{d}\ell(s)}{\mathrm{d}s} \right)^2, \tag{5.20}$$

where $\mathrm{d}\ell/\mathrm{d}s$ gives the relative change in length (the strain) along the filament. The stretch modulus μ is of order Ea^2. The effective (mechanical) spring constant of a segment of length ℓ is thus $k_{\mathrm{M}} \sim \mu/\ell \sim Ea^2/\ell$, compared with the effective (thermal) spring constant $k_{\mathrm{T}} \sim \kappa\ell_{\mathrm{p}}/\ell^4 \sim a^2\ell_{\mathrm{p}}/\ell^3$. Since the system will respond primarily according to the softer effective spring constant, the dominant response will be thermal if $\ell \gtrsim a^2\ell_{\mathrm{p}}$, and will be mechanical only if $\ell \lesssim a^2\ell_{\mathrm{p}}$. Thus, even segments of length much less than ℓ_{p} can still respond according to the thermal response described above. For F-actin, for example, even filament segments as short as a fraction of 1 μm in length may be dominated in their longitudinal compliance by the thermal response arising from bending fluctuations. The extent of this compliance, however, will be quite limited. Thus, there is expected to be a crossover from a thermal compliance, characterized by linear regime followed by stiffening, to an enthalpic stretching regime (Odijk 1995; Storm *et al.* 2005).

In addition to the mechanical or enthalpic response to stretching, there is also another purely mechanical response under compression: Euler buckling (Landau and Lifshitz 1986). When an initially straight elastic rod is subject to a compressive load, it initially responds by compressing longitudinally along its axis. Above a well-defined threshold, however, it undergoes a buckling instability. This is an instability in which, to a good approximation, the rod can simply not bear any increase in compressive load beyond the threshold value. Any additional load will cause the rod to completely collapse. This threshold compressive force can be calculated as follows.

When a rod of length ℓ undergoes an oscillatory transverse deflection of amplitude $u(x) = \sum_q u_q \sin(qx)$, it contracts longitudinally by an amount given by eqn (5.12). The energy of this deflection is given by eqn (5.13), where τ is the (tensile) load. For $\tau > 0$, each term in the series contributes a positive energy, so that the system is (harmonically) stable against increasing amplitude u_q, for each q. For compressive loads, however, where $\tau < 0$, this is no longer the case if $\tau < -\kappa q^2$. For such a mode q, the system is unstable to transverse displacements. As the compressive load $-\tau$ increases from zero, such an instability first happens at the smallest q possible, which is determined by the length of the rod: $q = q_1 = \pi/\ell$. Thus, the buckling instability occurs for a compressive force given by

$$\kappa \left(\frac{\pi}{\ell} \right)^2$$

and this instability corresponds to the fundamental oscillatory mode of the rod, where the wavelength of the instability is twice the length of the rod.

Interestingly, if the elastic rod is embedded in a surrounding elastic material, as can be the case for microtubules embedded in the matrix formed by the rest of the cytoskeleton, then this classical Euler buckling problem and the corresponding threshold are altered (Landau and Lifshitz 1986). We can see this from eqn (5.12), since a given amount of longitudinal compression $\Delta\ell$ requires a larger amplitude u_q for smaller q. Such large-amplitude lateral deflections are suppressed by the surrounding matrix. More precisely, for a transverse deflection u, the elastic energy per unit length along the rod is approximately given by

$$\frac{1}{2}\alpha_\perp u^2,$$

where

$$\alpha_\perp = \frac{4\pi}{\ln(\lambda/b)} G.$$

Here, G is the shear modulus of the surrounding elastic medium, λ is the bending wavelength, and b is a microscopic length of the order of the radius of the rod. Thus, using the mode decomposition above, we obtain the elastic energy

$$H = \frac{1}{2}\int dx \left[\kappa\left(\frac{\partial^2 u}{\partial x^2}\right)^2 + \tau\left(\frac{\partial u}{\partial x}\right)^2 + \alpha_\perp u^2\right] = \frac{\ell}{4}\sum_q \left(\kappa q^4 + \tau q^2 + \alpha_\perp\right) u_q^2. \quad (5.21)$$

The elastic energy of the matrix has the effect of both increasing the threshold force and shifting the instability to a shorter wavelength. The first unstable mode corresponds to

$$q_* = \sqrt[4]{\alpha_\perp/\kappa}, \quad (5.22)$$

and the critical force is now

$$f_c = 2\kappa q_*^2\,, \quad (5.23)$$

which can be much larger than that for unconstrained buckling. Such constrained buckling has been reported for microtubules under compression in cells, where the surrounding cytoskeletal meshwork can greatly enhance the compressive load-bearing capability of microtubules by up to 100 times (Brangwynne *et al.* 2006).

5.1.2.3 Dynamics

In the above, we have considered only static properties at the single-chain level. The dynamics of individual chains exhibits a rich behavior that can have important consequences even at the level of bulk solutions and networks. The principal dynamic modes come from the transverse motion, i.e., the degrees of freedom u and v described above. Thus, we must consider the time dependence of these quantities. The transverse equation of motion of the chain can be found from H_{bend} above, together with the hydrodynamic drag of the filaments through the solvent. This is done via a Langevin equation describing the net force per unit length on the chain at position x,

$$0 = -\zeta \frac{\partial}{\partial t} u(x,t) - \kappa \frac{\partial^4}{\partial x^4} u(x,t) + \xi_\perp(x,t), \tag{5.24}$$

which is, of course, zero within the linearized, inertia-free (low-Reynolds-number) hydrodynamics that we assume here. The first term represents the hydrodynamic drag per unit length of the filament. Here, we have assumed a constant transverse drag coefficient that is independent of wavelength. In fact, the actual (low-Reynolds-number) drag per unit length on a rod of length L is $\zeta = 4\pi\eta/\ln(AL/a)$, where L/a is the aspect ratio of the rod, and A is a constant of order unity that depends on the precise geometry of the rod. For a filament fluctuating freely in solution, a weak logarithmic dependence on wavelength is thus expected. In practice, the presence of other chains in the solution gives rise to an effective screening of the long-range hydrodynamics beyond a length of the order of the separation between chains, which can then be taken in place of L above. The second term in the Langevin equation above is the restoring force per unit length due to bending. It has been calculated from $-\delta H_{\text{bend}}/\delta u(x,t)$ with the help of integration by parts. Finally, we include a random force $\xi_\perp$.

A simple force balance in the Langevin equation above leads us to conclude that the characteristic relaxation rate of a mode of wave vector q is (Farge and Maggs 1993)

$$\omega(q) = \frac{\kappa q^4}{\zeta}. \tag{5.25}$$

The fourth-order dependence of this rate on q is to be expected from the appearance of a single time derivative along with four spatial derivatives in eqn (5.24). This relaxation rate determines, among other things, the correlation time for the fluctuating bending modes. Specifically, in the absence of an applied tension,

$$\langle u_q(t) u_q(0) \rangle = \frac{2kT}{\ell \kappa q^4} e^{-\omega(q)t}. \tag{5.26}$$

The fact that the relaxation rate varies as the fourth power of the wave vector q has important consequences. For example, while the time it takes for an actin filament bending mode of wavelength 1 μm to relax is of order 10 ms, it takes about 100 s for a mode of wavelength 10 μm.

The very strong wavelength dependence of the relaxation rate in eqn (5.25) has important consequences for the imaging of the thermal fluctuations of filaments, for instance, as is done in order to measure ℓ_p and the filament stiffness (Gittes *et al.* 1993). This is the basis, in fact, of most measurements to date of the stiffness of DNA, F-actin, and other biopolymers. Using eqn (5.26), for instance, one can both confirm thermal equilibrium and determine ℓ_p by measuring the mean square amplitudes of the thermal modes of various wavelengths. However, in order both to resolve the various modes and to establish that they behave according to the thermal distribution, one must sample over times long compared with $1/\omega(q)$ for the longest wavelengths $\lambda \sim 1/q$. At the same time, one must be able to resolve fast motion on times of order $1/\omega(q)$ for the shortest wavelengths. Given the strong dependence of these relaxation times on the corresponding wavelengths, a range of the order of a factor of 10 in the wavelengths of the modes corresponds to a range of 10^4 in observation times, for instance.

Another way to look at the result of eqn (5.25) is that a bending mode of wavelength λ relaxes (i.e., fully explores its equilibrium conformations) in a time of order $\zeta\lambda^4/\kappa$. Since it is also true that the longest-wavelength (unconstrained) bending mode has by far the largest amplitude, and thus dominates the typical conformations of any filament (see eqns (5.15) and (5.26)), we can see that in a time t, the *typical* or dominant mode that relaxes is one of wavelength $\ell_\perp(t) \sim (\kappa t/\zeta)^{1/4}$. As we have seen above in eqn (5.17), the mean square amplitude of the transverse fluctuations increases with the filament length ℓ as $\langle u^2\rangle \sim \ell^3/\ell_\mathrm{p}$. Thus, in a time t, the expected mean square transverse motion is given by (Farge and Maggs 1993; Amblard *et al.* 1996)

$$\langle u^2(t)\rangle \sim \frac{(\ell_\perp(t))^3}{\ell_p} \sim t^{3/4}, \tag{5.27}$$

because the typical, dominant mode contributing to the motion at time t is of wavelength $\ell_\perp(t)$.

The dynamics of the longitudinal motion can be calculated similarly. Here, however, we must account for the fact that the mean square longitudinal fluctuations $\langle\delta\ell^2(t)\rangle$ of a long filament involve a sum (in quadrature) of independently fluctuating segments along a full filament of length ℓ. The typical size of such independently fluctuating segments at time t is $\ell_\perp(t)$, of which there are $\ell/\ell_\perp(t)$. As shown above, the mean square amplitude of the longitudinal fluctuations of a fully relaxed segment of length $\ell_\perp(t)$ is of order $\ell_\perp(t)^4/\ell_\mathrm{p}^2$. Thus, the longitudinal motion is given by (Granek 1997; Gittes and MacKintosh 1998)

$$\langle\delta\ell(t)^2\rangle \sim \frac{\ell}{\ell_\perp(t)} \times \frac{\ell_\perp(t)^4}{\ell_\mathrm{p}^2} \sim t^{3/4}, \tag{5.28}$$

where the mean square amplitude is smaller than that for the transverse motion by a factor of order ℓ/ℓ_p. By using the Fluctuation-Dissipation Theorem (FDT) (see the appendix in Section 5.2), it is also possible to determine the time- or frequency-dependent compliance of such a filament with respect to a longitudinal force applied at one end from these end-to-end fluctuations. This can also be expressed in terms of a (complex) spring constant $K_\mathrm{eff}(\omega)$:

$$K_\mathrm{eff}(\omega) = \kappa\ell_\mathrm{p}\left(\frac{-2i\zeta}{\kappa}\right)^{3/4}\omega^{3/4}. \tag{5.29}$$

Both for the short-time fluctuations and for the static fluctuations of a filament segment of length ℓ, a filament end explores a disk-like region with a longitudinal motion smaller than the perpendicular motion by this factor.

For the problem as stated above, i.e., for an isolated fluctuating filament in a quiescent solvent, there is a potential problem with the analysis above, which includes only the effect of drag for motion perpendicular to the filament (Everaers *et al.* 1999). In fact, there is a finite propagation of tension along a semiflexible filament, expressed by yet another length (Morse 1998*b*),

$$\ell_{\|}(t) \sim \sqrt{\ell_{\perp}(t)\ell_{\mathrm{p}}} \sim t^{1/8}. \tag{5.30}$$

This represents, for instance, the range along the filament over which the tension has spread from a point of disturbance. At very short times, it is possible to observe a $t^{7/8}$ motion in some cases rather than the $t^{3/4}$ in eqn (5.28) (Everaers *et al.* 1999). For the high-frequency rheology of semiflexible polymer networks, however, only a dynamical regime corresponding to eqn (5.28) is observed (Gittes *et al.* 1997; Koenderink *et al.* 2006) and expected (Morse 1998*a*; Gittes and MacKintosh 1998). In solutions, consequences of the dynamics represented in eqn (5.30) are expected (Pasquali *et al.* 2001) and have been reported experimentally (Koenderink *et al.* 2006), depending on the length or molecular weight of the polymer.

5.1.3 Solutions of semiflexible polymers

Given the rigidity of semiflexible polymers at scales shorter than their contour length, it is not surprising that in solution they interact with each other in very different ways than flexible polymers would at, for example, the same concentration. In addition to the important characteristic lengths of the molecular dimensions (say, the filament diameter $2a$), the material parameter ℓ_{p}, and the contour length of the chains, there is another important new length scale in a solution, the *mesh size*, or typical spacing between polymers in solution, ξ. A simple estimate (Schmidt *et al.* 1989) shows how ξ depends on the molecular size a and the polymer volume fraction ϕ. In the limit where the persistence length ℓ_{p} is large compared with ξ, we can approximate the solution on the scale of the mesh as one of rigid rods. Hence, within a cubical volume of size ξ, there is of order one polymer segment of length ξ and cross section a^2, which corresponds to a volume fraction ϕ of order $(a^2\xi)/\xi^3$. Thus,

$$\xi \sim \frac{a}{\sqrt{\phi}}. \tag{5.31}$$

While the mesh size characterizes the typical spacing between polymers within a solution, it does not entirely determine the way in which they interact with each other. For instance, for a random static arrangement of rigid rods, it is not hard to see that polymers will not touch each other on average except on a much larger length scale: imagine threading a random configuration of rods at small volume fraction through a thin needle. An estimate of the distance between typical interactions (entanglements) of semiflexible polymers must account for their thermal fluctuations (Odijk 1983). As we have seen above, the transverse range of fluctuations δu a distance ℓ away from a fixed point grows according to $\delta u^2 \sim \ell^3/\ell_{\mathrm{p}}$. Along this length, such a fluctuating filament explores a narrow cone-like volume of order $\ell\,\delta u^2$. An entanglement that leads to a constraint of the fluctuations of such a filament occurs when, with probability of order unity, another filament crosses through this volume, in which case it will occupy a volume of order $a^2\,\delta u$, since $\delta u \ll \ell$. Thus, the volume fraction and the contour length ℓ between constraints are of order $\phi \sim a^2/(\ell\,\delta u)$, taking the corresponding length as an entanglement length

$$\ell_e \sim (a^4 \ell_p)^{1/5} \phi^{-2/5}, \tag{5.32}$$

which is larger than the mesh size ξ in the semiflexible limit $\ell_p \gg \xi$.

These transverse entanglements, separated by a typical length ℓ_e, govern the elastic response of solutions, in a way first outlined by Isambert and Maggs (1996). A more complete discussion of the rheology of such solutions can be found in Morse (1998*a*) and Hinner *et al.* (1998). The basic result for the rubber-like plateau shear modulus for such solutions can be obtained by noting that the number density of entropic constraints (entanglements) is thus $n\ell/\ell_c \sim 1/(\xi^2 \ell_e)$, where $n = \phi/(a^2 \ell)$ is the number density of chains of contour length ℓ. In the absence of other energetic contributions to the modulus, the entropy associated with these constraints results in a shear modulus of order $G \sim kT/(\xi^2 \ell_e) \sim \phi^{7/5}$. This has been well established in a experiments such as those of Hinner *et al.* (1998).

With increasing frequency, or for short times, the macroscopic shear response of solutions is expected to show the underlying dynamics of the individual filaments. One of the main signatures of the frequency response of polymer solutions in general is an increase in the shear modulus with increasing frequency. In practice, for high-molecular-weight F-actin solutions of approximately 1 mg/ml, this is seen for frequencies above a few hertz. Initial experiments to measure this response by imaging the dynamics of small probe particles have shown that the shear modulus increases as $G(\omega) \sim \omega^{3/4}$ (Gittes *et al.* 1997; Schnurr *et al.* 1997), which has since been confirmed in other experiments and by other techniques.

This behavior can be understood in terms of the dynamic longitudinal fluctuations of single filaments, as shown above (Morse 1998*a*; Gittes and MacKintosh 1998). Much as the static longitudinal fluctuations $\langle \delta\ell^2 \rangle \sim \ell^4/\ell_p^2$ correspond to an effective longitudinal spring constant $\sim kT\ell_p^2/\ell^4$, the time-dependent longitudinal fluctuations shown above in eqn (5.28) correspond to a time- or frequency-dependent compliance or stiffness, in which the effective spring constant increases with increasing frequency, as shown in eqn (5.29). This is because, on shorter timescales, fewer bending modes can relax, which makes the filament less compliant. Accounting for the random orientations of filaments in solution results in a frequency-dependent shear modulus

$$G(\omega) = \frac{1}{15} \rho \kappa \ell_p \left(-\frac{2i\zeta}{\kappa} \right)^{3/4} \omega^{3/4} - i\omega\eta, \tag{5.33}$$

where ρ is the polymer concentration measured in length per unit volume. This shear modulus has been quantitatively confirmed in recent measurements on *in vitro* networks (Koenderink *et al.* 2006), and has also been reported for the high-frequency response of living cells (Deng *et al.* 2006).

5.1.4 Network elasticity

In a living cell, there are many different specialized proteins for binding, bundling, and otherwise modifying the network of filamentous proteins. In fact, more than 100 actin-associated proteins alone have been identified. Not only is it important to understand the mechanical roles of, for example, cross-linking proteins, but as we shall see, these can have a much more dramatic effect on the properties of the network than is the case for solutions and networks of flexible polymers.

The introduction of cross-linking agents into a solution of semiflexible filaments introduces yet another important and distinct length scale, which we shall call the cross-link distance ℓ_c. As we have just seen, in the limit where $\ell_p \gg \xi$, individual filaments may interact with each other only infrequently. That is to say that, in contrast with flexible polymers, the distance between the interactions of one polymer with its neighbors (ℓ_e in the case of solutions) may be much larger than the typical spacing between polymers. Thus, if there are biochemical cross-links between filaments, these may result in significant variation of the network properties even when ℓ_c is larger than ξ.

Given a network of filaments connected to each other by cross-links spaced an average distance ℓ_c apart along each filament, the response of the network to macroscopic strains and stresses may involve two distinct single-filament responses: (1) bending of filaments, and (2) stretching/compression of filaments. Models based on both of these effects have been proposed and analyzed. Bending-dominated behavior has been suggested both for ordered (Satcher and Dewey 1996) and for disordered (Kroy and Frey 1996) networks. That individual filaments bend under network strain is perhaps not surprising, unless one thinks of the case of uniform shear. In this case, only rotation and stretching/compression of individual rod-like filaments are possible. This is the basis of so-called *affine* network models (MacKintosh *et al.* 1995). In contrast, the bending of constituents involves *nonaffine* deformations.

It has recently been shown (Head *et al.* 2003*a*, *b*; Wilhelm and Frey 2003) that which of these behaviors is expected depends, for example, on filament length and cross-link concentration. Nonaffine behavior is expected either at low concentrations or for short filaments, while the deformation becomes increasingly affine at high concentrations or for long filaments. For the first of these responses, the network shear modulus is expected to be of the form

$$G_{\rm NA} \sim \frac{\kappa}{\xi^4} \sim c^2 \tag{5.34}$$

(the subscript "NA" denotes "nonaffine") when the density of cross-links is high (Kroy and Frey 1996; Satcher and Dewey 1996).

For affine deformations, the modulus can be estimated using the effective single-filament longitudinal spring constant for a filament segment of length ℓ_c between cross-links, $\sim \kappa\ell_p/\ell_c^4$, as derived above. Given a total polymer length per unit volume of ρ, together with an effective tension of order $(\kappa\ell_p/\ell_c^3)\epsilon$, where ϵ is the strain, the shear modulus is expected to be

$$G_{\rm AT} \sim \rho\frac{\kappa\ell_p}{\ell_c^3}. \tag{5.35}$$

This shows that the shear modulus is expected to be strongly dependent on the density of cross-links. Recent experiments on *in vitro* model gels consisting of F-actin with permanent cross-links, for instance, have shown that the shear modulus can vary from less than 1 Pa to over 100 Pa at the same concentration of F-actin, when the cross-link concentration is varied (Gardel *et al.* 2004).

In the preceding derivation, we have assumed a thermal/entropic (affine, thermal) response of the filaments to longitudinal forces. As we have seen above, however, for shorter filament segments (for example, for small enough ℓ_c), one may expect a mechanical response characteristic of rigid rods that can stretch and compress. This would lead to a different expression (affine, mechanical) for the shear modulus

$$G_{\mathrm{AM}} \sim \rho\mu, \tag{5.36}$$

which is simply proportional to the concentration. The expectations for the various mechanical regimes are shown in Fig. 5.4 (Head *et al.* 2003*b*). Recent theory has also shown that initially nonaffinely deforming networks tend to deform increasingly affinely at higher shear strains (Onck *et al.* 2005), as the filaments tend to align with the shear. Interestingly, this can also give rise to a nonlinear stiffening response, as one goes from a smaller nonaffine modulus determined primarily by bending to an increasingly stretch-dominated response, even for linear elastic elements in the network. However, the maximum degree of such stiffening will be limited by the (nonlinear) affine modulus, which represents an upper bound on the stiffness.

5.1.5 Nonlinear response

One of the characteristics of the elastic response of many biopolymer networks is their nonlinear behavior (Janmey *et al.* 1994, Storm *et al.* 2005). This is shown in Fig. 5.5. In particular, these networks exhibit stiffening by up to a factor of 10 or more under strain. In fact, many biological tissues show the same sort of strain stiffening. These materials are compliant, while being able to withstand a wide range of shear stresses.

This strain-stiffening behavior can be understood in terms of the characteristic force–extension behavior of an individual semiflexible filament, as described above. As can be seen in Fig. 5.3, for small extensions or strains, there is a linear increase in the

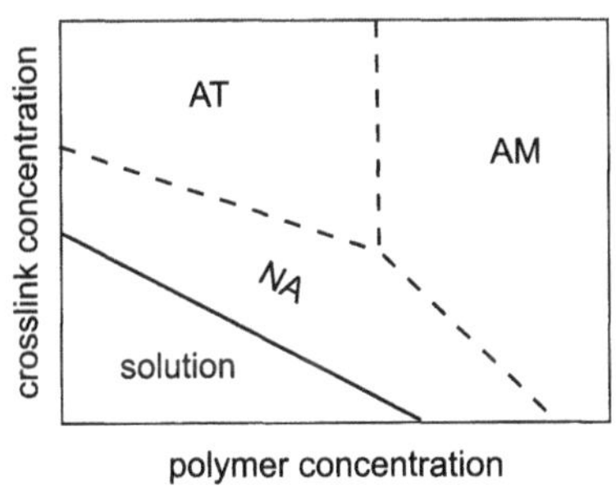

Fig. 5.4 A sketch of the expected diagram showing the various elastic regimes in terms of cross-link density and polymer concentration. The solid line represents the rigidity percolation transition, where rigidity first develops from a solution at a macroscopic level. The other, dashed lines indicate crossovers (not thermodynamic transitions). "NA" indicates the nonaffine regime, and "AT" and "AM" refer to the affine thermal (or entropic) and mechanical regimes, respectively.

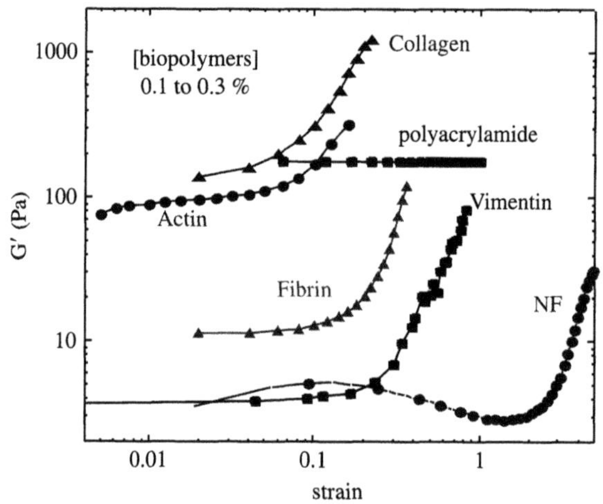

Fig. 5.5 A wide variety of biopolymers, including F-actin and intermediate filaments, form networks that stiffen dramatically under an applied shear strain. Figure adapted from Storm *et al.* (2005).

force. As the strain increases, however, the force is seen to grow more rapidly. In fact, in the absence of any compliance in the arc length of the filament, the force strictly diverges at a finite extension. This suggests that for a network, the macroscopic stress should diverge, while in the presence of high stress, the macroscopic shear strain is bounded and ceases to increase. In other words, after being compliant at low stress, such a material will be seen to stop responding, even under high applied stress.

This can be made more quantitative by calculating the macroscopic shear stress of a strained network, including random orientations of the constituent filaments. Specifically, for a given shear strain γ, the tension in a filament segment of length ℓ_c is calculated, based on the force–extension relation above. This is done within the affine approximation, in which the microscopic strain in any such filament segment is determined precisely by the macroscopic strain, as well as the filament's orientation with respect to the shear. The contribution of such a filament's tension to the macroscopic stress also depends on its orientation in space. Finally, the concentration or number density of such filaments crossing this horizontal plane is a function of the overall polymer concentration and the filament orientation.

The full nonlinear shear stress is calculated as a function of γ, the polymer concentration, and ℓ_c, by adding all such contributions from all (assumed random) orientations of filaments. This can then be compared with macroscopic rheological studies of cross-linked networks, such as were done recently by Gardel *et al.* (2004). These experiments measured the differential modulus $\partial\sigma/\partial\gamma$ versus the applied stress σ and found good agreement with the predicted increase in this modulus with increasing stress. In particular, given the quadratic divergence of the single-filament tension shown above (Fixman and Kovac 1973), this modulus is expected to increase as $\partial\sigma/\partial\gamma \sim \sigma^{3/2}$, which is consistent with the experiments of Gardel *et al.* (2004).

This provides a strong test of the underlying mechanism of network elasticity. In addition to good agreement between theory and experiment for densely cross-linked networks, these experiments have also shown evidence of a lack of strain-stiffening behavior in these networks at lower concentrations (of polymer or cross-links), which may provide evidence for the nonaffine regime of network response described above. It has also been shown that even initially nonaffinely deforming networks can exhibit a stretching regime, although this may occur at higher strains than those predicted by the affine model (Onck *et al.* 2005; Palmer and Boyce 2008; Huisman *et al.* 2008).

5.1.6 Active, nonequilibrium networks

Cells differ in many ways from artificial materials. Metabolism, for instance, keeps them far from equilibrium by means of motor proteins and other active processes that drive shape changes and motion. In cells, both the mechanical scaffold of the cytoskeleton and the force-generating elements that act on it form an interwoven composite. The result represents a new class of active materials that are currently the focus of intense research.

At the microscopic level, movement in cells is generated at the expense of chemical free energy, mostly in the form of adenosine triphosphate (ATP). Through ATP hydrolysis, conformational changes and resulting motions of motors are generated in polar protein filaments of F-actin and microtubules. The result is a highly dynamic and active composite material, very different from passive equilibrium materials, both natural and artificial. Understanding such active materials remains a challenge, but is necessary for understanding cellular dynamics, and also holds out the promise of designing synthetic active materials. Here, we discuss recent experimental results and theories, with a focus on nonequilibrium implications for both mechanics and dynamics.

The cytoskeletal motor proteins of the myosin (interacting with F-actin) and kinesin (interacting with microtubules) families have been studied extensively in single-molecule assays (Block 2007). Recent experiments combining reconstituted cytoskeletal networks with molecular motor activity have focused on both noncross-linked, fluid-like actin or microtubule solutions, in which motors generally cause large-scale structural changes and pattern formation, and cross-linked networks, in which motors can generate steady-state tensions in the network and can cause fluctuations that are much stronger than thermal fluctuations. Microtubules can be driven by kinesin motors to form astral patterns that are reminiscent of the poles of the mitotic spindle (Nedelec *et al.* 1997). In the case of actin solutions, the addition of myosin motors can cause network contraction, macroscopic phase separation, or even fluidization by driving filaments past each other to relax stress (Humphrey *et al.* 2002). By contrast, in permanently cross-linked networks, such fluidization is not expected, and stresses may build up within the elastic matrix.

In cells, the cytoskeleton is organized by various bundling and cross-linking proteins. The resulting structure is transient, and can re-form on timescales of minutes and longer. On shorter timescales, however, the solid-like or viscoelastic character will be predominant. It is thus also important to identify the effects of nonequilibrium force generators on cross-linked networks. In recent experiments, simplified model systems have been developed, consisting of cross-linked actin filaments together with muscle myosin II (Mizuno *et al.* 2007; Koenderink *et al.* 2009). In both cases, a very strong stiffening of the networks was found. Specifically, the shear modulus was found to increase by nearly 100-fold, in an ATP-dependent manner. This can be understood in terms of the nonlinear stiffening of equilibrium networks under external stress. The myosin motors induce tension in the actin filaments, as illustrated in Fig. 5.6. In a random network, this is expected to generate isotropic stress inside the network. It was shown that the effect of this isotropic stress could be mimicked by a shear stress imposed externally, which could then be used to estimate the tension generated by the motors.

These networks were also probed by a combination of micromanipulation and fluctuation-tracking techniques that were able, using basic principles of statistical physics, to unambiguously distinguish active/vital processes from passive/thermal fluctuations. To the networks were added micrometer-sized probe particles, whose

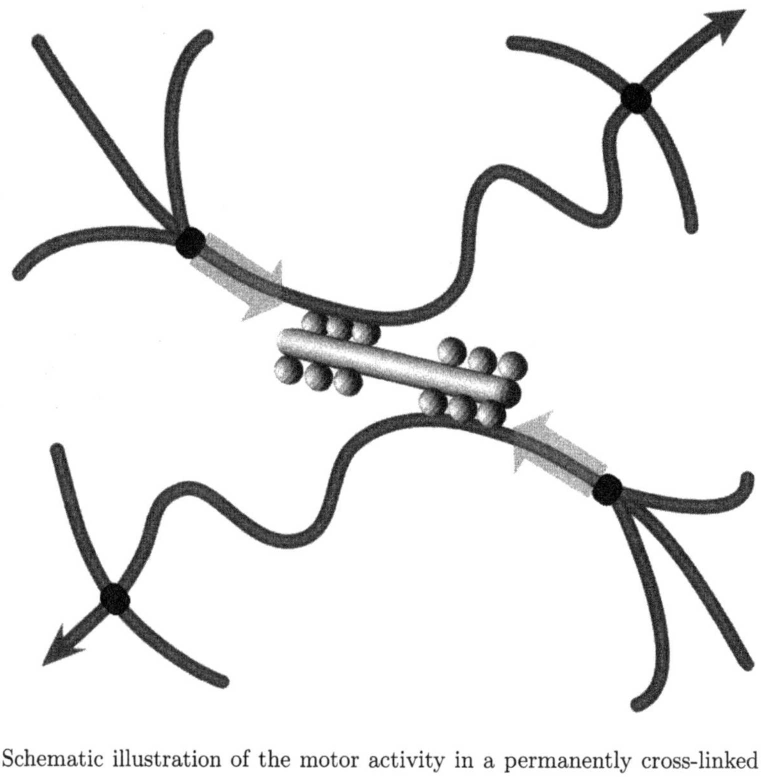

Fig. 5.6 Schematic illustration of the motor activity in a permanently cross-linked network. Forces indicated by the arrows generate tension in the bound filaments, which then couples to the surrounding filaments via the cross-links. For filaments such as F-actin, no significant compressive forces are expected, owing to Euler buckling. Figure adapted from Mizuno *et al.* (2007).

motion could be detected at the subnanometer level using interference microscopy (Gittes *et al.* 1997). This permitted a very sensitive measure of the fluctuations

$$C(\omega) = \int_{-\infty}^{\infty} \langle x(t)x(0)\rangle e^{i\omega t}\, dt \tag{5.37}$$

of the probe particles, whose origin could be either thermal or motor-driven. At the same time, calibrated forces f could be applied to the same particles, allowing a direct measurement of the response function $\chi(\omega)$ as a function of frequency. As discussed in the appendix in section 5.2, the response and fluctuations are fundamentally related in any equilibrium system. Specifically, according to the Fluctuation–Dissipation Theorem, the *equilibrium* fluctuations and the (imaginary part of the complex) response should agree to within a known factor depending on the frequency and temperature:

$$C_{\text{eq}}(\omega) = \frac{2kT}{\omega}\chi''(\omega). \tag{5.38}$$

As can be seen in Fig. 5.7, however, a strong violation of this equality at frequencies below about 10 Hz provides evidence of enhanced fluctuations due to motor activity. It was precisely when these nonequilibrium fluctuations were present that the network also stiffened by up to 100 times, apparently also as a result of motor activity (Mizuno *et al.* 2007). Although Koenderink *et al.* (2009) did not directly measure the fluctuations at a microscopic level as Mizuno *et al.* did, they observed a very similar level of nearly 100-fold motor-induced stiffening of the network at a macroscopic level.

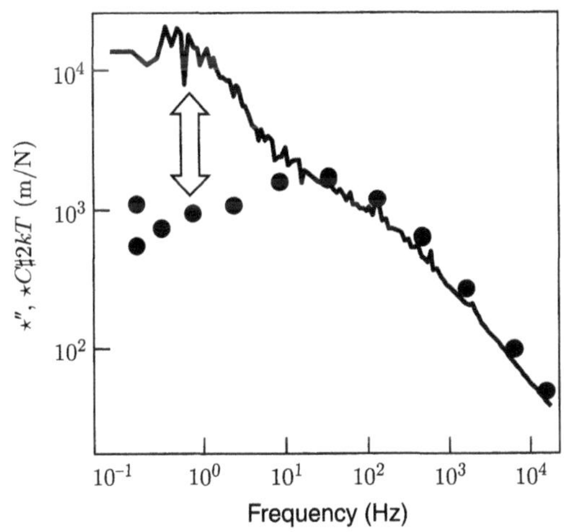

Fig. 5.7 The measured position fluctuations $C(\omega)$ (line) and directly measured response function $\chi''(\omega)$ (solid points) as functions of frequency. A clear departure from the predictions of the FDT is shown by the arrow. This clearly demonstrates the existence of nonequilibrium fluctuations that dominate over thermal fluctuations at frequencies below about 10 Hz. Figure adapted from Mizuno *et al.* (2007).

This provides a very powerful control mechanism for cells to react to external forces by controlling their rigidity, even before adapting their internal structure.

5.1.6.1 Models of active gels

Various models have been developed for nonequilibrium, or active, gels. Some of these efforts were aimed at understanding pattern formation in kinesin-driven microtubule arrays (Nedelec *et al.* 1997). Similar considerations were also applied to acto-myosin systems, for which enhanced fluidization was predicted owing to actin-driven sliding of filaments (Liverpool *et al.* 2001). This can also lead to the formation of bundled or oriented/polarized states of acto-myosin gels. A variety of hydrodynamic-like theories of active gels have been developed by identifying the relevant degrees of freedom and generalized fluxes or currents and forces (Liverpool *et al.* 2009 and references therein). Such models are covered in the chapter by Jülicher.

In contrast to these macroscopic or phenomenological approaches, attention has recently turned to the case of stable, elastic gels. Here flow and polarization are blocked, but internal stresses can develop, as in the experiments of Mizuno *et al.* (2007) and Koenderink *et al.* (2009). These stresses can lead to enhanced force fluctuations and stiffening of the network. Both the stiffening and the frequency dependence of the nonequilibrium, motor-induced fluctuations can be understood theoretically by means of a model based on the known properties and prior models of passive, equilibrium F-actin networks (MacKintosh and Levine 2008). To these were also added the contractile effect of motors. Specifically, the model assumes uncoordinated, isotropic dipolar forces generated by the motors, as sketched in Fig. 5.6. Such a model is expected to be valid provided that the density (both spatial and temporal) of contractions is not sufficiently great to cause interference between contractions within the network. In this case, the contractile events can be treated independently, and can be approximated by the effect of equal and opposite forces, as illustrated by the arrows in Fig. 5.6, which are expected to be separated by a distance of a few microns, at least for *in vitro* networks. In this way, strong tensile forces can be generated in the network, primarily along the line formed by the arrows. By contrast, compressive loads along the perpendicular direction, due to motors pushing the filaments, are not expected, since actin filaments buckle under very small compressive loads, at least for segments longer than a micron.

Although at low frequencies the elastic properties of the network are expected to dominate in a permanently cross-linked network, at the higher frequencies probed by experiments (Mizuno *et al.* 2007), strong viscous/viscoelastic effects due to the solvent are also expected. To capture this, a so-called two-fluid model (Brochard and de Gennes 1977) was used, in which the network displacement field $\vec{u}$ and fluid/solvent velocity $\vec{v}$ are coupled. For low Reynolds number, these fields satisfy the following coupled equations:

$$0 = \mu\, \nabla^2 \vec{u} + (\mu + \lambda)\vec{\nabla}(\vec{\nabla} \cdot \vec{u}) + \Gamma\left(\vec{v} - \frac{d\vec{u}}{dt}\right) + \vec{f} \tag{5.39}$$

and

$$0 = \eta \nabla^2 \vec{v} - \vec{\nabla} P - \Gamma \left(\vec{v} - \frac{d\vec{u}}{dt} \right), \tag{5.40}$$

where P is the pressure in the fluid, $\mu = G$ and λ are Lamé coefficients, and η is the solvent viscosity. Given a meshwork with a pore size ξ, the coupling Γ is expected to be of order η/ξ^2 (Gittes *et al.* 1997; Levine and Lubensky 2000). One must also include a model for the motor forces f, which couple to the network. Myosin motors in such systems form filaments or bundles, rather than acting alone. Such bundles are expected to bind to a filament and begin to move processively (without falling off) for many steps, perhaps moving several microns over a period of many seconds. Significant forces will only be generated in the network once the myosin filament binds to more than one actin. The resulting forces are expected only to build up over a few seconds, as the motors move. The motors may also stall, holding a nearly constant level of force within the network. By contrast, as soon as the motors unbind, the force vanishes, and the network relaxes over a period of time (Mizuno *et al.* 2007). Thus, the dominant feature is expected to be a rapid, step-like drop in the force once the motors unbind. This suggests a simple, general form of the force spectrum

$$C_{\mathrm{f}}(\omega) = \int \langle f(t) f(0) \rangle e^{i\omega t} \sim 1/\omega^2. \tag{5.41}$$

Here, the average $\langle \cdots \rangle$ is meant to represent a possibly nonequilibrium ensemble average, as opposed to a thermal average.

This simple picture already suggests an explanation for the observation that motor-induced fluctuations appear only at low frequencies: the spectrum of nonequilibrium forces decreases strongly at higher frequencies, eventually dropping below the thermal force fluctuations (MacKintosh and Levine 2008). More precisely, in a purely elastic limit, in which the response function of the gel is frequency-independent, the nonequilibrium forces drive position fluctuations in the network that directly reflect the force spectrum in eqn (5.41), which is equivalent to a mean square displacement (MSD)

$$\Delta x^2(t) = \langle (x(t) - x(0)) \rangle \sim t. \tag{5.42}$$

This is consistent with the low-frequency spectrum of position fluctuations in Mizuno *et al.* (2007).

This model can also account for the stiffening of permanently cross-linked actin networks by myosin motor activity. Phenomenologically, as the motors induce local stresses and strains in the network, regions near the motors can be expected to stiffen, in analogy with the macroscopic strain stiffening discussed in Section 5.1.5. More precisely, if the network is modeled by an isotropic distribution of semiflexible filaments, as the motors introduce tension in the actin filaments, the nonlinear force–extension curve in Fig. 5.3 means that filaments in the vicinity of a myosin filament will tend to become less compliant and stiffen. In a mean-field model, one can understand the relative degree of network stiffening in terms of the average tension per filament (MacKintosh and Levine 2008; Liverpool *et al.* 2009). As noted in Section 5.1.5, the stiffness is expected to increase with the force f as $f^{3/2}$. One can account for the approximately 100-fold stiffening observed by Mizuno *et al.* (2007) and Koenderink *et al.* (2009) by average filament tensions of the order of only a pN, which is easily achievable by motors such as myosin.

5.1.6.2 *Implications for cells*

The cytoplasm of living cells represents a highly dynamic system, undergoing continual remodeling driven by molecular-motor-generated forces. At a microscopic level, these nonequilibrium forces and resulting motions are usually directed, as in the transport of cargo by motors along cytoskeletal filaments and in retrograde flow during cell locomotion (WatermanStorer and Salmon 1997). By contrast, thermal/equilibrium dynamics is fundamentally nondirected, as illustrated by the ever-present Brownian motion. It is well established, nevertheless, that random thermal fluctuations play an important role in a variety of cellular processes. The combination of thermal fluctuations with nonequilibrium energy consumption can, for instance, lead to directed force generation by polymerization ratchets (Mogilner and Oster 1996).

In order to measure dynamics in the cytoskeleton of living cells, a number of experimental techniques have been developed. One approach is to embed small particles (ranging from nanometers to micrometers in size) (Lubyphelps *et al.* 1987). The motion of these tracer particles is often interpreted as thermal diffusion, with the MSD of the particle growing linearly with time, and with diffusion coefficients D often measured to be comparable to that of a freely diffusing particle in water. Other experiments have shown deviations from apparently diffusive behavior, and have recorded so-called superdiffusive motion with MSDs growing in proportion to t^{α}, where $\alpha > 1$, which indicates partly directed motion (Caspi *et al.* 2000). This is to be expected if a particle is dragged along by moving motors. For thermally driven motion in the viscoelastic cytoskeleton, strictly subdiffusive motion characterized by $\alpha < 1$ is expected. Thus subdiffusive and superdiffusive motion are indications of thermal and active processes, respectively. In spite of the microscopically directed nature of motor activity, however, such activity can also give rise to random motion within the cytoplasm, which may dominate over the ever-present thermal motion and yet may appear deceptively to be diffusive (Brangwynne *et al.* 2008). Experiments by Lau *et al.* (2003), for instance, have shown evidence for nonthermal fluctuations in living cells that can appear approximately diffusive. Recently, Brangwynne *et al.* (2008) have used analysis of the shape and motion of microtubules to probe cytoplasmic mechanics and dynamics. This was done by resolving the various bending amplitudes $a_q(t)$ of the microtubules and then tracking the evolution of these amplitudes over time, on a scale of 0.1–10 s. This approach has an added benefit of being sensitive to motion on different length scales within the cytoplasm, by tracking the time dependence of bending modes of different wavelengths $\lambda = 2\pi/q$. In these studies, it was found that the MSD of the bending amplitudes increased in an ATP-dependent manner as

$$\Delta a_q^2(t) = \langle (a_q(t) - a_q(0))^2 \rangle \sim t\,, \tag{5.43}$$

which is consistent with the expected spectrum of the acto-myosin fluctuations described above in eqns (5.41) and (5.42). The interpretation of these experiments on cells, combined with parallel work on *in vitro* model systems of actin, myosin, and stabilized microtubules, was that the inert microtubules moved in a way that reflected the contractile fluctuations of the surrounding acto-myosin matrix, and thus the observed time dependence of the displacement fluctuations in an elastic medium.

5.1.7 Discussion

Cytoskeletal filaments are not only essential structural elements in cells but also have proven to be remarkable model systems for the study of semiflexible polymers. Their size alone makes it possible to visualize individual filaments directly. They are also unique in the extreme separation of two important lengths, the persistence length ℓ_{p} and the size of a single monomer. In the case of F-actin, ℓ_{p} is more than a thousand times the size of a single monomer. This makes for not only quantitative but also qualitative differences from most synthetic polymers. We have seen, for instance, that the way in which semiflexible polymers entangle is very different. This makes for a surprising variation of the stiffness of these networks with only changes in the density of cross-links, even at the same concentration.

In spite of the molecular complexity of filamentous proteins as compared with conventional polymers, a quantitative understanding of the properties of single filaments provides a quantitative basis for modeling solutions and networks of filaments. In fact, the macroscopic response of cytoskeletal networks quite directly reflects, for example, the underlying dynamics of an individual semiflexible chain fluctuating in its Brownian environment.

In developing our current understanding of cytoskeletal networks, a crucial role has been played by *in vitro* model systems. Major challenges, however, remain for understanding the cytoskeleton of living cells. In the cell, the cytoskeleton is hardly a passive network. Among other differences from the model systems studied to date is the presence of active contractile or force-generating elements such as motors that work in concert with filamentous proteins. Nevertheless, *in vitro* systems are beginning to be used for quantitative studies of, for example various actin-associated proteins for cross-linking, bundling, and force generation. The active gel systems, in particular, are likely to be very fruitful in the coming years.

5.2 Appendix: Thermal fluctuations and linear responses

5.2.1 Brownian motion and thermal diffusion

In an effort to understand equilibrium fluctuations, we begin with arguably the simplest problem illustrating these fluctuations: Brownian motion in a liquid. This term derives from the botanist Robert Brown, who observed random motions of pollen grains in a fluid in 1828. This kind of motion is general for sufficiently small particles suspended in a fluid. We now understand this motion as being due to the ever-present thermal fluctuations in the fluid itself. In a series of papers beginning in 1905, Einstein gave a theoretical basis for Brownian motion and the closely related phenomenon of diffusion. It is hard to overstate the importance of this seemingly mundane result for Brownian motion, as it played a key role in firmly establishing the atomic or discrete nature of matter—something that was definitely not the prevailing view prior to 1905. This connection to the discrete nature of matter was made possible by the so-called Stokes–Einstein expression for the diffusion constant D for a spherical particle of radius a,

$$D = \frac{kT}{6\pi\eta a}, \tag{5.44}$$

where η is the solvent viscosity. Here, $\mu = 1/(6\pi\eta a)$ is the mobility of the particle, a result due to Stokes. The novel result of Einstein, which was actually derived slightly earlier by Sutherland, connected diffusion of an observable micrometer-sized particle with the molecular-scale thermal energy kT. This was used by Perrin to measure Avagadro's number N_{A}, in essence, identifying the atomic size.

This Stokes–Einstein relation also represents a special case of a much more general relationship between thermal fluctuations (e.g., diffusion) and response to a force (e.g., the mobility). This more general result is known as the Fluctuation–Dissipation Theorem. As we now know, the drag force on a particle (or, equivalently, its mobility) when subject to an external force is fundamentally related to the strength of the random forces that it experiences in the fluid. The proof of this theorem can be found in various standard statistical-physics texts. We will not derive it here, but rather describe the motivation behind it by an example and state the general form of the theorem.

The diffusion constant is defined by the time evolution of the mean square displacement of a particle. Specifically, for a particle beginning initially at the origin, the motion along a single coordinate direction x is given by

$$\langle x^2(t)\rangle = 2Dt. \tag{5.45}$$

We obtain similar results for $\langle y^2(t)\rangle$ and $\langle z^2(t)\rangle$. By adding these, we obtain

$$\langle r^2(t)\rangle = \langle x^2(t)\rangle + \langle y^2(t)\rangle + \langle z^2(t)\rangle = 6Dt \tag{5.46}$$

for motion in three dimensions. In reality, for the example of a particle suspended in a fluid, we expect that the kicks that the particle experiences in the fluid are random. Even so, since the particle has a mass, there will be some effect of its inertia. So, we expect that the motion will not be truly random, but will be correlated for short times. For such short times, the particle will move ballistically, and we expect that the displacement will grow linearly with time. Thus, we expect that $\langle x^2\rangle \propto t^2$, rather than $\langle x^2\rangle \propto t$, for short enough times. Actually, this must be true. We can already see that there is a problem with eqn (5.45), because it would imply that $x \simeq \sqrt{t}$, which leads to an unphysical divergence of the velocity at short times. Thus, only for times long compared with some microscopic correlation time (which we can be more quantitative about below) will diffusion, characterized by eqn (5.45), be valid.

The diffusion constant is also related to the velocity fluctuations of a particle. If $x(t)$ is the distance that the particle has traveled in the x-direction in a time interval t, this is related to the (x-component of the) velocity $v(t)$ by

$$x(t) = \int_0^t dt' \; v(t'). \tag{5.47}$$

We can then write the mean square displacement as

$$\langle x^2(t) \rangle = \int_0^t dt' \int_0^t dt'' \; \langle v(t')v(t'') \rangle. \tag{5.48}$$

The quantity $\langle v(t')v(t'') \rangle$ is an example of a *time correlation function*—in this case, the 'velocity autocorrelation function' (VACF). This quantity measures the degree of correlation between the particle velocity at two different times t' and t''. When $t' = t''$, the value of the velocity correlation function is simply $\langle v^2 \rangle$. At sufficiently long times, when the particle loses all memory of its original velocity, $\langle v(t')v(t'') \rangle = \langle v(t') \rangle \times \langle v(t'') \rangle = 0$, because the average velocity of a particle in equilibrium is zero. Time correlation functions such as the VACF are properties of a system in equilibrium. Their value can therefore never depend on the times t' and t'' individually, but only on the difference $t' - t''$. This is because the properties of a system in equilibrium are time invariant. Furthermore, because we can permute $v(t')$ and $v(t'')$, we must have that

$$\langle v(t')v(t'') \rangle = K(t' - t''), \tag{5.49}$$

where $K(t)$ is a symmetric function. Thus,

$$\langle x^2(t) \rangle = \int_0^t dt' \int_0^t dt'' \; \langle v(t')v(t'') \rangle = 2 \int_0^t dt' \int_0^{t'} dt'' K(t' - t''), \tag{5.50}$$

and

$$\begin{aligned} \frac{\partial}{\partial t} \langle x^2(t) \rangle &= 2 \frac{\partial}{\partial t} \int_0^t dt' \int_0^{t'} dt'' K(t' - t'') \\ &= 2 \int_0^t dt'' K(t - t'') = 2D. \end{aligned} \tag{5.51}$$

We only expect the diffusive behavior of eqn (5.45) for times much longer than the decay time of correlations in the velocity, and hence we may replace the upper limit in the last integral of eqn (5.51) by $+\infty$. This gives us a simple relationship between D and the VACF:

$$D = \int_0^\infty d\tau \; \langle v(\tau)v(0) \rangle \, . \tag{5.52}$$

As noted above, we expect $K(t) = \langle v(t)v(0) \rangle$ to be a symmetric function of t that must vanish for times t greater than some relaxation time τ, which represents the timescale over which a particle can be said to forget its initial velocity. Thus, we might expect the correlation function K to be represented by an exponential dependence, in which τ is the time constant:

$$K(t) = \langle v(t)v(0)\rangle = \langle v^2\rangle e^{-|t|/\tau}. \tag{5.53}$$

In fact, we are making an additional implicit assumption that $\langle v(t)v(0)\rangle$ is a maximum at $t = 0$, which can be shown. It is also possible to derive the assumed exponential dependence directly from the approach in the next section. But, if we simply take eqn (5.53) as a hypothesis, we find that

$$D = \frac{kT}{m}\tau \tag{5.54}$$

and

$$\tau = \frac{m}{6\pi\eta a}, \tag{5.55}$$

since the average kinetic energy is

$$\frac{1}{2}m\langle v^2\rangle = \frac{kT}{2}. \tag{5.56}$$

For a micron-size particle in a liquid such as water, the relaxation time τ turns out to be of the order of a microsecond.

5.2.2 Langevin theory

In order to see at a deeper level the connection between the random fluctuations in a fluid and the dissipative nature of the resulting particle motion, we turn now to an approach due to Langevin. The approach here will be to account more directly for the fluctuating forces acting on the particles by the surrounding fluid. We shall see how thermal equilibrium (e.g., the expected uniform spatial distribution of particles) can be established by these fluctuations.

We begin with the simple equation of motion (in one dimension) satisfied by a particle of mass m,

$$m\frac{dv}{dt} = \mathcal{F}(t), \tag{5.57}$$

where $\mathcal{F}$ is the force acting on the particle. We might be tempted at this point to assume that this force is random, and independent of the particle motion. After all, these forces are due to the fluctuating solvent, right? In fact, there are two physically distinct parts to this force: (i) the viscous drag on the particle, which depends on v, and (ii) the randomly fluctuating forces $\zeta(t)$. We assume here that the first of these can be described by the Stokes formula for the drag on a spherical particle in a viscous fluid, $-6\pi\eta av$, where η is the fluid viscosity and a is the radius of the sphere. This is valid for small particles and small velocities. It is important to note that this force is random and has zero mean only to the extent that the particle velocity is random with zero mean. Thus, we can say that, in the absence of external forces, it has a zero *time* average.

It is important also to distinguish another kind of average that we have been a bit sloppy about so far: the *ensemble* average. This represents an average over a large number of similarly prepared systems (an *ensemble*). We can consider, for instance, the motion of a particle with some initial condition, say a velocity $v(0)$ at time $t = 0$. We expect that this initial velocity will change over time, and that it will become uncorrelated with the velocity at much later times t. Let τ denote the correlation time, after which memory of the initial conditions is lost. As a specific example, if the initial velocity were large compared with the expected thermal velocities, we would expect that it would decay toward zero for times large compared with τ. But the precise way in which it will do this depends on the details of the fluctuating force ζ due to the fluid. We are not interested in the exact trajectory that an individual particle follows owing to each and every collision with individual fluid molecules. Rather, we want to know what to expect for the way the particle velocity decays in a typical sample of a certain kind of solvent (e.g., with known viscosity η) at a certain temperature. Thus, we want to know how the particle velocity behaves on the average, i.e., the ensemble average. However, the ensemble average of the random force is zero, i.e., $\langle\zeta(t)\rangle = 0$, while the ensemble average of the particle velocity is not zero, at least for times smaller than τ. Mathematically, we could say that the ensemble corresponds to a large set of possible functions $\zeta(t)$ consistent with a given fluid at a given temperature. The properties of these functions are yet to be determined, but the fact that the average $\langle\zeta(t)\rangle$ over this set is zero for all times t is clear by construction. At any time t, ensemble average quantities can be obtained by averaging over the possible functions $\zeta(t)$.

We rewrite the equation of motion as follows:

$$m\frac{dv(t)}{dt} = -6\pi\eta a v(t) + \zeta(t). \tag{5.58}$$

If we take the average of this over the ensemble, we obtain

$$m\frac{d\langle v(t)\rangle}{dt} = -6\pi\eta a\langle v(t)\rangle,$$

which implies that

$$\langle v(t)\rangle = v(0)e^{-t/\tau} \quad \text{for } t > 0, \tag{5.59}$$

where we can now identify $\tau = m/(6\pi\eta a)$ as the relaxation time for effects of the initial condition to die out. Interestingly, this is consistent with the relaxation time we obtained from our hypothesis in eqn (5.53) above. This is actually an example of a very important observation at the heart of the FDT: the way a system fluctuates about equilibrium (in this case, the VACF) also characterizes the way it returns to equilibrium after an external perturbation (here, an initial nonzero velocity $v(0)$, for example in response to an initial force on the particle). Also, the essentially dissipative nature of this result is evident from the fact that this result cannot be reversed in time. Equation (5.59) is only true for $t > 0$, while eqn (5.53) turns out to be valid for both positive and negative times t.

Using the Langevin approach and the equation of motion above, one can also derive the mean square displacement of the particle for times $t < \tau$. We will not do this here, but by multiplying eqn (5.58) by $x(t)$ before taking the ensemble average, and using the fact that $\langle x(t)\zeta(t)\rangle = 0$, it is possible to show that

$$\langle x^2 \rangle = \frac{kT}{3\pi\eta a}\left[t - \tau\left(1 - e^{-t/\tau}\right)\right]. \tag{5.60}$$

At long times, for which $t \gg \tau$, we find, as expected, that

$$\langle x^2 \rangle = 2Dt,$$

where D is given by eqn (5.44). By contrast, however, for short times, $\langle r^2 \rangle \propto t^2$, thus resolving the problem with our earlier analysis for short times. Incidentally, although $\zeta(t)$ is a random variable, it can be correlated with other variables of interest. Thus, for instance, although it is correct to assume that $\langle x(t)\zeta(t)\rangle = 0$, as we have done above, it can also be shown that $\langle v(t)\zeta(t)\rangle \neq 0$.

5.2.3 Linear response and the Fluctuation–Dissipation Theorem

The Stokes–Einstein relation connects thermal diffusion in the absence of any applied forces to the mobility, which expresses the way the velocity responds to an applied force. This response is assumed to be linear and instantaneous, i.e.,

$$v(t) = \mu f(t).$$

More generally, given that we expect finite relaxation times in most systems, we consider a more general form of response, in which a physical observable (here, the velocity v) depends linearly on the history of the prior forces acting to create the response:

$$v(t) = \int_{-\infty}^{t} \chi(t,t')f(t')\,dt'. \tag{5.61}$$

This should actually be considered to be the ensemble or average response of many similar systems to the same force history $f(t')$. It is very important to emphasize that the response at time t can depend only on forces acting at times $t' < t$. This causality has many implications. Furthermore, if we are interested in a system that is characterized by equilibrium in the absence of any applied forces, then the origin of time should not matter. In other words, the response should be time-translation invariant, provided the force is similarly translated in time. Thus, the response function $\chi(t,t') = \chi(t-t')$ should be a function only of the time difference $t-t'$. In view of the causal relationship between force and response, it is convenient to consider $\chi(t)$ to be a function that simply vanishes for negative arguments t, in which case we can consider the integral in eqn (5.61) to extend over the full real line in t'.

In fact, for the problem at hand, we have already effectively calculated the response function χ in eqn (5.59). To see this, consider a force $f(t)$ given by

$$f(t) = \begin{cases} f \text{ for } t < 0, \\ 0 \text{ otherwise.} \end{cases} \tag{5.62}$$

Then, we know that $v(0) = \mu f$, and this initial condition relaxes according to eqn (5.59). Thus, for $t > 0$,

$$v(t) = \mu f e^{-t/\tau} = f \int_t^\infty \chi(t')\, dt', \tag{5.63}$$

from which we find that

$$\chi(t) = \frac{\mu}{\tau} e^{-t/\tau} = \frac{1}{m} e^{-t/\tau} \quad \text{for } t > 0 \tag{5.64}$$

by differentiation with respect to t. Thus,

$$\tau = m\mu = \frac{m}{6\pi\eta a}, \tag{5.65}$$

as above.

By comparing eqn (5.64) with eqns (5.53)–(5.56), it would appear that

$$\langle v(t)v(0)\rangle = kT\chi(t) \quad \text{for } t > 0. \tag{5.66}$$

This expresses a relationship between the equilibrium fluctuations of a system in the absence of any applied force f, in the specific form of the VACF, and the way the system responds linearly to such forces. In other words, if we know how a system fluctuates about equilibrium, we also know how it will respond when perturbed away from equilibrium. This is the essence of the FDT. But eqn (5.66) is only a special case of the much more general FDT. We will not prove this theorem, but merely state it and show how the above is indeed a special case. In doing so, it is important to think a bit more abstractly about the way a system may respond to perturbations such as the force f above.

There may be many physical observables of our system (e.g., a coordinate $x(t)$, a velocity $v(t)$, or even the magnetization $M(t)$ of a magnet) that respond to a number of possible perturbing, force-like variables (e.g., a mechanical force f, a magnetic field H, or an electric field E). Thus, we can consider a matrix of response functions χ_{AB} that relate variables A and B. For instance, in the above treatment we have considered v as depending linearly on f, for which we might want to define a response function χ_{vf}. Then, eqn (5.61) has a matrix-like form:

$$v = \chi_{vf} * f,$$

where a convolution is implied. This is actually not the way things are usually defined. Instead of defining the response function χ in terms of the perturbing field or force-like (intensive) variable f directly, one usually identifies it by the (extensive) variable that is thermodynamically conjugate to f. In thermodynamics we learn, for instance, that the volume V of a system is conjugate to the pressure P applied to the system. Here, the work done in a volume change dV can be described by $P\, dV$. Similarly, the force f and the displacement x are conjugate, as are the magnetic field H and the magnetization M. The general statement of the FDT is as follows: if a physical observable $A(t)$ responds linearly to a field or force-like perturbation h that is thermodynamically conjugate to B, for which we write

$$A(t) = \int_{-\infty}^{t} \chi_{AB}(t - t')h(t')\,dt', \tag{5.67}$$

then

$$kT\chi_{AB}(t) = -\frac{d}{dt}\langle \Delta A(t + t')B(t')\rangle \quad \text{for } t > 0, \tag{5.68}$$

where $\Delta A(t) = A(t) - \langle A\rangle$. Again, this tells us that the way a system will respond to a perturbing force or field is completely determined by the thermal/equilibrium average properties of the system in the absence of any such perturbation. It is important to note that all the averages $\langle\cdots\rangle$ here refer to the unperturbed state.

In example above, we were interested in the response of the velocity v to the force f that was conjugate to x. Thus, the correlation function above becomes $\langle v(t + t')x(t')\rangle$, since $\langle v\rangle = 0$. If we take the derivative of this expression with respect to t', we find that

$$\langle \dot{v}(t + t')x(t')\rangle + \langle v(t + t')v(t')\rangle = 0,$$

since the equilibrium correlation function must be time-translation invariant. Thus,

$$-\frac{d}{dt}\langle \Delta A(t + t')B(t')\rangle = \langle v(t + t')v(t')\rangle,$$

and eqn (5.66) becomes a special case of the FDT.

While the statement of the FDT in eqn (5.68) may seem strange, it turns out to be very powerful and useful. It is, however, most frequently seen and applied in the frequency domain, rather than in the time domain. A common application of this is to a damped harmonic oscillator. A concrete example of this is a micrometer-size particle trapped by a harmonic potential, such as that produced by optical tweezers, in a viscous liquid. Here, the particle position $x(t)$ responds to a force f that is also conjugate to x:

$$x(t) = \int_{-\infty}^{t} \chi_{xx}(t - t')f(t')\,dt'.$$

The corresponding correlation function $K_{xx}(t) = \langle x(t)x(0)\rangle$ has a Fourier transform

$$C(\omega) = \tilde{K}_{xx}(\omega) = \int_{-\infty}^{\infty} \langle x(t)x(0)\rangle e^{i\omega t}\,dt. \tag{5.69}$$

This is often referred to as the power-spectral Density, or PSD, of the particle position. The FDT in this case can be expressed as

$$C(\omega) = \frac{2kT}{\omega}\chi''_{xx}(\omega), \tag{5.70}$$

where χ'' refers to the imaginary part of the (complex) response function $\tilde{\chi}(\omega)$.

References

Alberts, B., *et al.* (1994). *Molecular Biology of the Cell* (3rd edn). Garland Press, New York.

Amblard, F., Maggs, A. C., Yurke, B., Pargellis, A. N., and Leibler, S. (1996). *Physical Review Letters* **77** 4470.

Block, S. M. (2007). *Biophysical Journal* **92** 2986.

Boal, D. (2002). *Mechanics of the Cell.* Cambridge University Press, Cambridge.

Brangwynne, C. P., *et al.* (2006). *Journal of Cell Biology* **173** 733.

Brangwynne, C. P., *et al.* (2008). *Journal of Cell Biology* **183** 583.

Brochard, F. and de Gennes, P. G. (1977). *Macromolecules* **10** 1157.

Bustamante, C., Marko, J. F., Siggia, E. D., and Smith, S. (1994). *Science* **265** 1599.

Caspi, A., Granek, R., and Elbaum, M. (2000). *Physical Review Letters* **85** 5655.

Deng, L. H., *et al.* (2006). *Nature Materials* **5** 636.

Everaers, R., Jülicher, F., Ajdari, A., and Maggs, A. C. (1999). *Physical Review Letters* **82** 3717.

Farge, E. and Maggs, A. C. (1993). *Macromolecules* **26** 5041.

Fixman, M. and Kovac, J. (1973). *Journal of Chemical Physics* **58**, 1564.

Gardel, M., *et al.* (2004). *Science* **304** 1301.

Gittes, F. and MacKintosh, F. C. (1998). *Physical Review E* **58** R1241.

Gittes, F., Mickey, B., Nettleton, J., and Howard, J. (1993). *Journal of Cell Biology* **120** 923.

Gittes, F., Schnurr, B., Olmsted, P. D., MacKintosh, F. C., and Schmidt, C. F. (1997). *Physical Review Letters* **79** 3286.

Granek, R. (1997). *Journal de Physique II (France)* **7** 1761.

Head, D. A., Levine, A. J., and MacKintosh, F. C. (2003a). *Physical Review Letters* **91** 108102.

Head, D. A., Levine, A. J., and MacKintosh, F. C. (2003b). *Physical Review E* **68** 061907.

Hinner, B., Tempel, M., Sackmann, E., Kroy, K., and Frey, E. (1998). *Physical Review Letters* **81** 2614.

Howard, J. (2001). *Mechanics of Motor Proteins and the Cytoskeleton.* Sinauer Press, Sunderland, MA.

Huisman, E. M., Storm, C., and Barkema, G. T. (2008). *Physical Review E* **78** 051801.

Humphrey, D., Duggan, C., Saha, D., Smith, D., and Kas, J. (2002). *Nature* **416** 413.

Isambert, H. and Maggs, A. C. (1996). *Macromolecules* **29** 1036.

Janmey P. A., *et al.* (1994). *Journal of Biological Chemistry* **269**, 32503.

Koenderink, G. H., Atakhorrami, M., MacKintosh, F. C., and Schmidt, C. F. (2006). *Physical Review Letters* **96** 138307.

Koenderink, G. H., *et al.* (2009). *Proceedings of the National Academy of Sciences of the USA* **106** 15192.

Kroy, K., and Frey, E. (1996). *Physical Review Letters* **77** 306.

Landau, L. D. and Lifshitz, E. M. (1986). *Theory of Elasticity.* Pergamon, Oxford.

Lau, A. W. C., *et al.* (2003). *Physical Review Letters* **91** 198101.

Levine, A. J. and Lubensky, T. C. (2000). *Physical Review Letters* **85** 1774.

Liverpool, T. B., Maggs, A. C., and Ajdari, A. (2001). *Physical Review Letters* **86** 4171.

Liverpool, T. B., Marchetti, M. C., Joanny, J. F., and Prost, J. (2009). *Europhysics Letters* **85** 18007.

Lubyphelps, K., Castle, P. E., Taylor, D. L., and Lanni, F. (1987). *Proceedings of the National Academy of Sciences of the USA* **84** 4910.

MacKintosh, F. C. and Levine, A. J. (2008). *Physical Review Letters* **100** 018104.

MacKintosh, F. C., Käs, J., and Janmey, P. A. (1995). *Physical Review Letters* **75** 4425.

Mizuno, D., Tardin, C., Schmidt, C. F., and MacKintosh, F. C. (2007). *Science* **315** 370.

Mogilner, A. and Oster, G. (1996). *Biophysical Journal* **71** 3030.

Morse, D. C. (1998*a*). *Physical Review E* **58** R1237.

Morse, D. C. (1998*b*). *Macromolecules* **31** 7044.

Nedelec, F. J., Surrey, T., Maggs, A. C., and Leibler, S. (1997). *Nature* **389** 305.

Odijk, T. (1983). *Macromolecules* **16** 1340.

Odijk, T. (1995). *Macromolecules* **28** 7016.

Onck, P. R., Koeman, T., van Dillen, T., and van der Giessen, E. (2005). *Physical Review Letters* **95** 178102.

Palmer, J. S. and Boyce, M. C. (2008). *Acta Biomaterialia* **4** 587.

Pasquali, M., Shankar, V., and Morse, D. C. (2001). *Physical Review E* **64** 020802(R).

Satcher, R. and Dewey, C. (1996). *Biophysical Journal* **71** 109.

Schmidt, C. F., Baermann, M., Isenberg, G., and Sackmann, E. (1989). *Macromolecules* **22**, 3638.

Schnurr, B., Gittes, F., MacKintosh, F. C., and Schmidt, C. F. (1997). *Macromolecules* **30** 7781.

Storm, C., *et al.* (2005). *Nature* **435** 191.

WatermanStorer, C. M. and Salmon, E. D. (1997). *Journal of Cell Biology* **139** 417.

Wilhelm, J. and Frey, E. (2003). *Physical Review Letters* **91** 108103.

Part III

Biomechanics

6
Morphoelasticity: A theory of elastic growth

Alain GORIELY and Derek MOULTON

Mathematical Institute, University of Oxford

6.1 Introduction

Growth is the process by which a body increases in size through the addition of mass. In biological systems, growth can serve a number of different purposes, and occurs in many different forms. Growth may be restricted to particular locations in the body. In particular, tip growth is often found in microscopic filamentary systems (Gooday and Trinci 1980; Howard and Valent 1996; Goriely and Tabor 2003). Surface growth and accretion refers to the deposition of material on the surface of a body—this type of growth is found in the formation of horns, teeth, and seashells (Skalak and Hoger 1997; Thompson 1992). In volumetric growth, on the other hand, growth occurs throughout the bulk of the body. This is common in the growth of hearts, tumors, and arteries (Taber and Humphrey 2001; Cowin 2004).

Continuum mechanics and nonlinear elasticity provide a natural framework to study growth. Of foremost importance is capturing the correct relationship between growth and elasticity. The basic idea is that in a growth process, the deformation of a body can be due to both a change in mass and an elastic response. This concept, first put in mathematical terms by Rodriguez *et al.* (1994), states that the deformation tensor can be decomposed into a growth tensor, which instructs different parts of the body how to add mass locally, and an elastic tensor, which captures the elastic reorganization necessary to ensure integrity of the body.

A key benefit to studying growth through continuum mechanics is that it provides direct access to stresses. The relationship between stress and growth is a key concept in the theory of elastic growth. Stress may induce growth; for instance, a body may grow until a homeostatic "target" stress is reached. On the other hand, growth can induce residual stresses, in particular when the body is inhomogeneous or growing differentially. Residual stresses are stresses that persist in the body even after all body forces have been removed. They are known to play an important role in the functioning of bodies. For instance, residual stress is key to the regulation of the transmural stress in arteries (Humphrey 2003), and is also known to occur in tree and plant tissues (Vandiver and Goriely 2008) leading to tension and compression woods. A particularly spectacular example of these internal stresses is found in the wapas trees in French Guiana. Owing to residual stress, these trees explode when cut, giving them the nickname "killing trees" (Détienne and Thiel 1988). More generally, the generation of residual stress is connected to the question of how growth can alter the properties of a material (Goriely and Ben Amar 2005; Ben Amar and Goriely 2005).

This chapter is structured as follows. We begin with a 1D theory of growth, providing several examples to illustrate the basic concepts and some of the questions that arise in morphoelasticity. Then, we develop the full 3D theory from first principles. The fundamental concept, the decomposition of the deformation tensor, is derived in a manner different from any appearing in the literature. We conclude with several applications and future challenges.

6.2 1D growth

In this section, we develop the theory of elastic growth in one dimension. Consider a rod of length L_0 in its initial reference configuration which undergoes a deformation

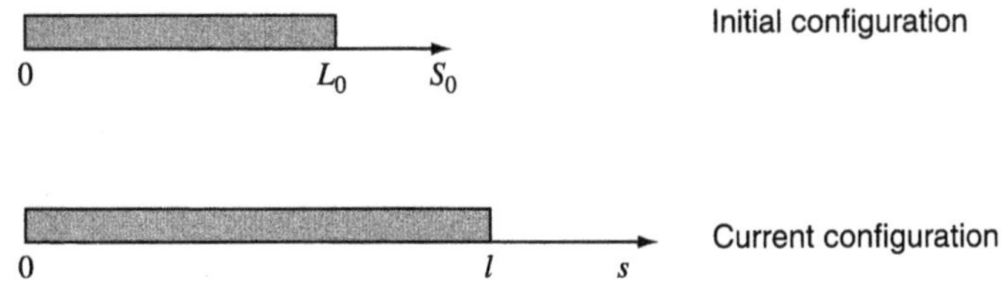

Fig. 6.1 One-dimensional deformation of a rod.

so that it is of length l in the current configuration. Let S_0 describe position along the rod in the initial configuration and s position in the current configuration (Fig. 6.1). We define the geometric stretch

$$\lambda = \frac{\partial s}{\partial S_0}. \tag{6.1}$$

In general, $\lambda = \lambda(S_0)$, and λ is constant for a uniform stretch; $\lambda > 1$ corresponds to elongation and $\lambda < 1$ to a reduction. We next consider several classes of deformation.

6.2.1 Pure elastic deformation

In a pure elastic deformation, $\lambda = \alpha$, where α is the *elastic stretch.* It we let σ denote the stress, the characteristics of the deformation are determined by a constitutive relationship between σ and α. For example, Hooke's Law for a linear material gives $\sigma = E(\alpha - 1)$, where E is the Young's modulus. Solving for α and inserting the result in eqn (6.1), the deformation is described by

$$s = \left(\frac{\sigma}{E} + 1\right) S_0, \tag{6.2}$$

from which we can write $l = (\sigma/E + 1)\, L_0$. In this case, the length increases linearly with the stress. A more accurate description is given by the nonlinear neo-Hookean relationship $\sigma = \mu(\alpha^2 - \alpha^{-1})$, where $\mu > 0$ is the elastic modulus.

6.2.2 Pure growth

A pure growth is characterized by having no elastic response. Here, $\lambda = \gamma$, where γ describes the growth. Whereas in the previous section α described a pure elastic *stretch*, γ describes the *addition* or *removal* of material. If $\gamma > 1$, there is growth, while $\gamma < 1$ indicates shrinking or resorption of material. In general, the material may undergo growth or resorption at different times and at different rates, and a constitutive relation is needed. For pure growth, the constitutive law is of the form

$$\frac{\partial \gamma}{\partial t} = G(\gamma, s, S_0). \tag{6.3}$$

In the simple case of uniform growth, $G = 1$, which gives $\gamma = t$ and $s = S_0 t$. Exponential growth is captured by $G = k\gamma$. Then $s = S_0 e^{kt}$ and the length of the rod grows as $l(t) = L_0 e^{kt}$. Alternatively, suppose that growth occurs only towards the tip. For an exponentially growing tip, we take

$$\frac{\partial \gamma}{\partial t} = G(\gamma, s) = \begin{cases} k\gamma, 0 < s < a, \\ 0, a < s < l. \end{cases} \tag{6.4}$$

Equation (6.4) is coupled with $\partial s/\partial S_0 = \gamma$ and the initial conditions $s(0) = S_0$ and $\gamma(0) = 1$. The difficulty computationally is that $s = s(S_0, t)$ is a moving interface. To understand the behavior of the growing tip, consider a discrete numeric process. At time t_i, the rod will grow according to $\gamma_i = e^{kt_i}$. How much does the rod grow in one time step, from t_i to t_{i+1}? Letting s_n be the current configuration at the nth step, we have

$$e^{kt_i} = \gamma_i = \frac{\partial s_{i+1}}{\partial S_0} = \frac{\partial s_{i+1}}{\partial s_i}\frac{\partial s_i}{\partial S_0} = \frac{\partial s_{i+1}}{\partial s_i} e^{kt_{i-1}}. \tag{6.5}$$

This implies that $\partial s_{i+1}/\partial s_i = e^{k(t_i - t_{i-1})} = e^{k\Delta t}$. Thus the region $0 < s_i < a$ grows by a factor $e^{k\Delta t}$, meaning that the rod extends by an amount $\Delta l = a(e^{k\Delta t} - 1)$. The important thing to note is that this is independent of the actual time, i.e. the rod extends by the same amount at every time step. Even though the tip is growing exponentially, the length of the rod increases only *linearly*. As another example, we present a toy model for tumor growth (more models of this type are presented by Preziosi and Vitale in chapter 7 of this volume). The setup is depicted in Fig. 6.2. Let $r(R_0, t)$ give the radial position in a solid sphere in the current configuration, where R_0 is the radial position in the initial configuration, and let $b(t)$ be the radius of the sphere in the current configuration. The tumor is surrounded by a bath of nutrient, which is transmitted to the tumor through diffusion with uptake. We suppose that the growth is exponential but proportional to the concentration of the nutrient. Hence, the growth is modeled by

$$\begin{aligned} \frac{\partial \gamma}{\partial t} &= k\gamma u(r), \\ \frac{\partial r}{\partial R_0} &= \gamma. \end{aligned} \tag{6.6}$$

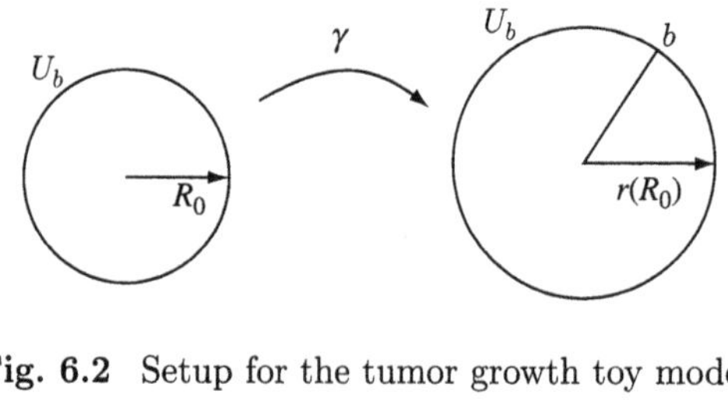

Fig. 6.2 Setup for the tumor growth toy model.

Here $u(r)$ is the concentration of the nutrient and satisfies

$$\frac{\partial u}{\partial t} = \frac{D}{r^2}\frac{\partial}{\partial r}\left(r^2 \frac{\partial u}{\partial r}\right) - Q, \tag{6.7}$$

where D is the diffusion coefficient and Q is the nutrient uptake (assumed to be constant). The outer boundary condition for $u(r,t)$ is $u(b,t) = U_b$, that is, the bath provides a constant nutrient supply to the outer surface. The other boundary condition, and the growth behavior, depends on the size of the tumor. If b is small enough that $u(r)$ can diffuse all the way to the center of the sphere, the boundary condition is $0 \le u(0) < \infty$. For some b_{cr}, the diffusion will satisfy $u(0) = 0$. For $b > b_{\text{cr}}$, the boundary condition is $u(a) = 0$, where $a > 0$ is the smallest radius to which the nutrient can diffuse. This is shown schematically in Fig. 6.3.

For $b < b_{\text{cr}}$, i.e. when the tumor is "small", the entire sphere grows and so growth is exponential. After this initial phase, growth is restricted to a spherical shell and so is equivalent to the tip growth problem discussed above. Thus, as a function of time, the size of the tumor $b(t)$ increases exponentially at first and then transitions to a linear rate.

We have provided only a qualitative description of the behavior of the growing tumor. As an exercise left for the reader, it is informative to derive a differential equation for the evolution of $b(t)$ and determine, *as* $t \to \infty$, *the size of the region* $b - a$. This corresponds to the so-called "penetration length" of the nutrient.

6.2.3 Growth with elastic response

In an elastic body, growth is subject to an elastic response. The fundamental assumption of morphoelasticity, to be justified later, is that the geometric stretch $\lambda = \partial s/\partial S_0$ is the product of an elastic term α and a growth term γ. That is,

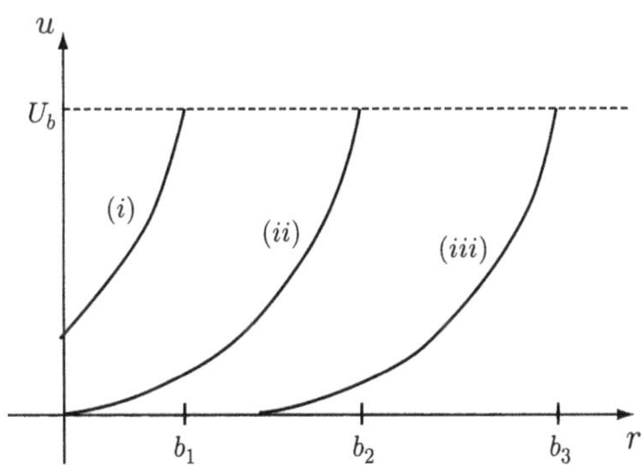

Fig. 6.3 The concentration of the nutrient $u(r)$ as a function of r for different-sized tumors. In (i), the tumor radius b_1 is small enough for the nutrient to diffuse all the way to the center, and growth is exponential. In (ii), the radius has a critical value b_{cr} such that $u(0) = 0$. In (iii), $b > b_{\text{cr}}$—here, growth occurs only in the region where $u(r) > 0$ and so the growth is linear.

$$\lambda = \alpha\gamma, \tag{6.8}$$

where α satisfies some constitutive relationship with the stress, $H(\alpha, \sigma) = 0$, and γ satisfies a growth equation $\partial_t \gamma = G(\alpha, \gamma, \sigma)$.

To give an example, consider a rod growing between two walls. We take the growth to be linear and uniform, so that $\gamma = 1 + t$, and use the linear Hookean model for the elastic relationship, $\sigma = E(\alpha - 1)$. Owing to the wall constraint, we must have $\lambda = 1$. This implies $\alpha = 1/\gamma$, from which we obtain

$$\sigma = -E\left(\frac{t}{1+t}\right). \tag{6.9}$$

Essentially, as the rod grows, a compressive stress builds up to contain the rod in the fixed space. Observe from eqn (6.9) that as $t \to \infty$, $\sigma \to -E$. This is problematic in that it states that only a finite stress is developed in compressing an infinite rod. The problem is due to the linearity assumption. If instead the nonlinear neo-Hookean relationship $\sigma = \mu\left(\alpha^2 + \alpha^{-1}\right)$ is used, we obtain

$$\sigma = \mu\left(\frac{1}{1+t^2} - t + 1\right). \tag{6.10}$$

For the nonlinear model, there is nothing that slows down the growth, and $\sigma \to -\infty$ as $t \to \infty$.

To better understand the relationship between growth and elasticity, it is useful to consider a two-step process, pictured in Fig. 6.4. Starting from the initial, stress-free configuration, in Step 1 the rod *grows* to a configuration which is stress-free but not compatible because it does not fit within the walls. Hence this configuration is called a virtual configuration. Step 2 is the elastic response, which may be thought of as the

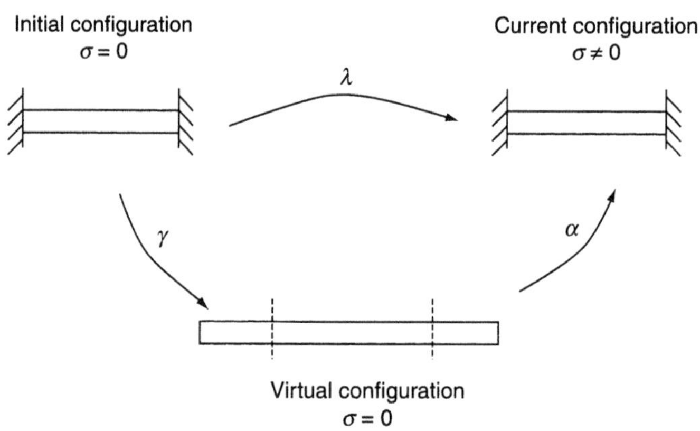

Fig. 6.4 Decomposition of the deformation into a growth step, which leads to an incompatible, virtual configuration, and an elastic-response step, which fits the rod back into a physically compatible state and introduces stress.

problem of fitting the rod back in between the walls. This step introduces stress, and maps the rod from the virtual configuration to the current configuration.

In many situations, the growth of a body depends on the stress level within the body. For instance, the body may function best at a particular stress, and grow until this target stress is reached. This situation can be modeled by

$$\frac{\partial \gamma}{\partial t} = k\gamma(\sigma - \sigma^*), \tag{6.11}$$

where σ^* is the target stress, or *homeostatic stress.* Consider the geometry of a rod between two walls and a neo-Hookean material with $\sigma = \mu\left(\alpha^2 - \alpha^{-1}\right)$. As before, $\gamma = 1/\alpha$, and eqn (6.11) may be recast in terms of α as

$$\frac{\partial \alpha}{\partial t} = \alpha\sigma^* - \mu(\alpha^3 - 1). \tag{6.12}$$

Setting the right-hand side to zero, we define α^* as the solution to $\alpha\sigma^* = \mu(\alpha^3 - 1)$, the equilibrium elastic stretch.

Assuming that the initial configuration is stress-free, $\alpha(0) = 1$. If $\sigma^* > 0$, then $\alpha^* > 1$ and so α will increase, tending to a limit of α^* as $t \to \infty$. This corresponds to γ decreasing from 1 to $\gamma^* = 1/\alpha^*$. In this case, the target stress is tensile, and the rod contracts until the target stress is reached. On the other hand, if $\sigma^* < 0$, the target stress is compressive. Here, $\alpha^* < 1$, and so α decreases to α^*; that is, the rod grows until the proper compression is reached.

6.3 3D growth

In this section, we develop the full 3D theory of morphoelasticity. Let $\mathbf{X}$ be the position vector for a body $\mathcal{B}_0$ before deformation, and $\mathbf{x}$ the position vector for the body in the current configuration at time t, denoted by $\mathcal{B}$. This is pictured in Fig. 6.5. The deformation of the body is described by a function χ, so that $\mathbf{x}(t) = \chi(\mathbf{X}, t)$. Then the deformation gradient tensor is $\mathbf{F} = \text{Grad}\,\chi$, the gradient taken with respect to the reference variables.

Without growth, the motion of the body is governed by the following equations:
Conservation of mass:

$$\dot{\rho} + \rho \text{ div } \mathbf{v} = 0; \tag{6.13}$$

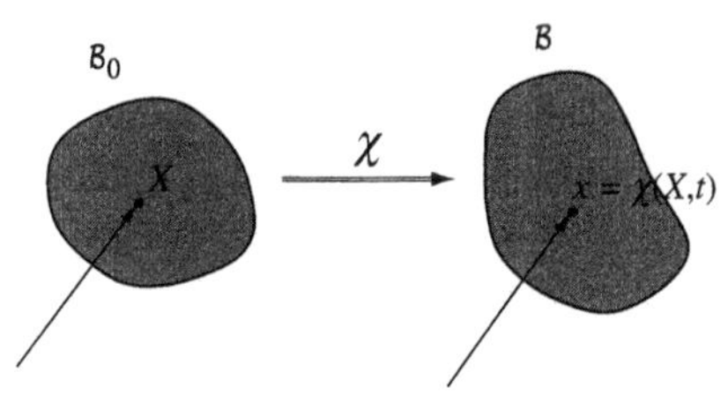

Fig. 6.5 Deformation χ from reference configuration $\mathcal{B}_0$ to current configuration $\mathcal{B}$.

Balance of linear momentum:

$$\rho\dot{\mathbf{v}} = \text{div } \mathbf{T} + \rho\mathbf{b}; \tag{6.14}$$

Balance of angular momentum:

$$\mathbf{T}^{\mathrm{T}} = \mathbf{T}. \tag{6.15}$$

Here ρ is the density, $\mathbf{v} \equiv \dot{\mathbf{x}} = \partial\mathbf{x}/\partial t$ is the velocity of a material point, $\mathbf{T}$ is the Cauchy stress, and $\mathbf{b}$ represents any body forces. Note that div is the divergence operator in the current configuration.

We now derive the corresponding equations with added growth. To do so, we will need to make use of the following transport formulas (derived in, e.g., Ogden 2003):

$$\frac{d}{dt}\int_{\Omega} \phi \, dv = \int_{\Omega} (\dot{\phi} + (\text{div } \mathbf{v})\phi) \, dv, \tag{6.16}$$

$$\frac{d}{dt}\int_{\Omega} \mathbf{u} \, dv = \int_{\Omega} (\dot{\mathbf{u}} + (\text{div } \mathbf{v})\mathbf{u}) \, dv, \tag{6.17}$$

where ϕ and $\mathbf{u}$ are scalar and vector fields associated with the moving body. Assume that growth can occur through volumetric growth, captured by the growth rate function $\gamma(\mathbf{x})$, or by the flux of material through the boundary, described by the vector $\mathbf{\Pi}(\mathbf{x})$. The balance of mass, applied to an arbitrary volume element Ω, yields

$$\frac{d}{dt}\int_{\Omega} \rho \, dv = \int_{\Omega} \gamma \, dv + \int_{\partial\Omega} \mathbf{\Pi} \cdot d\mathbf{s}. \tag{6.18}$$

By application of the divergence theorem and the transport formula (6.16), and using the fact that Ω is arbitrary, we localize the mass equation

$$\dot{\rho} + \rho \text{ div } \mathbf{v} = \gamma + \text{div } \mathbf{\Pi}. \tag{6.19}$$

The balance of linear momentum is expressed as

$$\frac{d}{dt}\int_{\Omega} \rho\mathbf{v} \, dv = \int_{\partial\Omega} \mathbf{T} \cdot d\mathbf{a} + \int_{\Omega} \rho\mathbf{b} \, dv + \int_{\Omega} \gamma\mathbf{v} \, dv + \int_{\partial\Omega} \mathbf{\Pi} \otimes \mathbf{v} \, d\mathbf{s}. \tag{6.20}$$

Use of the divergence theorem on the terms integrated over $\partial\Omega$, and the transport formula (6.17), leads to

$$\int_{\Omega} (\rho\dot{\mathbf{v}} + \dot{\rho}\mathbf{v} + (\text{div } \mathbf{v})\rho\mathbf{v}) \, dv = \int_{\Omega} (\text{div } \mathbf{T} + \rho\mathbf{b} + \gamma\mathbf{v} + \text{div}(\mathbf{\Pi} \otimes \mathbf{v})) \, dv. \tag{6.21}$$

If we insert the expression for $\dot{\rho}$ obtained from (6.19) and use the arbitrary nature of Ω, eqn (6.21) simplifies to

$$\rho\dot{\mathbf{v}} = \text{div } \mathbf{T} + \rho\mathbf{b} + (\mathbf{\Pi} \cdot \nabla)\mathbf{v}. \tag{6.22}$$

Lastly, defining $\tilde{\mathbf{T}} := \mathbf{T} + \mathbf{\Pi} \otimes \mathbf{v}$, angular momentum is balanced if $\tilde{\mathbf{T}}$ is symmetric.

To simplify the situation, it is assumed that growth processes take place on slow timescales so that we can neglect fluxes through the boundary and their momenta. This is the slow-growth assumption of morphoelasticity. It enables us to set $\mathbf{\Pi} = \mathbf{0}$. If we drop all terms containing $\mathbf{\Pi}$ and compare the result with eqns (6.13)–(6.15), the only apparent difference in the governing equations with added growth is the addition of the function γ in eqn (6.19); that is, the mass equation becomes

$$\dot{\rho} + \rho \operatorname{div}\mathbf{v} = \gamma. \tag{6.23}$$

Note also that $\tilde{\mathbf{T}} = \mathbf{T}$, so that the Cauchy stress is symmetric. To close the system, a constitutive law is needed to relate the stress to the deformation, i.e. a relation of the form $\mathbf{T} = H(\mathbf{F})$. We will assume that the material is hyperelastic. In the absence of growth, the relation is

$$\mathbf{T} = J^{-1}\mathbf{F}\frac{\partial W}{\partial \mathbf{F}} - p\mathbf{1}, \tag{6.24}$$

where W is the strain energy function and $J = \det \mathbf{F}$. If the material is incompressible, $J = 1$ and p is the hydrostatic pressure; for a compressible material, $p = 0$. However, this relation requires the initial body $\mathcal{B}_0$ to be *unstressed.* This raises the question: what if we don't know the unstressed initial configuration? Suppose we only know the body at time t in the configuration $\mathcal{B}$, a stressed state. How do we develop a response function theory for a stressed material? In answering this question, we will develop the theory of morphoelasticity.

As a first step, we remove any external or body forces from $\mathcal{B}$, that is, *unload the body.* Let $\mathcal{B}_\mathrm{r}$ be this unloaded configuration, and denote the map from $\mathcal{B}_\mathrm{r}$ to $\mathcal{B}$ by χ_l. It is important to note that $\mathcal{B}_\mathrm{r}$ is not stress-free, since residual stresses may still exist within the body. There are several possible approaches at this point. One is to develop a constitutive relation that includes the residual stress field. That is, if $\mathbf{T}^\mathrm{r}$ is the stress in $\mathcal{B}_\mathrm{r}$, one could try to find a relation $\mathbf{T} = H(\mathbf{F}, \mathbf{T}^\mathrm{r})$. Another possibility, outlined in Ogden (2003), is to add the residual stress field, i.e. define the stress to be the sum of the residual stress and the stress due to loading: $\mathbf{T} = \mathbf{T}^\mathrm{r} + \mathbf{T}^\mathrm{l}$.

A third option, and the one we will expand upon here, is to find a description of the residually stressed material in terms of volume element growth. The approach is to define a stress-free reference configuration $\mathcal{B}_0$ and build a map from $\mathcal{B}_0$ to $\mathcal{B}_\mathrm{r}$ where residual stress is induced in the body by growth. This is depicted in Fig. 6.6.

Conceptually, the idea is to repeatedly *cut* the body $\mathcal{B}_\mathrm{r}$, thereby relieving the residual stress, until the body consists of a (possibly infinite) collection of stress-free "components". These components together comprise a stress-free *virtual configuration.* By keeping track of the deformation due to the "cuts" and reversing the directionality, a deformation tensor can be defined which takes the unstressed virtual configuration to the fully stressed current configuration. The final step is then to introduce a tensor which takes the connected, stress-free state $\mathcal{B}_0$ to the virtual configuration.

Mathematically, the idea relies upon Signorini's Theorem, which states that the *mean stress* in a simply connected body which is in mechanical equilibrium and without external loads vanishes identically, i.e.

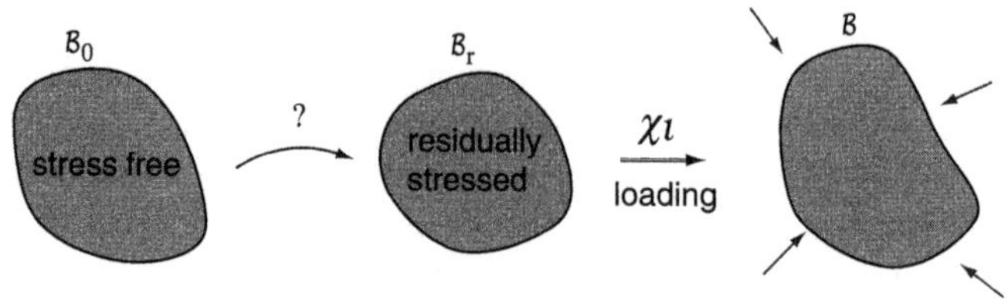

Fig. 6.6 Conceptual schematic illustration of the theory developed here. A stress-free reference configuration $\mathcal{B}_0$ is sought, such that in the deformation from $\mathcal{B}_0$ to $\mathcal{B}_r$, residual stress develops because of growth of volume elements.

$$\bar{\mathbf{T}} := \frac{1}{v}\int_{\Omega} \mathbf{T}\, dv = \mathbf{0} \tag{6.25}$$

for any volume element Ω of volume v. Signorini's Theorem, along with the continuity of the stress $\mathbf{T}$, implies that there exists a point $\mathbf{x}^* \in \Omega$ such that $\mathbf{T}(\mathbf{x}^*) = \mathbf{0}$. Now, consider an arbitrary point $\mathbf{x} \in \mathcal{B}_r$. We start to cut the body, and follow the portion of the body which contains $\mathbf{x}$, denoted by $\mathcal{B}_r^n$ (see Fig. 6.7). We define

$$d_n := |\mathbf{x} - \mathbf{x}_n^*|, \tag{6.26}$$

where $\mathbf{T}(\mathbf{x}_n^*) = \mathbf{0}$. Thus, d_n is the distance between $\mathbf{x}$ and the (not necessarily unique) stress-free point $\mathbf{x}_n^* \in \mathcal{B}_r^n$ after the nth cut. As $n \to \infty$, $d_n \to 0$, and thus

$$\lim_{n\to\infty} \mathbf{T}(\mathbf{x}) = \mathbf{0}.$$

The body is now completely unstressed at $\mathbf{x}$. We repeat this process for all points in $\mathcal{B}_r$ and we are left with an *unstressed virtual configuration*, which we call $\mathcal{B}_v$. To make sense of $\mathcal{B}_v$, consider that for each point $\mathbf{x}$ in $\mathcal{B}_r$ there is a map which takes the tangent space at that point to the tangent space in $\mathcal{B}_v$; in this sense, $\mathcal{B}_v$ is not really a "configuration" but rather is defined through a collection of tangent spaces, that is, a

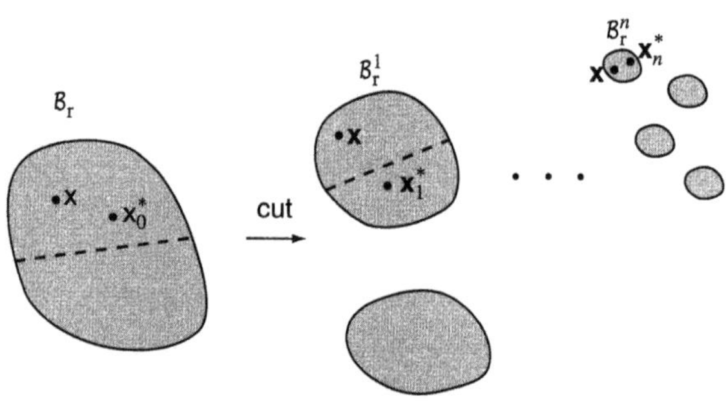

Fig. 6.7 The residually stressed body $\mathcal{B}_r$ is repeatedly cut, relieving stress. As the number of cuts $n \to \infty$, the distance between the point $\mathbf{x}$ and the stress-free point $\mathbf{x}^*$ goes to zero.

tangent bundle. This tangent bundle is the tangent bundle of a manifold, the "virtual configuration". Reversing this map, we define $\mathbf{F}_\mathrm{r}$ as the tensor that takes $\mathcal{B}_\mathrm{v}$ to the residually stressed body $\mathcal{B}_\mathrm{r}$. Letting $\mathbf{F}_\mathrm{l}$ be the deformation tensor from $\mathcal{B}_\mathrm{r}$ to $\mathcal{B}$, i.e. the deformation that loads $\mathcal{B}_\mathrm{r}$, we define the elastic deformation tensor

$$\mathbf{A} := \mathbf{F}_\mathrm{l}\mathbf{F}_\mathrm{r}.$$

We now have a tensor $\mathbf{A}$ from a stress-free configuration to the current configuration. However, $\mathcal{B}_\mathrm{v}$ is not Euclidean—this is the price to pay for being unstressed—and so $\mathbf{A}$ is not in general compatible: there is no function χ_r such that $\mathbf{A} = \operatorname{grad} \chi_r$. To resolve this, the final step is to describe a deformation from the stress-free reference configuration $\mathcal{B}_0$ to the stress-free virtual configuration $\mathcal{B}_\mathrm{v}$. The fundamental assumption of morphoelasticity, due to Rodriguez *et al.* (1994), is to introduce the *growth deformation tensor* $\mathbf{G}$, which takes the tangent space of $\mathcal{B}_0$ to the tangent space of $\mathcal{B}_\mathrm{v}$. The full deformation tensor from $\mathcal{B}_0$ to $\mathcal{B}$, denoted by $\mathbf{F}$, is defined as the composition of the growth deformation $\mathbf{G}$ followed by the elastic deformation $\mathbf{A}$,

$$\mathbf{F} = \mathbf{A} \cdot \mathbf{G}. \tag{6.27}$$

Then $\mathbf{F}$ is defined from a stress-free Euclidean reference configuration, and so the constitutive relation (6.24) applies. The complete theory is sketched in Fig. 6.8.

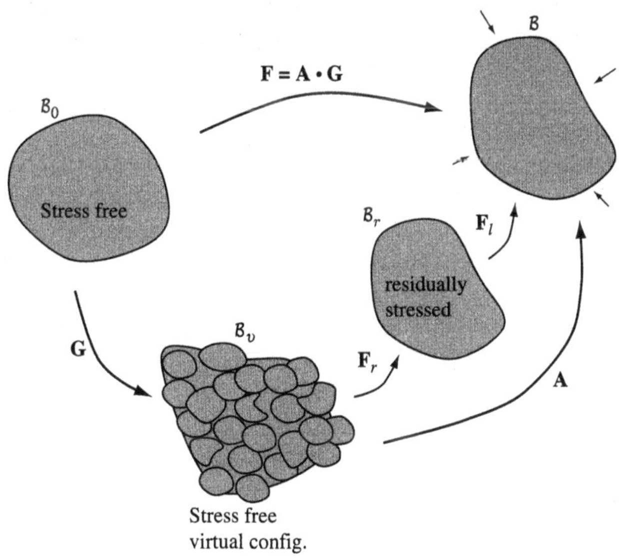

Fig. 6.8 Schematic illustration of the decomposition of the deformation tensor $\mathbf{F}$ into the growth tensor $\mathbf{G}$ and the elastic deformation tensor $\mathbf{A}$. The tensor $\mathbf{A}$, in turn, is composed of the reorganization deformation $\mathbf{F}_\mathrm{r}$, in which residual stress is introduced, and the deformation due to loading, $\mathbf{F}_\mathrm{l}$.

Physically, the theory is as follows: $\mathbf{G}$ describes the local increase or decrease of mass in $\mathcal{B}_0$. Since $\mathbf{G}$ only informs us about the change of mass locally, this growth step can lead to an incompatible (but still stress-free) configuration, where, for instance, neighboring cells overlap. Thus the growth is accompanied by an elastic response—the deformation $\mathbf{A}$—in which material is reorganized to ensure compatibility, and which includes the response to any applied loads. It is through the elastic response that residual stress is introduced.

As far as practical problems and solving equations are concerned, there are two ways to view the theory. We can either define a certain growth tensor and external loads and compute the new stressed configuration, or, starting with a stressed configuration, obtain the corresponding G. In the first case, suppose the reference configuration $\mathcal{B}_0$ is known and the growth tensor $\mathbf{G}$ is provided. The constitutive equation (6.24) transforms to

$$\mathbf{T} = \mathbf{A}\frac{\partial W}{\partial \mathbf{A}} - p\mathbf{1}. \tag{6.28}$$

In the case of mechanical equilibrium and in the absence of body forces, the governing equation is div $\mathbf{T} = \mathbf{0}$. By solving this along with appropriate boundary conditions specifying any applied loads, the deformation, as well as the stress field in the current configuration, can be completely determined. In Section 6.4, we demonstrate the process with an example.

In the second case, the question we began with was how to determine a response theory for a residually stressed material. Suppose that the current configuration $\mathcal{B}$ is known, but the reference configuration $\mathcal{B}_0$ is not. Equation (6.28) may again be used in conjunction with div $\mathbf{T} = 0$, in this case to solve for the deformation in the opposite direction, given that the growth deformation $\mathbf{G}$ is known. That is, if a stress-free reference configuration is unknown but the form of growth which induced the residual stress is known and the current configuration is also known, the reference state can be determined. Once the stress-free reference state is found, the response theory is well defined. Hence, if external loading is applied to a residually stressed material, the resulting deformation can be determined in two steps: first, ignore the loading by letting $\mathbf{A} = \mathbf{F}_\mathrm{r}$ only and solve for the "reverse deformation" to determine the stress-free reference configuration $\mathcal{B}_0$. Once this is known, the external loading can be added back into the elastic deformation tensor by letting $\mathbf{A} = \mathbf{F}_\mathrm{l}\mathbf{F}_\mathrm{r}$, and then the deformation $\mathbf{F} = \mathbf{A} \cdot \mathbf{G}$ encompasses the induced residual stress due to growth as well as deformation due to external loading. Solving for the deformation in the forward direction enables one to determine the desired configuration with loading, $\mathcal{B}$.

6.3.1 Continuous growth

Now consider the situation where a material evolves and is residually stressed. This material can be subject to further growth. In such a case, growth is applied to a residually stressed material rather than a stress-free state. This is equivalent to a subsequent growth deformation of a previously grown body, and is tied to the problem of a continuously or incrementally growing body. More broadly, this is the subject of morphodynamics. In a continuously growing body, the growth can be modeled by a relation of the form

$$\dot{\mathbf{G}} = \mathcal{G}(\mathbf{G}, \mathbf{A}, \mathbf{T}, \ldots; \mathbf{X}, t), \tag{6.29}$$

where the rate of growth is not necessarily constant, and might in general depend on the growth and strain tensors, stress, position, time, and potentially other factors. This can be approximated by a discrete growth process, where eqn (6.29) is replaced by $\mathbf{G}(t + \Delta t) = \mathbf{G}(t) + \Delta t\ \mathcal{G}$. This relation may be seen to define an incremental growth, $\mathbf{G}_{\text{inc}} := \mathbf{G}(t + \Delta t) - \mathbf{G}(t)$. At each step, the material grows according to some incremental growth law, followed by an incremental elastic response necessary for compatibility. The problem is that beyond the first step, the body is not stress-free, and so the constitutive relation (6.28) does not apply. One solution, given in detail in Goriely and Ben Amar (2007), is as follows: for each incremental step, define a total growth tensor, defined from the initial, stress-free configuration $\mathcal{B}_0$, which takes into account all previous growth increments. In this way, the ith incremental step becomes equivalent to a single step defined from a stress-free state.

Consider Fig. 6.9, which illustrates the idea for two incremental steps. The problem is that the virtual state $\mathcal{B}_{\text{v}2}$, after the second incremental step $\mathbf{G}_{\text{inc}}^{(2)}$, is not stress-free. The key is to unload $\mathcal{B}_{\text{v}2}$ by defining a stress-free virtual configuration $\mathcal{V}_2$ and finding a deformation from $\mathcal{V}_2$ to $\mathcal{B}_{\text{v}2}$. In brief, since the growth steps do not introduce stress, the stress in $\mathcal{B}_{v2}$ arises from the elastic response $\mathbf{A}_{\text{inc}}^{(1)}$ and a possible rotation of the principal axis due to $\mathbf{G}_{\text{inc}}^{(2)}$. Thus, the deformation from $\mathcal{V}_2$ to $\mathcal{B}_{v2}$ is of the form $\mathbf{R} \cdot \mathbf{A}_{\text{inc}}^{(1)}$, where $\mathbf{R}$ is a rotation. By inverting this map, the growth tensor $\mathbf{G}_2$ can be defined. In particular, if the growth and elastic tensors commute, as is the case for

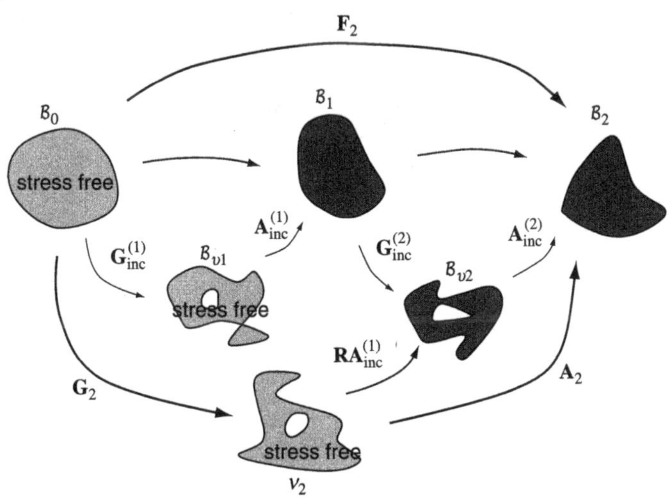

Fig. 6.9 Two incremental steps. The total growth tensor $\mathbf{G}_2$ is defined after creating a map from the stress-free virtual configuration $\mathcal{V}_2$ to the stressed virtual configuration $\mathcal{B}_{\text{v}2}$.

radial deformations in a spherical or cylindrical geometry, $\mathbf{G}_2$ takes the convenient form

$$\mathbf{G}_2 = \mathbf{G}_{\text{inc}}^{(2)} \cdot \mathbf{G}_{\text{inc}}^{(1)}. \tag{6.30}$$

Since $\mathbf{G}_2$ is defined from the stress-free state $\mathcal{B}_0$, the state $\mathcal{B}_2$ can be found using the constitutive relation (6.28) for the elastic deformation $\mathbf{A}_2$ and the decomposition $\mathbf{F}_2 = \mathbf{A}_2 \cdot \mathbf{G}_2$. Extending this idea to an arbitrary number of steps is relatively straightforward, in which case the total growth becomes a product of all subsequent growth increments. In essence, the process consists of *evolving* the virtual configuration as the body grows.

6.3.2 Relationship between γ and G

To complete the basic theory of morphoelasticity, we need to relate the growth tensor $\mathbf{G}$ to the growth rate function $\gamma(\mathbf{x})$ appearing in the mass balance equation (6.19). To do this, we first note that when expressed in the reference frame, the mass equation is

$$\frac{d}{dt}(J\rho) = J\gamma. \tag{6.31}$$

Next, we write $J = J_A J_G$, so that J_A measures the change in volume during the elastic deformation, and J_G the change during the growth deformation. Let ρ_G be the density with respect to the virtual configuration, and consider an element of mass in the virtual state, $dM = \rho_G\, dV$. Since mass is not added in the elastic deformation, the mass in the current state is

$$dm = \rho\, dv = dM = \rho_G\, dV, \tag{6.32}$$

and since $dv = J_A\, dV$, we have $\rho J_A = \rho_G$. Using this relationship along with the fact that $J_G = \det \mathbf{G}$, eqn (6.31) can be written, after some tensor algebra, as

$$\dot{\rho} + \rho_G \operatorname{tr}(\mathbf{G}^{-1}\dot{\mathbf{G}}) = J_A\gamma. \tag{6.33}$$

In particular, if the density does not change with growth, $\dot{\rho} = 0$, and the relation is

$$\rho \operatorname{tr}(\mathbf{G}^{-1}\dot{\mathbf{G}}) = \gamma. \tag{6.34}$$

In the case of constant volume, $J_G = 1$, and the relationship can be written $\dot{\rho_G} = J_A\gamma$, in which case we can interpret γ as the rate of densification.

6.4 Sample problem: A growing cylindrical tube

In this section, we present as a sample problem the growth and deformation of an incompressible, hyperelastic cylindrical tube. We restrict ourselves to radial deformations uniform along the tube axis. Therefore, we focus only on a cross section of the tube. The geometry is sketched in Fig. 6.10. The radius of the undeformed tube is defined by $A \le R \le B$. The deformation is described by the function $r(R)$, which gives the location of circles of radius R after deformation. Let the inner radius in the

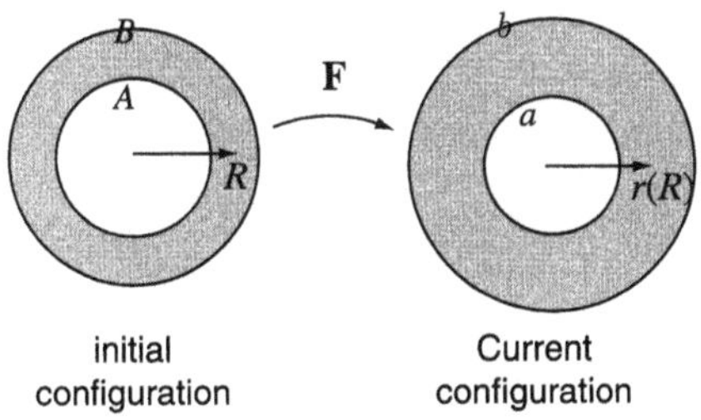

Fig. 6.10 Geometry for the growing cylindrical tube before and after deformation.

current configuration be $r(A) = a$ and let the outer radius be $r(B) = b$, so that the tube after deformation is given by $a \leq r \leq b$.

The deformation gradient tensor, expressed in cylindrical coordinates, is

$$\mathbf{F} = \text{diag}(r'(R), r/R, 1). \tag{6.35}$$

The elastic strain tensor is $\mathbf{A} = \text{diag}(\alpha_1, \alpha_2, 1)$, and the growth tensor for the symmetric growth is $\mathbf{G} = \text{diag}(\gamma_1, \gamma_2, 1)$. Here, γ_1 corresponds to radial growth and γ_2 to circumferential growth. Thus if $\gamma_2 = 1$ and $\gamma_1 \neq 1$, the growth is purely in the radial direction, with each radial fiber increasing if $\gamma_1 > 1$ and decreasing if $\gamma_1 < 1$. The case $\gamma_1 = 1$, $\gamma_2 \neq 1$ corresponds to circumferential fibers gaining or losing mass (see Fig. 6.11). If the γ_i are functions of R, the growth is inhomogeneous. Isotropic growth occurs when $\gamma_1 = \gamma_2$.

The decomposition $\mathbf{F} = \mathbf{A} \cdot \mathbf{G}$ implies

$$r' = \alpha_1 \gamma_1, \quad r/R = \alpha_2 \gamma_2. \tag{6.36}$$

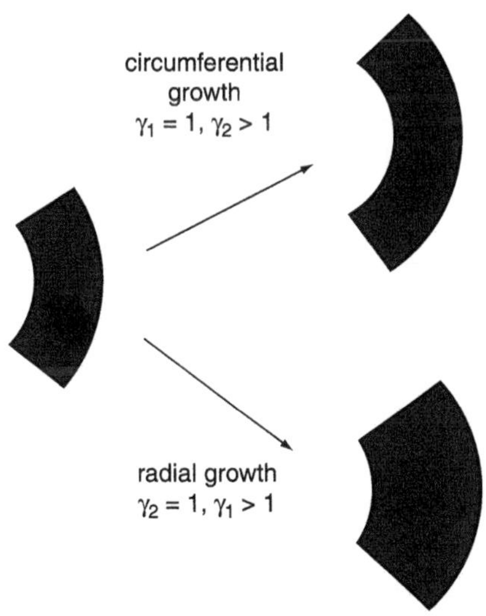

Fig. 6.11 Transformation of an area element under circumferential and radial growth.

The elastic incompressibility condition is det $\mathbf{A} = 1$. If we define $\alpha := \alpha_2$, this implies $\alpha_1 = 1/\alpha$. Together with eqn (6.36), the deformation is thus determined by the equation $r' = \gamma_1\gamma_2 R/r$, which after integration yields

$$r^2 - a^2 = 2\int_A^R \gamma_1\gamma_2 R\, dR. \tag{6.37}$$

Once the value of the inner radius a is known, the deformation is completely determined. Let t_1 and t_2 be the radial and circumferential (hoop) stresses of the Cauchy stress tensor. In the absence of body forces, the equilibrium linear-momentum equation is div $\mathbf{T} = \mathbf{0}$, and the only nonvanishing component is

$$\frac{\partial t_1}{\partial r} + \frac{1}{r}(t_1 - t_2) = 0. \tag{6.38}$$

The constitutive equation (6.28) has components

$$t_1 = \alpha_1 W_1 - p, \quad t_2 = \alpha_2 W_2 - p, \tag{6.39}$$

where $W = W(\alpha_1, \alpha_2)$ is the strain energy function and $W_i = \partial W/\partial\alpha_i$. Using incompressibility, we define the auxiliary function $\hat{W}(\alpha) = W(\alpha^{-1}, \alpha)$. Eliminating p in eqn (6.39), eqn (6.38) gives the following closed equation for the stress:

$$\frac{\partial t_1}{\partial r} = \frac{\alpha\hat{W}'(\alpha)}{r}. \tag{6.40}$$

Integrating eqn (6.40) from $r = a$ to b and defining $-P = t_1(b) - t_1(a)$, we have

$$-P = \int_a^b \frac{\alpha\hat{W}'(\alpha)}{r}\, dr. \tag{6.41}$$

Note that P is the applied load on the cylindrical tube, such that $P > 0$ in the case of external pressure and $P < 0$ for internal pressure. For given growth functions γ_i and a fixed pressure P, eqn (6.41) is an equation for the unknown parameter a, since b is related to a through eqn (6.37). Equation (6.41) may also be recast as an integral either in R or in α. Once the value of a is known, the deformation is completely determined. Then the stress functions t_1 and t_2 are found via

$$t_1(r) = \int_a^r \frac{\alpha\hat{W}'(\alpha)}{r}\, dr, \quad t_2 = t_1 + \alpha\hat{W}'(\alpha). \tag{6.42}$$

As we have discussed, growth can create residual stress even in the absence of external forcing. To demonstrate this, we consider a neo-Hookean material, characterized by a strain energy $W = (\mu/2)(\alpha_1^2 + \alpha_2^2 - 2)$, $\mu > 0$. We set the external pressure $P = 0$, and consider anisotropic homogeneous growth, that is, γ_i is constant with $\gamma_1 \neq \gamma_2$. In Fig. 6.12, we plot the radial and hoop stresses as functions of R for $\gamma_1 = 1.5$, $\gamma_2 = 1$. The radial growth of the tube induces a compressive residual radial stress in the material. In terms of the hoop stress, the inner radius is in compression, while the outer radius is in tension. Circumferential growth or, equivalently, radial resorption creates a tensile radial stress; this is the case $\gamma_1 < 1$ (not plotted).

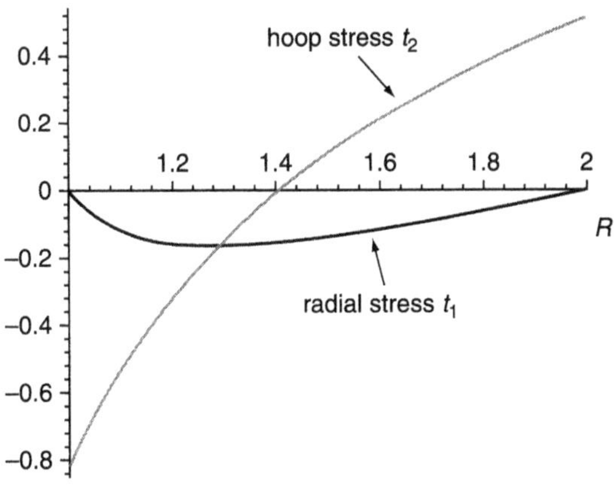

Fig. 6.12 Residual stress in a deformed neo-Hookean tube after anisotropic growth. ($A = 1$, $B = 2$, $\mu = 1$.)

6.4.1 Stability

To make progress in morphoelasticity, it is usually necessary to make assumptions about the class of allowable deformations. For instance, it is typical to restrict one's attention to symmetric deformations, as we did in the previous example. It is reasonable to expect an initially symmetric body subjected to symmetric growth and loading to remain symmetric, and such a restriction enables analytical progress to be made. However, it also raises the question of stability—when can we claim that a symmetric solution of a mechanical equilibrium is stable?

The approach to the question of stability is to add a perturbation to a finite deformation. The perturbation comes in the form of an imposed incremental deformation from a wider class of deformations, i.e. without the symmetry of the finite deformation. Let $\chi^{(0)}$ be a known finite deformation and let $\mathbf{T}^{(0)}$ be the related Cauchy stress, also known. We introduce an incremental deformation $\chi^{(1)}$ and consider

$$\chi = \chi^{(0)} + \epsilon\chi^{(1)}, \tag{6.43}$$

where ϵ is a small parameter characterizing the size of the imposed perturbation. The stability analysis proceeds by taking expansions of the deformation gradient and the elastic strain tensor (the growth tensor remains constant). That is, we write

$$\mathbf{F} = \text{Grad}\ \chi = \mathbf{F}^{(0)} + \epsilon\mathbf{F}^{(1)}\mathbf{F}^{(0)}, \quad \mathbf{A} = \mathbf{A}^{(0)} + \epsilon\mathbf{A}^{(1)}\mathbf{A}^{(0)}. \tag{6.44}$$

This form of the expansions is chosen so that the incremental deformation is expressed in terms of the current configuration. The relation $\mathbf{F} = \mathbf{A}\cdot\mathbf{G}$ implies $\mathbf{F}^{(0)} = \mathbf{A}^{(0)}\mathbf{G}$ and $\mathbf{F}^{(1)} = \mathbf{A}^{(1)}$. Incompressibility implies $\text{tr}(\mathbf{F}^{(1)}) = 0$. Next, we expand the Cauchy stress tensor as $\mathbf{T} = \mathbf{T}^{(0)} + \epsilon\mathbf{T}^{(1)} + O(\epsilon^2)$ and the hydrostatic pressure as $p = p^{(0)} + \epsilon p^{(1)}$. The constitutive relation (6.28) reads, to zeroth order,

$$\mathbf{T}^{(0)} = \mathbf{A}^{(0)}\cdot(W^{(0)}_{\mathbf{A}} - p^{(0)}\mathbf{1}) \tag{6.45}$$

and to first order

$$\mathbf{T}^{(1)} = \mathbf{A}^{(1)}(\mathbf{T}^{(0)} + p^{(0)}) + \mathcal{L} : \mathbf{A}^{(1)} - p^{(1)}\mathbf{1}, \tag{6.46}$$

where $\mathcal{L} : \mathbf{A}^{(1)} = \mathbf{A}^{(0)} W^{(0)}_{\mathbf{AA}} : \mathbf{A}^{(1)}\mathbf{A}^{(0)}$, and $W^{(0)}_{\mathbf{A}}$, $W^{(0)}_{\mathbf{AA}}$ are the first and second derivatives of W with respect to $\mathbf{A}$, evaluated at $\mathbf{A}^{(0)}$. The components of the fourth-order tensor $\mathcal{L}$ are the instantaneous elastic moduli, these are functions of the strain components α_i and the derivatives of the strain energy $W(\alpha_1, \alpha_2)$ (see Ogden 1984 for a derivation and formulas).

Setting div $\mathbf{T} = \mathbf{0}$ gives at $O(1)$ div$(\mathbf{T}^{(0)}) = \mathbf{0}$, which is already satisfied by the finite deformation. At $O(\epsilon)$, the incremental equilibrium equation div$(\mathbf{T}^{(1)}) = \mathbf{0}$ may be written (Ben Amar and Goriely 2005) as

$$\mathbf{A}^{(1)}\text{grad } p^{(0)} + \text{div}(\mathcal{L} : \mathbf{A}^{(1)}) - \text{grad } p^{(1)} = 0. \tag{6.47}$$

The boundary conditions consist of prescribing either $\chi^{(1)}$ or $\mathbf{n} \cdot \mathbf{T}^{(1)}$; that is, either the deformation or the traction is prescribed on the boundary.

For a given finite deformation, all terms at order zero are known. Then, eqn (6.47) plus the boundary conditions forms the system for the linear stability analysis. If a solution to the boundary value problem exists, a bifurcation from the set of symmetric deformations is detected. To demonstrate, this, in the following subsection we return to the example of a growing cylindrical tube and consider its stability.

6.4.2 Stability of a growing cylindrical tube

As before, we consider only a cross section of the cylindrical tube, and so all quantities are assumed to be independent of z; instability is characterized by a bifurcation to a solution with angular dependence. At a point of instability, the tube "buckles" circumferentially to a noncircular cross section. This is the phenomenon of mucosal folding, which is seen to occur in a number of biological tissues, including airways (Lambert *et al.* 1994), the esophagus (Liao *et al.* 2007), and the gastrointestinal tract (Gregersen 2003).

Working in polar coordinates, the general form for the incremental deformation is

$$\chi^{(1)} = [u(r,\theta), v(r,\theta)]. \tag{6.48}$$

Then

$$\mathbf{F}^{(1)} = \mathbf{A}^{(1)} = \begin{bmatrix} u_r & \dfrac{u_\theta - v}{r} \\ v_r & \dfrac{u + v_\theta}{r} \end{bmatrix}. \tag{6.49}$$

The incompressibility condition is

$$\text{tr}(\mathbf{A}^{(1)}) = u_r + (u + v_\theta)/r = 0. \tag{6.50}$$

Inserting eqn (6.49) into the incremental equilibrium equation (6.47) yields two differential equations involving the unknown functions u, v, and $p^{(1)}$. To simplify the problem, these functions are assumed to be of the form

$$\begin{aligned} u(r,\theta) &= f(r)\sin(n\theta), \\ v(r,\theta) &= g(r)\cos(n\theta), \\ p^{(1)}(r,\theta) &= h(r)\sin(n\theta). \end{aligned} \tag{6.51}$$

In this manner, instability is examined as a function of the buckling mode n, the number of folds in the buckled state. By making these substitutions and using the incompressibility condition (6.50) to solve for $g(r)$ in terms of $f(r)$, the system can be simplified to a single fourth-order differential equation for $f(r)$:

$$B_4 f^{(4)}(r) + B_3 f'''(r) + B_2 f''(r) + B_1 f'(r) + B_0 f(r) = 0, \tag{6.52}$$

where

$$B_4 = \gamma_2^2 R^2, \quad B_3 = \frac{2\gamma_2\left(\gamma_1\gamma_2 R^2 + 2r^2\right)}{r\gamma_1},$$

$$B_2 = -\frac{3{\gamma_2}^4 R^4\gamma_1 - 8{\gamma_2}^3 R^2 r^2 + n^2{\gamma_2}^4 R^4\gamma_1 + r^4 n^2\gamma_1}{r^2{\gamma_2}^2 R^2\gamma_1},$$

$$B_1 = \frac{-3r^4 n^2\gamma_1\gamma_2 R^2 + 2\,r^6 n^2 + 3{\gamma_2}^5 R^6\gamma_1 - 4{\gamma_2}^4 R^4 r^2 + n^2{\gamma_2}^5 R^6\gamma_1 - 2{\gamma_2}^4 R^4 n^2 r^2}{r^3\gamma_1{\gamma_2}^3 R^4},$$

$$B_0 = \frac{-3{\gamma_2}^4 R^4\gamma_1 + 4{\gamma_2}^3 R^2 r^2 + 3n^2{\gamma_2}^4 R^4\gamma_1 - 4{\gamma_2}^3 R^2 n^2 r^2 - r^4 n^2\gamma_1 + r^4 n^4\gamma_1}{r^4{\gamma_2}^2 R^2\gamma_1}.$$

Note that $R = R(r)$ is determined from the finite deformation via eqn (6.37). The system is closed with the addition of boundary conditions—a typical choice is to set both the shear and the normal stresses equal to zero on the inner and outer edges. Alternatively, part of the tube may be held fixed, in which case one might require the perturbation to vanish at one boundary.

With given boundary conditions, a solution to the boundary value problem indicates the onset of instability. The approach is to leave one parameter in the system free and search for a critical value of that parameter at which the boundary value problem has a solution. This parameter might be the strain α at the inner boundary $r = a$, the applied pressure P, or the rate of growth. Here we briefly consider the case of homogeneous anistropic growth: we set $\gamma_2 = 1$, $\gamma_1 \neq 1$, and let the external pressure P be the bifurcation parameter.

For thin shells, an asymptotic analysis may be used to solve the ordinary differential equation (6.52) directly, but in the general case numerical methods are needed. One approach is to assume

$$f(r) = a_1\xi_1(r) + a_2\xi_2(r). \tag{6.53}$$

We then treat the system as an initial value problem at $r = a$, and solve two copies of the system with linearly independent initial conditions for the $\xi_i(r)$, such that both ξ_1 and ξ_2 satisfy the two boundary conditions at $r = a$ (note that eight initial conditions must be given in total, i.e. $\xi_i(a), \xi_i'(a), \xi_i''(a), \xi_i'''(a),\ \ i = 1, 2$, whereas the boundary conditions at $r = a$ provide only four total conditions). We integrate both of the ξ_i forward to $r = b$ and form the determinant of boundary values at $r = b$,

$$D(P) = \begin{vmatrix} c_1(\xi_1(b); P) & c_1(\xi_2(b); P) \\ c_2(\xi_1(b);)) & c_2(\xi_2(b); P) \end{vmatrix}, \tag{6.54}$$

where $c_1(f(b)) = 0,\ c_2(f(b)) = 0$ are the two boundary conditions at $r = b$. Owing to linearity, if the determinant $D(P) = 0$, then there exist values a_1 and a_2 for which $f(r)$ given by eqn (6.53) solves the boundary value problem. The parameter P is iterated until the determinant vanishes.

In Fig. 6.13, the critical value of P is plotted as a function of the radial growth γ_1, for various modes n and for fixed tube thickness $A/B = 0.5$. For a given γ_1, the smallest value of P_{cr} over all modes is the critical external pressure, denoted P_{cr}^*, at which the tube becomes unstable and buckles, and the corresponding mode n_{cr} is the critical buckling mode. For $P < P_{\text{cr}}^*$, the tube remains circular.

The value of P_{cr}^* for $\gamma_1 = 1$ is the critical pressure without growth, which occurs for the mode $n_{\text{cr}} = 2$. Growth can have both stabilizing and destabilizing effects. For

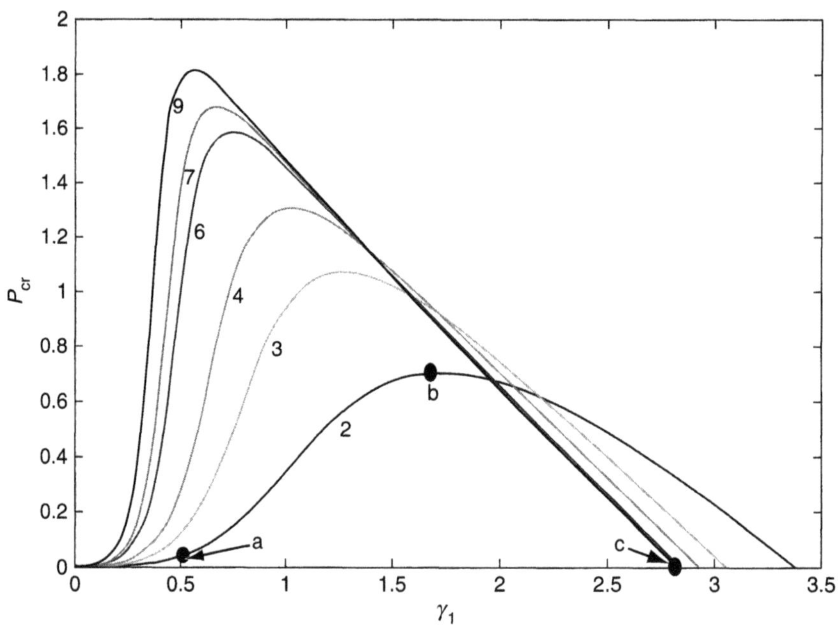

Fig. 6.13 Critical buckling pressure as a function of radial growth γ_1 for modes $n = 2, 3, 4, 6, 7, 9$.

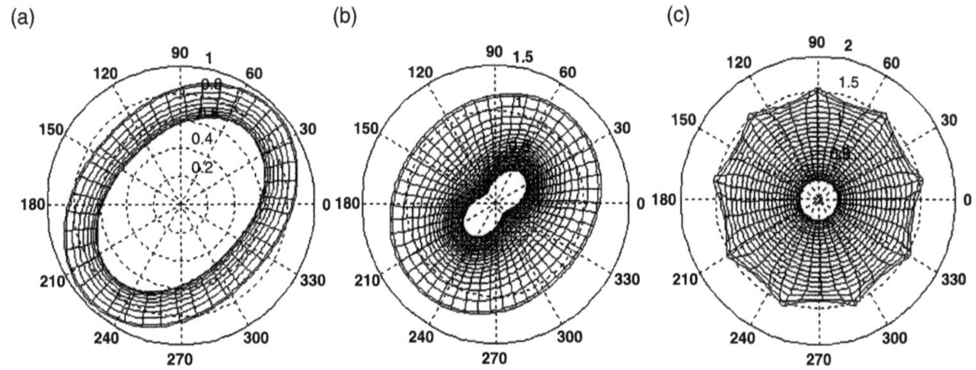

Fig. 6.14 Tube deformation after bifurcation corresponding to the points marked in Fig. 6.13. Note that the amplitude has been chosen to illustrate the structure—it has no physical significance, since the shapes were produced in a linear stability analysis.

this thickness, radial resorption ($\gamma_1 < 1$) is destabilizing—P^*_{cr} is smaller for all values of $\gamma_1 < 1$ than for $\gamma_1 = 1$. On the other hand, radial growth ($\gamma_1 > 1$) has a stabilizing effect initially but is destabilizing for large γ_1. Observe that the "strongest" tube possible, i.e. highest value of P^*_{cr}, is for $\gamma_1 \approx 1.72$, the point marked b in Fig. 6.13. With this in mind, one could imagine an application where one uses differential growth and bifurcation curves such as those in Fig. 6.13 in the *design* of biomaterials in which fine-tuning of the critical pressure is desired.

Along with the change in critical pressure, growth has a dramatic effect on the shape after bifurcation as well. Samples of the tube after deformation are shown in Fig. 6.14 for the three points marked a, b, c in Fig. 6.13. Note that point c corresponds to buckling due solely to differential growth (since $P^*_{\text{cr}} = 0$). It should not be surprising that differential growth can cause instability without any external loading, since it was demonstrated earlier that growth can induce residual stress. However, growth also changes the geometry of the body, which can have a stabilizing effect (as stubbier bodies are less likely to suffer buckling than thin ones). Further results can be established regarding the effect of tube thickness on buckling. Also, a greater range of possibilities opens up when considering a tube with two connected layers of material with different stiffness properties. For a detailed exploration, see Moulton and Goriely (2011).

6.5 Conclusions

Growth processes are abundant in nature, yet the rules governing them are inherently complex. Moreover, growth occurs in different forms, and the varying types of growth may be fundamentally different both on a biological and on a mathematical level. Theoretical and experimental analyses to understand growth processes on a mechanical level have been developed over a number of years and a good deal of understanding has been achieved. Of particular interest is the residual stress that can be induced in a body by growth—exactly how and why it arises, and the role that it plays in elastic materials are key issues.

We have presented a summary of some of the key ideas and challenges underlying morphoelasticity in this chapter. We illustrated the basic concept by considering one-dimensional growth. Even in one dimension, interesting and challenging questions arise. We then developed the full three-dimensional theory, where the question of how to think about a residually stressed body culminated in the fundamental idea of morphoelasticity, the decomposition of the deformation tensor into growth and elastic deformations. We then illustrated the process by considering a growing cylindrical tube, demonstrating how growth can alter material properties through a stability analysis.

Many problems remain unsolved in morphoelasticity, and the theory itself is far from complete. One of the big future challenges in the field is in morphodynamics. In particular, we discussed a continuously growing body as satisfying a relation

$$\dot{\mathbf{G}} = \mathcal{G}(\mathbf{G}, \mathbf{A}, \mathbf{T}, \ldots; \mathbf{X}, t); \tag{6.55}$$

however, finding appropriate biologically derived growth laws $\mathcal{G}$ presents an important challenge. Moreover, the incremental growth theory presented here largely becomes impractical when one is studying nonsymmetric deformations or geometries, and stability in a continuous framework presents a much greater difficulty.

Acknowledgments

This work was supported by the National Science Foundation under grant DMS-0907773 (AG). This publication is also based on work supported by Award No. KUK-C1-013-04, made by the King Abdullah University of Science and Technology (KAUST).

References

Ben Amar, M. and Goriely, A. (2005). Growth and instability in elastic tissues. *J. Mech. Phys. Solids*, **53**, 2284–2319.

Cowin, S. C. (2004). Tissue growth and remodeling. *Annu. Rev. Biomed. Eng.*, **6**, 77–107.

Détienne, P. and Thiel, J. (1988). Monograph on wapas in French Guiana. *Bois et Forêts des Tropiques*, **216**, 43–68.

Gooday, G. W. and Trinci, A. P. J. (1980). Wall structure and biosynthesis in fungi. In: Gooday, G. W., Lloyd, D., and Trinci, A. P. J. (eds.), *The Eukaryotic Microbial Cell*, vol. 30, pp. 207–251. Cambridge University Press, London.

Goriely, A. and Ben Amar, M. (2005). Differential growth and instability in elastic shells. *Phys. Rev. Lett.*, **94**(19), 198103.

Goriely, A. and Ben Amar, M. (2007). On the definition and modeling of incremental, cumulative, and continuous growth laws in morphoelasticity. *Biomech. Model. Mechanobiol.*, **6**(5), 289–296.

Goriely, A. and Tabor, M. (2003). Biomechanical models of hyphal growth in actinomycetes. *J. Theor. Biol.*, **222**, 211–218.

Gregersen, H. (2003). *Biomechanics of the Gastrointestinal Tract: New Perspectives in Motility Research and Diagnostics.* Springer, London.

Howard, R. J. and Valent, B. (1996). Breaking and entering: Host penetration by the fungal rice blast pathogen, *Magnaporthe grisea. Annu. Rev. Microbiol.*, **50**, 491–512.

Humphrey, J. D. (2003). Continuum biomechanics of soft biological tissues. *Proc. Roy. Soc. Lond. A*, **459**, 3–46.

Lambert, R. K., Codd, S. L., Alley, M. R., and Pack, R. J. (1994). Physical determinants of bronchial mucosal folding. *J. Appl. Physiol.*, **77(3)**, 1206–1216.

Liao, D., Zhao, J., Yang, J., and Gregersen, H. (2007). The oesophageal zero-stress state and mucosal folding from a giome perspective. *World J. Gastroenterol.*, **13**(9), 1347–1351.

Moulton, D. E. and Goriely, A. (2011). Circumferential buckling instability of a growing cylindrical tube. *J. Mech. Phys. Solids*, **59**, 525–537.

Ogden, R. W. (1984). *Non-linear Elastic Deformation.* Dover, New York.

Ogden, R. W. (2003). *Nonlinear Elasticity with Application to Material Modelling.* Polish Academy of Sciences, Institute of Fundamental Technological Research, warsaw.

Rodriguez, E. K., Hoger, A., and McCulloch, A. (1994). Stress-dependent finite growth in soft elastic tissues. *J. Biomech.*, **27**, 455–467.

Skalak, R. and Hoger, A. (1997). Kinematics of surface growth. *J. Math. Biol.*, **35**, 869–907.

Taber, L. A. and Humphrey, J. D. (2001). Stress-modulated growth, residual stress, and vascular heterogeneity. *J. Biomech. Eng.*, **123**, 528–535.

Thompson, D. W. (1992). *On Growth and Form: The Complete Revised Edition.* Dover, New York.

Vandiver, R. and Goriely, A. (2008). Tissue tension and axial growth of cylindrical structures in plants and elastic tissues. *Europhys. Lett. (EPL)*, **84**(58004).

7

Mechanics of tumor growth: Multiphase models, adhesion, and evolving configurations

Luigi Preziosi and Guido Vitale

Dipartimento di Matematica, Politecnico di Torino

7.1 Introduction

In the last ten years, mixture theory has been applied with success to describe several problems related to tumor growth, as reviewed by Ambrosi and Preziosi (2002), Araujo and McElwain (2005*a*), Graziano and Preziosi (2007), Lowengrub *et al.* (2010), and Tracqui (2009). This chapter will give the basics required to deduce multiphase models for a few but essential constituents (cells, extracellular matrix, extracellular liquid, and possibly blood and lymphatic vasculature). All the steps of the modeling procedure are explained in detail, special attention being paid to the meaning of all the different terms involved in the model. We will also show how to take account of the presence of several subpopulations of cells, the different components of the extracellular matrix, and possibly a homogenized vascular network.

Though the general model can be quite complicated, there are two basic assumptions that allow one to obtain more manageable models. The first one consists in assuming that the components of the extracellular matrix and possibly of the vasculature form such an intricate network that they all move together, so that the same deformation and velocity describe their evolution. The second one consists in assuming that the pressure gradient and the interaction forces between the liquid and the other constituents are very small when compared, for instance, with the internal stresses in the solid constituents or, say, the adhesion forces acting between the cells and the extracellular matrix.

We will also focus on other mechanical aspects related to tumor growth, due to the activation of mechanotransduction pathways inside the cell, contact inhibition phenomena, and the formation of stiffer fibrotic tissues.

In order to focus on the aspects above, we will forget about the role that nutrients and growth factors have on growth. This would require the introduction of reaction–diffusion equations in a multicomponent model. The reader interested in such extensions can look at Astanin and Preziosi (2008), and Bear and Bachmat (1990).

Specifically, Sections 2 and 3 will describe how to derive the balances of mass and momentum, respectively, for multicomponent systems. Sections 4 through 6 will focus on the interaction forces acting between different constituents, on the relative importance of the interaction with the extracellular liquid with respect to the interactions between solid-like constituents, on how to transfer microscopic information about the forces exerted by the adhesion molecules up to the continuum level, and on how to model the interaction force between cells and the extracellular matrix. Section 7 will specialize the model to the case in which the ensemble of cells is treated as a liquid, with two applications having to do with two different mechanical aspects related to tumor growth: contact inhibition and the formation of fibrotic tissues. Section 8 will describe how to model the ensemble of cells as a solid. Finally, Sections 9 and 10 will give a comparison with several experimental tests on cell aggregates and multicellular spheroids.

7.2 Mass balance for a multicomponent system

Tumors, like their surrounding tissues, are mainly made of cells, extracellular matrix (ECM), blood and lymphatic vessels, and physiological liquids. The growth and

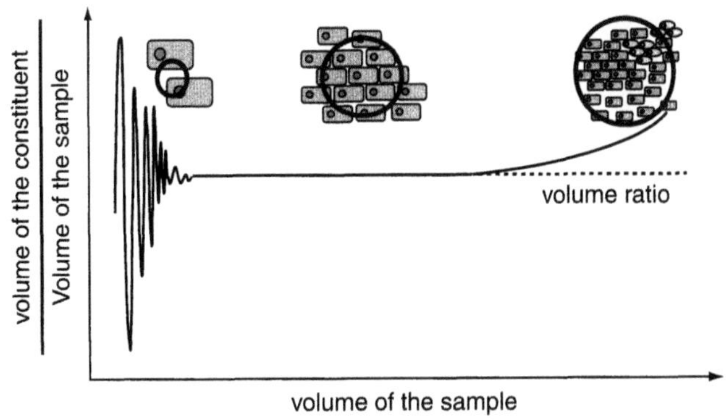

Fig. 7.1 Plot of the volume ratio of a constituent as a function of sample volume.

behavior of cells depend on many chemical compounds, for example nutrients, growth factors, and chemotactic factors. Of course, the typical size of molecules is extremely small compared with cells, vascular structures, and ECM fibers, so that the space they occupy can be neglected and they can be considered as part of the other constituents, i.e., dissolved in the extracellular liquid, within the vasculature, attached to the extracellular matrix, or within the cells.

For all the other constituents of the mixture, it is useful to refer to their volume ratios, which can be defined as follows. Given a point in the mixture, consider a sequence of spheres centered at that point. By measuring the ratio of the volume occupied by the constituent of interest inside the sphere to the volume of the sphere, one may observe the dependency shown in Fig. 7.1. For sample volumes of the size of the microscopic constituent, the ratio is likely to oscillate very strongly owing to microscopic inhomogenity. At the other extreme, macroscopic inhomogenities may affect the ratio for large sample volumes. However, for sample volumes in between, which in this case means at scales larger than the cell size and smaller than the typical tissue scale, the ratio is nearly constant and allows us to define a quantity called the volume ratio of the constituent.

Let us then denote by $\phi_c, \phi_m, \phi_\ell, \phi_v \in [0,1]$ the volume ratios occupied by the cells, extracellular matrix, extracellular liquid, and vascular structures, respectively. Of course,

$$\sum_{\alpha=c,m,\ell,v} \phi_\alpha \leq 1, \tag{7.1}$$

because the above structures can at most occupy the entire space, and the model must be such that the solution satisfies this unilateral constraint. If, instead, one has considered all the constituents occupying the region of interest, then the mixture is said to be *saturated* and

$$\sum_{\alpha=c,m,\ell,v} \phi_\alpha = 1. \tag{7.2}$$

Actually, in some cases the upper constraint on the volume ratio (i.e., 1 in eqn (7.2)) is replaced by a constant value $\overline{\phi} < 1$, possibly space-dependent, allowing some fixed portion of space to be occupied by other constituents not considered in the mixture. To be specific, if we are fixing the volume occupied by vessels or by a general stroma, eqn (7.2) can be replaced by

$$\phi_c + \phi_m + \phi_\ell = \overline{\phi} \leq 1. \tag{7.3}$$

From the mathematical viewpoint, this geometrical constraint still plays the same role as the saturation assumption (7.2).

The mass balance equations for the various constituents can be obtained starting from a global mass balance of a constituent over a general control volume V fixed in space, with boundary ∂V and external normal to the boundary $\mathbf{n}$. As an example, let us focus on the cellular constituent of a tissue. Denoting by ρ the density of single cells, the mass of the constituent contained in V is

$$M = \int_{\mathrm{V}} \rho\phi_c \, dV$$

and can change owing to the flux caused by the motion of the constituent through the boundary ∂V and to growth or death of cells. One then has

$$\frac{dM}{dt} = -\int_{\partial\mathrm{V}} \rho\phi_c \mathbf{v}_c \cdot \mathbf{n} \, d\Sigma + \int_{\mathrm{V}} \rho\Gamma_c \, dV,$$

where $\mathbf{v}_c$ is the cell velocity and Γ_c is the growth/death rate for the cellular mass.

Using the divergence theorem, one can write

$$\int_{\mathrm{V}} \left[\frac{\partial}{\partial t}(\rho\phi_c) + \nabla \cdot (\rho\phi_c \mathbf{v}_c) - \rho\Gamma_c\right] dV = 0.$$

If the integrands are smooth, the arbitrariness of the volume of integration V ensures that the integrand must vanish, i.e.,

$$\frac{\partial}{\partial t}(\rho\phi_c) + \nabla \cdot (\rho\phi_c \mathbf{v}_c) = \rho\Gamma_c, \tag{7.4}$$

where ρ can be taken constant and equal to the density of water.

The same procedure can be repeated for all the volume ratios, so that one can write the mass balance equations as

$$\frac{\partial \phi_\alpha}{\partial t} + \nabla \cdot (\phi_\alpha \mathbf{v}_\alpha) = \Gamma_\alpha \qquad \text{for} \quad \alpha = c, m, \ell, v. \tag{7.5}$$

In the case of a saturated mixture, summing the above equations gives and using (7.2) or (7.3) gives

$$\nabla \cdot \sum_{\alpha=c,m,\ell,v} (\phi_\alpha \mathbf{v}_\alpha) = \sum_{\alpha=c,m,\ell,v} \Gamma_\alpha. \tag{7.6}$$

If vessels are not taken into account, as for instance happens in the avascular phase of growth, and mass exchange occurs only between the constituents taken into consideration, the mixture is said to be *closed*; and, if for the sake of simplicity one takes all constituents to have the same density ρ, one can write

$$\Gamma_c + \Gamma_m + \Gamma_\ell = 0 \tag{7.7}$$

(otherwise it is the sum of the $\rho_\alpha \Gamma_\alpha$ that needs to vanish).

On the other hand, the presence of a homogenized vascular or lymphatic structure makes the system open, because blood and lymphatic fluids are flowing in the vessels and perfusing through their walls. This allows the sum of the growth rates to be nonvanishing. In some models, for the sake of simplicity, the presence of blood and lymphatic vessels and even of the extracellular liquid is neglected, though their influence is reflected in the presence of external mass source/sink terms not satisfying eqn (7.7). However, in this case also one needs to ensure that during its evolution, the solution is such that the geometrical condition (7.1) is never violated. This approach was used, for instance, by Franks *et al.* (2003*a*,*b*), and Franks and King (2003).

Depending on the details needed to describe the phenomenon of interest, modeling the cellular population as a single constituent might not be enough. It may be necessary to distinguish different cell populations, for example tumor cells, endothelial cells, epithelial cells, fibroblasts, macrophages, and lymphocytes, or to distinguish different clones within the same population characterized by relevant differences in their behavior, or to distinguish the cells according to their phase in the cell cycle because, for instance, their response to therapies depends on whether the cell is resting, growing, or undergoing mitosis.

If this is the case, the same procedure can be repeated for each subpopulation, so that actually eqn (7.4) is split in to I equations

$$\frac{\partial \phi_{c_i}}{\partial t} + \nabla \cdot (\phi_{c_i} \mathbf{v}_{c_i}) = \Gamma_{c_i}, \qquad i = 1, \ldots, I, \tag{7.8}$$

where I is the number of subpopulations, ϕ_{c_i} is the volume ratio of the subpopulation i, $\mathbf{v}_{c_i}$ is its velocity, and Γ_{c_i} is its growth/death rate, which in this case can also take account of transition phenomena, for example between clones or between phases of the cell cycle. Of course,

$$\sum_{i=1}^{I} \phi_{c_i} = \phi_c, \qquad \sum_{i=1}^{I} \phi_{c_i} \mathbf{v}_{c_i} = \phi_c \mathbf{v}_c, \qquad \sum_{i=1}^{I} \Gamma_{c_i} = \Gamma_c, \tag{7.9}$$

and therefore summing all of eqn (7.8) over i gives eqn (7.4) back.

In a similar way, it might be necessary to distinguish the ECM in to its main constituents, for example collagen, elastin, fibronectin, vitronectin, and proteoglycans, because of their different mechanical behavior and chemical properties or because of different production/degradation phenomena. One then has

$$\frac{\partial \phi_{m_j}}{\partial t} + \nabla \cdot \left(\phi_{m_j} \mathbf{v}_m\right) = \Gamma_{m_j}, \qquad j = 1, \ldots, J, \tag{7.10}$$

where m_j is the volume ratio of the jth component and Γ_{m_j} is its remodeling rate. We stress that in eqn (7.10), the ECM velocity is taken to be the same for all ECM components, which means describing them as an intricate network of fibers that have to move all together. This is called a *constrained submixture assumption*, which, as we shall see in the following, will call for a single force balance equation for the ECM as a whole. As before,

$$\sum_{j=1}^{J} m_j = m, \qquad \sum_{j=1}^{J} \Gamma_{m_j} = \Gamma_m, \tag{7.11}$$

and summing eqn (7.10) over j gives

$$\frac{\partial \phi_m}{\partial t} + \nabla \cdot (\phi_m \mathbf{v}_m) = \Gamma_m \tag{7.12}$$

back. The same procedure could be repeated for the vascular networks because lymphatic, arterial, and venous flows play different roles. For these constituents, however, it is plausible to assume that the vessels move together with the ECM structure, which is entangled in the network of vessels. In this case, $\mathbf{v}_v = \mathbf{v}_m$.

7.3 Force balance for a multicomponent system

In order to determine the velocity fields appearing in the mass balance equations in the theory of mixtures, one can write a momentum balance equation for each constituent. As we shall see in the following, in dealing with growth phenomena, inertial terms can be neglected, luckily, so that the equations simplify to force balance equations. A further simplification is related to the fact that the ensemble of cells and ECM constituents forms a porous material in which the extracellular liquid flows, so it is possible to use classical tools and results from the theory of porous media. On the other hand, the behavior of the porous material is complicated by its deformability and by the presence of adhesion bonds between the constituents and within individual constituents, which, because of their weakness, are easily breakable, as occurs during growth or remodeling.

In order to clarify the origin of all terms appearing in the momentum equations, we focus here on the cellular constituent of the tissue.

Starting again with the integral formulation, the variation of the momentum of the cellular constituent in the fixed volume V (the term on the left-hand side of eqn (7.13) below) is due to

- the momentum flux caused by the motion of the cells through the boundary ∂V (the first term on the right-hand side of eqn (7.13));
- contact forces with other cells through the boundary ∂V, which are codirected with $\mathbf{n}$ (the second term on the right-hand side of eqn (7.13));
- contact forces due to interaction with other constituents within the domain through the interface separating the constituents, say the part of the cell membrane wetted by

the extracellular liquid or in contact with the extracellular matrix through adhesion sites (the third term on the right-hand side of eqn (7.13));

- momentum supply related to mass exchange among the constituents, for example fluid absorbed by a growing cell or ECM production (the fourth term on the right-hand side of eqn (7.13));
- body forces, for example chemotaxis or haptotaxis, which can be modeled in this way, though they actually involve the activation of subcellular mechanisms rather than an external action Chauviere and Preziosi (2010) (fifth term on the right-hand side of eqn (7.13)).

One then has

$$\frac{d}{dt}\int_V \rho\phi_c \mathbf{v}_c \, dV = -\int_{\partial V} \rho\phi_c \mathbf{v}_c(\mathbf{v}_c \cdot \mathbf{n})\, d\Sigma + \int_{\partial V} \widetilde{\mathbf{T}}_c^T \mathbf{n}\, d\Sigma$$
$$+ \int_V \widetilde{\mathbf{m}}_c \, dV + \int_V \rho\Gamma_c \mathbf{v}_c \, dV + \int_V \rho\phi_c \mathbf{b}_c \, dV, \tag{7.13}$$

where $\widetilde{\mathbf{T}}_c$ is called the partial stress and $\widetilde{\mathbf{m}}_c$ is called the interaction force.

So, using the divergence theorem, we can write

$$\int_V \left[\frac{\partial}{\partial t}(\rho\phi_c \mathbf{v}_c) + \nabla \cdot (\rho\phi_c \mathbf{v}_c \otimes \mathbf{v}_c - \widetilde{\mathbf{T}}_c) - \rho\phi_c \mathbf{b}_c - \widetilde{\mathbf{m}}_c - \rho\Gamma_c \mathbf{v}_c\right] dV = 0. \tag{7.14}$$

If the integrands are smooth, the arbitrariness of the volume of integration V ensures that the integrand must vanish, i.e.,

$$\frac{\partial}{\partial t}(\rho\phi_c \mathbf{v}_c) + \nabla \cdot (\rho\phi_c \mathbf{v}_c \otimes \mathbf{v}_c) = \nabla \cdot \widetilde{\mathbf{T}}_c + \rho\phi_c \mathbf{b}_c + \widetilde{\mathbf{m}}_c + \rho\Gamma_c \mathbf{v}_c. \tag{7.15}$$

Actually, using the mass balance equation (7.5), eqn (7.15) can be simplified to

$$\rho\phi_c \left(\frac{\partial \mathbf{v}_c}{\partial t} + \mathbf{v}_c \cdot \nabla \mathbf{v}_c\right) = \nabla \cdot \widetilde{\mathbf{T}}_c + \rho\phi_c \mathbf{b}_c + \widetilde{\mathbf{m}}_c. \tag{7.16}$$

As already stated, in growth phenomena the inertial term on the left-hand side can be neglected, so that eqn (7.16) simplifies even further into

$$\nabla \cdot \widetilde{\mathbf{T}}_c + \rho\phi_c \mathbf{b}_c + \widetilde{\mathbf{m}}_c = \mathbf{0}. \tag{7.17}$$

Proceeding in a similar way for the other constituents, one can write

$$\nabla \cdot \widetilde{\mathbf{T}}_m + \widetilde{\mathbf{m}}_m = \mathbf{0}, \tag{7.18}$$

$$\nabla \cdot \widetilde{\mathbf{T}}_\ell + \widetilde{\mathbf{m}}_\ell = \mathbf{0}, \tag{7.19}$$

where we have dropped the body force term because, apart from gravity, which plays no role in the phenomena of interest in this chapter, it is difficult to imagine relevant body forces acting on the ECM and the liquid.

We observe explicitly that the equation for the cellular constituent can be specialized to any subpopulation. On the other hand, even if it is found necessary to specify different ECM components, the constrained submixture hypothesis ensures that a single force balance equation (7.26) is enough to determine the common velocity $\mathbf{v}_m$. Of course, all the ECM components will contribute to the constitutive equation for the stress tensor according to their relative proportions.

If vascular structures are considered and they move together with the ECM, eqn (7.18) will still hold if the contribution due to vessels is included in $\mathbf{T}_m$ and the interaction of the vessels with the other constituents is included in $\mathbf{m}_m$.

If the mixture is closed and then momentum is exchanged only between the constituents taken into consideration, then one can enforce (Bowen 1976; Rajagopal and Tao 1995)

$$\sum_{\alpha=c,m,\ell} (\widetilde{\mathbf{m}}_\alpha + \rho\Gamma_\alpha \mathbf{v}_\alpha) = \mathbf{0}\,; \tag{7.20}$$

or, coherently with the fact that inertia can be neglected, the contribution due to mass exchange can be dropped with respect to that due to the interaction forces (Preziosi and Farina 2001), so that

$$\sum_{\alpha=c,m,\ell} \widetilde{\mathbf{m}}_\alpha = \mathbf{0}. \tag{7.21}$$

Hence, one can say that the interaction forces add up to zero, as might be expected, since they act as internal forces between the constituents of the whole mixture.

Actually, each term $\mathbf{m}_\alpha$ contains all forces acting on the constituent α owing to its interaction with the other constituents β. We assume here that an action–reaction principle holds for each interaction pair, i.e., $\mathbf{m}_{\alpha\beta} = -\mathbf{m}_{\beta\alpha}$, where $\mathbf{m}_{\alpha\beta}$ is the force acting on the constituent α owing to β. This, however, is an approximation, for instance because of the presence of exchanges of mass and other effects; but, as stated above, these can be considered negligible in growth phenomena.

7.4 Liquid–solid interactions in a saturated mixture

In a saturated and closed mixture, the geometrical constraint (7.2) describing saturation implies that not all motions are possible, but only those satisfying eqn (7.6) with a vanishing right-hand side, or

$$0 = \sum_{\alpha=c,m,\ell,v} \nabla\cdot(\phi_\alpha \mathbf{v}_\alpha) = \sum_{\alpha=c,m,\ell,v} (\phi_\alpha \mathbf{I} : \mathbf{L}_\alpha + \mathbf{v}_\alpha \cdot \nabla\phi_\alpha), \tag{7.22}$$

where $\mathbf{I}$ is the identity tensor and the colon indicates tensor scalar product. Without going into technical details, this means that it is possible to test the stress response only for those deformation gradients which satisfy the above constraint, or that the validity of the second law of thermodynamics is checked only for all those processes which satisfy the saturation constraint.

A similar reasoning is used classically in the study of the flow of incompressible fluids or the deformation of composite materials that can be assumed to be inextensible in a given direction. Enforcing the above conditions implies that one can only study those deformations which satisfy the constraints, and the second law of thermodynamics should hold only for these classes of processes. The consequence is that in the former case the isotropic part of the stress tensor for an incompressible fluid cannot be determined constitutively but is a reaction that is adjusted so that the incompressibility constraint is satisfied. Similarly, in the latter case, in which a material is inextensible along a direction $\mathbf{N}$ in the reference configuration, one cannot determine the component $\mathbf{n} \otimes \mathbf{n}$ (where $\mathbf{n} = \mathbf{FN}$ and $\mathbf{F}$ is the deformation gradient) of the stress tensor.

In the case of mixtures, the saturation constraint implies that the constitutive equations for the partial stresses and the interaction forces are characterized by the presence of a Lagrange multiplier, classically identified with the interstitial pressure of the extracellular liquid (Bowen 1976; Rajagopal and Tao 1995). The equation for the cellular constituent is written as

$$\widetilde{\mathbf{T}}_c = -\phi_c P \mathbf{I} + \phi_c \mathbf{T}_c, \tag{7.23}$$

$$\widetilde{\mathbf{m}}_c = P \nabla \phi_c + \mathbf{m}_c, \tag{7.24}$$

where $\mathbf{T}_c$ is called the excess stress and $\mathbf{m}_c$ the excess interaction force. Similar conditions hold for the other constituents.

Neglecting for the sake of simplicity the presence of vessels, one can then write

$$-\phi_c \nabla P + \nabla \cdot (\phi_c \mathbf{T}_c) + \mathbf{m}_c + \rho \phi_c \mathbf{b}_c = \mathbf{0}, \tag{7.25}$$

$$-\phi_m \nabla P + \nabla \cdot (\phi_m \mathbf{T}_m) + \mathbf{m}_m = \mathbf{0}, \tag{7.26}$$

$$-\phi_\ell \nabla P + \mathbf{m}_\ell = \mathbf{0}. \tag{7.27}$$

In the force balance equation for the liquid, when dealing with porous materials, the excess stress tensor for the extracellular liquid is usually assumed to be negligible. This allows one to obtain a Darcy-like law (Bowen 1976; Rajagopal and Tao (1995)).

In fact, the interaction forces of all the constituents with the liquid can be taken to be proportional to the velocity difference between the liquid and the other constituents through invertible positive matrices $\mathbf{M}_c$ and $\mathbf{M}_m$, so that

$$\mathbf{m}_{\ell c} = -\mathbf{M}_c(\mathbf{v}_\ell - \mathbf{v}_c), \tag{7.28}$$

$$\mathbf{m}_{\ell m} = -\mathbf{M}_m(\mathbf{v}_\ell - \mathbf{v}_m), \tag{7.29}$$

Therefore, eqn (7.27) can be written as

$$\mathbf{M}_m(\mathbf{v}_\ell - \mathbf{v}_m) + \mathbf{M}_c(\mathbf{v}_\ell - \mathbf{v}_c) = -\phi_\ell \, \nabla P, \tag{7.30}$$

or

$$\mathbf{v}_\ell = \mathbf{M}^{-1}\left(\sum_{\alpha=c,m} \mathbf{M}_\alpha \mathbf{v}_\alpha - \phi_\ell \, \nabla P\right), \tag{7.31}$$

where $\mathbf{M} = \mathbf{M}_c + \mathbf{M}_m$, which explicitly gives the liquid velocity in terms of the other velocities and the pressure gradient.

In particular, in the case of a rigid porous material, $\mathbf{v}_c = \mathbf{v}_m = \mathbf{0}$, if one defines $\mathbf{M}^{-1} = \mathbf{K}/\mu\phi_\ell^2$, where $\mathbf{K}$ is the permeability tensor and μ is the viscosity of the extracellular liquid, one has the classical Darcy's law,

$$\mathbf{v}_\ell = -\frac{\mathbf{K}}{\mu\phi_\ell}\nabla P. \tag{7.32}$$

Using eqn (7.6), one can also determine the following equation relating the pressure to the other variables:

$$\nabla \cdot \left(\frac{\mathbf{K}}{\mu}\nabla P\right) = \nabla \cdot \left[\sum_{\alpha=c,m}\left(\phi_\alpha \mathbf{I} + \frac{\mathbf{K}}{\phi_\ell \mu}\mathbf{M}_\alpha\right)\mathbf{v}_\alpha\right] - \sum_{\alpha=c,m,\ell} \Gamma_\alpha. \tag{7.33}$$

By the action–reaction principle assumed at the beginning of this section, the same terms as in eqn (7.29) are found, with the opposite sign, in the equations for the cells, the ECM constituents, and possibly the vessels. These equations also contain the pressure gradient terms, which, because of eqn (7.27), are of the same order as the interaction force with the extracellular liquid. It can be argued that these terms are much smaller than the interaction forces acting between the cells and between cells and the ECM, so that the momentum equations can be simplified to

$$\nabla \cdot (\phi_c \mathbf{T}_c) + \mathbf{m}_{cm} + \rho\phi_c \mathbf{b}_c = \mathbf{0}, \tag{7.34}$$

$$\nabla \cdot (\phi_m \mathbf{T}_m) - \mathbf{m}_{cm} = \mathbf{0}. \tag{7.35}$$

Under these hypotheses, the equations above do not depend either on the interstitial pressure or on the liquid velocity. Therefore, they can be solved in principle without solving eqns (7.31) and (7.33), which is required only if we want to describe the evolution of either the interstitial pressure or the liquid velocity. Equations (7.31) and (7.33) are solvable in cascade after integrating eqns (7.34) and (7.35).

We observe that summing eqns (7.34) and (7.35) gives the force balance equation for the tissue,

$$\nabla \cdot (\phi_c \mathbf{T}_c + \phi_m \mathbf{T}_m) + \rho\phi_c \mathbf{b}_c = \mathbf{0}, \tag{7.36}$$

with the pressure gradient term and the interaction forces with the liquid neglected, compatibly with the assumptions made before writing eqns (7.34) and (7.35). This might be a more convenient equation to use, because the mixture stress present in the divergence is the stress that can be actually measured experimentally.

We also observe that if one considers an unsaturated mixture, for instance if one neglects the interstitial liquid and its mechanical interactions with the other constituents, one can obtain eqns (7.34) and (7.35) again. Of course, the unilateral geometrical condition on the volume ratio (7.1) need still to be ensured by the model.

7.5 Modeling adhesion forces

In order to properly model the adhesion forces that cells can exert, it is useful to report what is known from the experimental point of view. In fact, several researchers have tried to measure the strength of single or clustered adhesion bonds formed by a cell (see, for instance, Baumgartner *et al.* (2000), Canetta *et al.* (2005), and Sun *et al.* (2005)). A typical experiment is done using an atomic-force-microscopy cantilever with a tip that can possibly be functionalized with adhesion molecules to check the specific interaction of the cell adhesion molecules with the molecules placed on the tip of the cantilever. After the tip has been put in contact with the cell for some time, either the cantilever or the plate carrying the cell is pulled away at a constant speed, typically in the range 0.2–5 μm/s. If the tip of the cantilever does not attach to the cell, there is no deflection of the cantilever when they are separated. This can be achieved experimentally, for instance, by the addition of an antibody attached to the external domain of the adhesion molecule (Baumgartner *et al.* 2000), by interfering with the links between the adhesion molecules and the cell cytoskeleton (Canetta *et al.* 2005), or by disrupting the actin cytoskeleton (Sun *et al.* 2005). On the other hand, adhesion gives rise to a deflection of the cantilever, which can be related to the stretching force exerted by the cell. Of course, with time the distance between the cell and the cantilever increases, increasing the deflection angle and the stretching force. It is then observed that after some time the adhesive bond breaks, causing a characteristic jump in the deflection of the cantilever, which returns to its undeformed configuration. In this way, it is possible to evaluate the maximum force of an adhesion bond.

Baumgartner *et al.* (2000) measured the strength of the adhesion bonds to be in the range of 35–55 pN, giving a distribution function for the critical unbinding force such as the one shown in Fig. 7.2(a).

Similar results were obtained by Canetta *et al.* (2005) and Sun *et al.* (2005). In particular, Sun *et al.* did not functionalize the microsphere and allowed a longer resting period on the cell surface, ranging from 2 to 30 s. Again, pulling the cantilever away at a constant speed in the range 3–5 μm/s caused the rupture of one or more adhesive bonds. Sun *et al.* used several different cell types (Chinese hamster ovary cells, endothelial cells, and human brain tumor cells), all showing an adhesive strength of a single bond slightly below 30 pN (Fig. 7.2(b)).

Unfortunately, all these measurements were done at the cellular scale. In order to use this information in the continuum model introduced in the previous section, one needs to upscale the results of the above experiments to the macroscopic scale.

A way to do that was suggested by Olz and Schmeiser (2010) and Olz *et al.* (2008), who solved a similar problem when dealing with the actin cytoskeleton. In order to

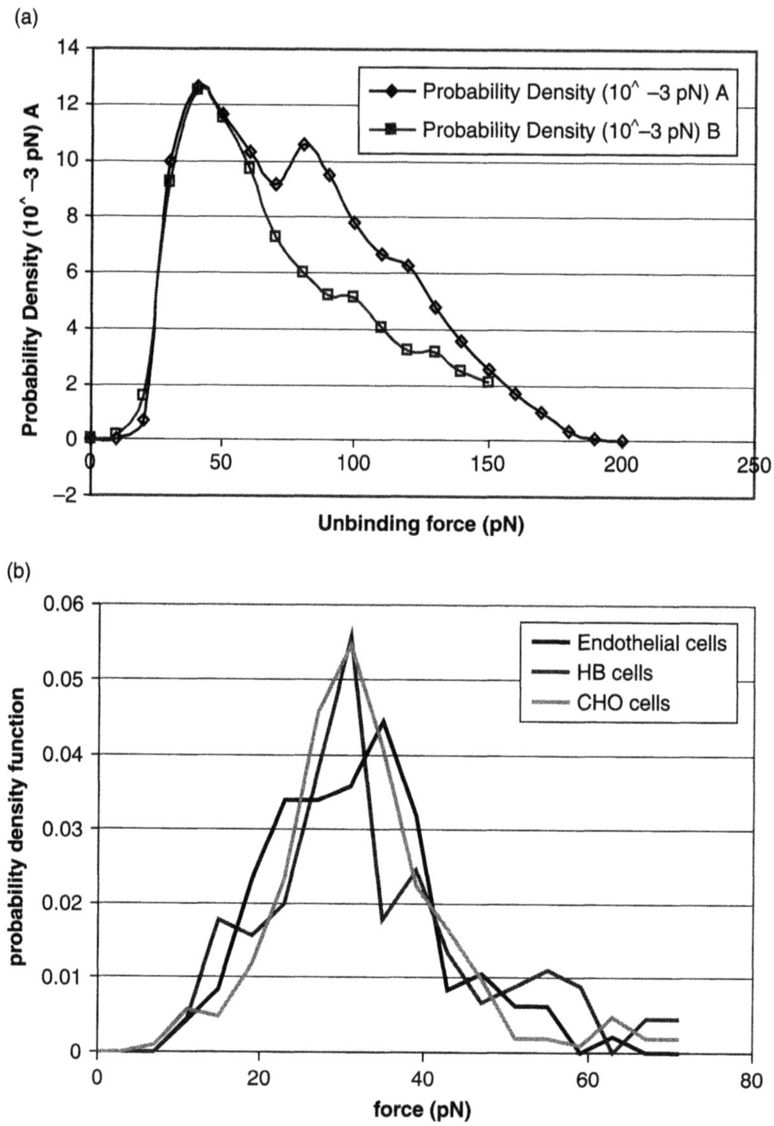

Fig. 7.2 Distribution function for the force of unbinding events: (a) when a single bond is acting (squares) and when several adhesion bonds are clustered together (diamonds) (data from Baumgartner *et al.* 2000), and (b) for different types of cells (data from Sun *et al.* 2005).

describe the solution, that, as a starting point let us focus on a single bond that is formed at a certain time between a cell and the extracellular fiber the cell is moving on. While the cell is moving, this bond, initially unstretched, will stretch and exert a force on the cell that depends on the distance traveled from the instant of formation of the bond. The length of time during which this has occured is the age of the bond

a, and the force is $\mathbf{F}_{\text{mic}}\big([\mathbf{x}(t) - \mathbf{x}(t-a)]\big)$. For instance, if we can assume Hooke's law for the bond, $\mathbf{F}_{\text{mic}} = -k_{\text{mic}}[\mathbf{x}(t) - \mathbf{x}(t-a)]$, and if the velocity of the cell is constant, $\mathbf{F}_{\text{mic}} = -k_{\text{mic}}\dot{\mathbf{x}}a$. However, from Fig. 4 in Baumgartner *et al.* (2000) it is evident that there is a hardening effect, i.e., the unbinding force increases with the relative velocity of the cantilever with respect to the cell.

Of course, at time t the cell does not have a single bond active but an ensemble of bonds, and, during motion, bonds continuously form and are broken. Following Olz and Schmeiser (2010) and Olz *et al.* (2008) and focusing on a one-dimensional situation, we introduce a probability distribution $f(t,a)$ over the age of the bonds, so that the number of bonds formed at time t is

$$N(t) = \int_0^{+\infty} f(t,a)\,da.$$

If, as time goes on, the bonds break at a rate ζ, which in general is not constant but is a function of F_{mic}, then

$$\frac{df}{dt} = -\zeta f,$$

where d/dt is the material derivative following the bond during its evolution. Taking into account the obvious fact that after a time interval Δt the bond ages by the same amount, i.e., $da/dt = 1$, we find that the evolution equation for the bond distribution is

$$\frac{\partial f}{\partial t} + \frac{\partial f}{\partial a} = -\zeta f, \tag{7.37}$$

as in classical structured-population models (Iannelli *et al.* 2005) and in transport theory.

The above partial differential equation needs a boundary condition for $a = 0$ related to the law of bond formation. We could take it to be constant, but as will be shown in the following, this would give rise in some cases to unreasonable results. A better boundary condition should take into account the fact that the cell can form a maximum number of bonds N_{max}, so that one can assume that the formation of new bonds is proportional to the bonds that can still be formed, i.e.,

$$f(t, a=0) = \beta\left(N_{\text{max}} - \int_0^{+\infty} f(t,a)\,da\right). \tag{7.38}$$

Referring to Preziosi and Vitale (2011) for more details and 3D version. The final output of the computation is the total force, which is given by

$$F = -k_{\text{mic}} \int_0^{+\infty} \xi(x,t,a)(t,a)\,da. \tag{7.39}$$

where ξ is the stretch of the bond formed at time $t-a$ when x was superimposed to another point of the ECM.

Before proceeding, it is worthwhile to write the equations in dimensionless form in order to justify the quasi-stationary version of the problem used later on. We scale lengths by the typical cell size L and time by $T = L/\dot{x}$, i.e., the time needed by a cell to move across a cell length. It is useful to introduce the ratio $\epsilon = A/T$, where A is the typical age of a bond. We assume that the characteristic time related to cell motion is much larger than the typical age of a bond, so that $\epsilon \ll 1$. We also scale f by $N_{\max}/A$.

The dimensionless form of eqn (7.37) is then

$$\epsilon \frac{\partial \tilde{f}}{\partial \tilde{t}} + \frac{\partial \tilde{f}}{\partial \tilde{a}} = -\tilde{\zeta}(\tilde{F}_{\text{mic}})\tilde{f}, \tag{7.40}$$

where $\tilde{\zeta} = A\zeta$, which needs to be joined to the boundary condition

$$\tilde{f}(\tilde{t}, \tilde{a} = 0) = \tilde{\beta}\left(1 - \int_0^{+\infty} \tilde{f}(\tilde{t}, \tilde{a})\, d\tilde{a}\right), \tag{7.41}$$

where $\tilde{\beta} = A\beta$. Finally, we set $L\tilde{x} = x$; so that (7.39) writes

$$F = -k_{\text{mic}} N_{\max} L \int_0^{+\infty} [\tilde{x}(T\tilde{t}) - \tilde{x}(T(\tilde{t} - \epsilon\tilde{a}))]\tilde{f}(\tilde{t}, \tilde{a})\, d\tilde{a}. \tag{7.42}$$

In the limit $\epsilon \to 0$,

$$F_{\text{mic}} \approx -k_{\text{mic}}\dot{x}a \tag{7.43}$$

and the problem reduces to its quasi-stationary version

$$\begin{cases} \dfrac{\partial \tilde{f}}{\partial \tilde{a}} = -\tilde{\zeta}\tilde{f}, \\ \tilde{f}(\tilde{a} = 0) = \tilde{\beta}\left(1 - \displaystyle\int_0^{+\infty} \tilde{f}(\tilde{a})\, d\tilde{a}\right), \end{cases} \tag{7.44}$$

with a macroscopic force that can be approximated by

$$F = -\epsilon k_{\text{mic}} N_{\max} \dot{\tilde{x}} \int_0^{+\infty} \tilde{a} f(\tilde{a})\, d\tilde{a}, \tag{7.45}$$

where $\dot{\tilde{x}} = T\dot{x}/L$.

We prefer, however, to work with dimensional variables, and therefore go back to the dimensional quasi-stationary problem,

$$\begin{cases} \dfrac{\partial f}{\partial a} = -\zeta(k_{\text{mic}}\dot{x}a) f, \\ f(a = 0) = \beta\left(1 - \displaystyle\int_0^{+\infty} f(a)\, da\right), \\ F = -k_{\text{mic}}\dot{x} \displaystyle\int_0^{+\infty} a f(a)\, da. \end{cases} \tag{7.46}$$

Before proceeding, we observe that in the case of a non-Hookean microscopic force of the bond, the last equation in eqn (7.46) is replaced by

$$F = \int_0^{+\infty} F_{\text{mic}}(\dot{x}a) f(a)\, da. \tag{7.47}$$

For any ζ, the partial differential equation in eqn (7.46) can be solved giving

$$f(a) = C \exp\left[-\int_0^a \zeta(k_{\text{mic}}\dot{x}\alpha)\, d\alpha\right], \tag{7.48}$$

where C can be determined from the boundary condition; we obtain

$$C = \frac{\beta N_{\max}}{1 + \beta \displaystyle\int_0^{+\infty} \exp\left[-\int_0^a \zeta(k_{\text{mic}}\dot{x}\alpha)\, d\alpha\right] da}. \tag{7.49}$$

Hence,

$$f(a) = \frac{\beta N_{\max} \exp\left[-\displaystyle\int_0^a \zeta(k_{\text{mic}}\dot{x}\alpha)\, d\alpha\right]}{1 + \beta \displaystyle\int_0^{+\infty} \exp\left[-\int_0^a \zeta(k_{\text{mic}}\dot{x}\alpha)\, d\alpha\right] da} \tag{7.50}$$

and

$$F = -k_{\text{mic}}\beta N_{\max}\dot{x} \frac{\displaystyle\int_0^{+\infty} a \exp\left[-\int_0^a \zeta(k_{\text{mic}}\dot{x}\alpha)\, d\alpha\right] da}{1 + \beta \displaystyle\int_0^{+\infty} \exp\left[-\int_0^a \zeta(k_{\text{mic}}\dot{x}\alpha)\, d\alpha\right] da}. \tag{7.51}$$

For instance, if $\zeta = \zeta_0$ (constant), then

$$f(a) = \frac{\beta N_{\max}\zeta_0}{\beta + \zeta_0} e^{-\zeta_0 a} \tag{7.52}$$

and

$$F = -k_{\text{mic}} \frac{\beta N_{\max}}{\zeta_0(\beta + \zeta_0)} \dot{x}. \tag{7.53}$$

One then finds the classical drag law, asserting that the interaction force acting on a cell moving with a velocity $\dot{x}$ is proportional to the velocity.

However, more general forces can be obtained. For instance, if ζ is taken to be proportional to the microscopic force (i.e., $\zeta = ck_{\text{mic}}\dot{x}a$), then

$$f(a) = \beta N_{\max} \frac{\exp[-ck_{\text{mic}}a^2/2]}{1 + \beta/\sqrt{2ck_{\text{mic}}\dot{x}\pi}} \tag{7.54}$$

and

$$F = -\frac{\beta}{c}\frac{N_{\max}}{1+\beta/\sqrt{2ck_{\text{mic}}\dot{x}\pi}}. \tag{7.55}$$

The cases in which the bonds break only if F_{mic} exceeds a threshold F_m are more interesting, because from the data of Baumgartner *et al.* (2000), and Sun *et al.* (2005) it seems that there is a nonvanishing F_m ranging between 5 pN and 10 pN. If, for instance,

$$\zeta(F_{\text{mic}}) = \zeta_0 H(F_{\text{mic}} - F_m), \tag{7.56}$$

where H is the Heaviside function, then

$$F = -\beta N_{\max}\frac{F_0^2 + F_m F_0 + \frac{1}{2}F_m^2}{(\beta+\zeta_0)F_0 + \beta F_m}, \tag{7.57}$$

where

$$F_0 = \frac{k_{\text{mic}}\dot{x}}{\zeta_0}. \tag{7.58}$$

For small velocities, i.e. for small F_0 $|F|$ tends to $F_m/2$, while for large velocities it tends to $\beta N_{\max}k_{\text{mic}}\dot{x}/[(\beta+\zeta_0)\zeta_0]$. This behavior, shown in Fig. 7.3, is compatible with that proposed by Preziosi and Tosin (2009), where it was argued that if cells are not pulled strongly enough to detach them from the ECM, they remain attached to it. If they detach, the excess force can be assumed initially to be proportional to the relative velocity of the cell with respect to the ECM (see also Fig. 7.4).

It should be noticed here that if the rate of formation of bonds were simply constant, i.e., if the integral in the boundary condition were absent, then the above procedure would yield a force that blew up for small velocity. This corresponds biologically to the fact that if the cell barely moves, bonds always form but never break. So, in the limit, an infinite number of bonds form, corresponding to an infinite force. This is of course unphysical and justifies the presence of a saturation term in the boundary condition (7.38).

It is also known that adhesion bonds have a limited strength and that all bonds break for large enough microscopic forces (e.g., 200 pN in Baumgartner *et al.*'s experiments). In the model above, this means that ζ goes to infinity above a critical value F_M, or, better, that the breaking time $(1/\zeta)$ goes to zero. This assumption leads to

$$F = -\beta N_{\max}\frac{F_0^2 + F_m F_0 + \frac{1}{2}F_m^2 - (F_M + F_0)\,F_0 \exp\left[-(F_M - F_m)/F_0\right]}{\zeta_0 F_0 + \beta[F_m + F_0 - F_0\exp\left[-(F_M - F_m)/F_0\right]]}, \tag{7.59}$$

which is characterized by a behavior similar to eqn (7.57) for small velocities but it goes back to zero for large velocities owing to the breaking of all the bonds.

Different breaking distributions $b(F_{\text{mic}})$ would give rise to different macroscopic forces, taking into account the fact that the breaking distribution (e.g., those in Fig. 7.2) is related to ζ by

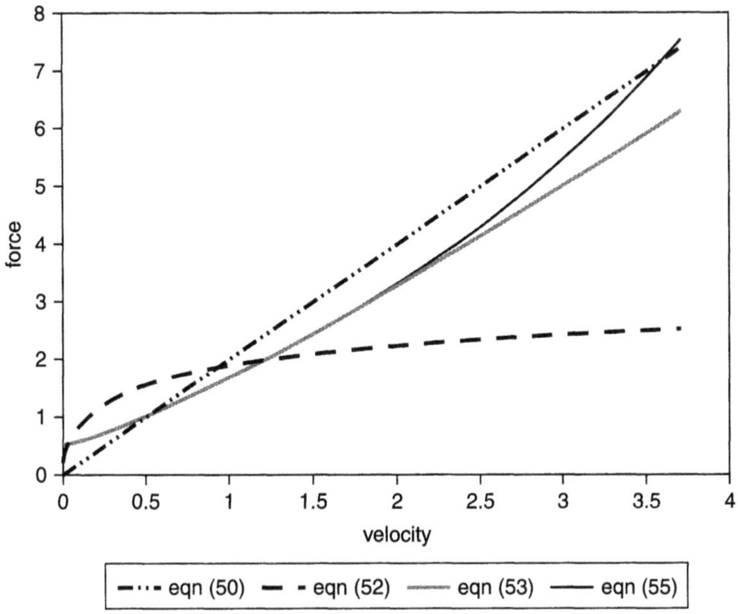

Fig. 7.3 Macroscopic adhesion laws for several different cell–ECM microscopic interaction laws.

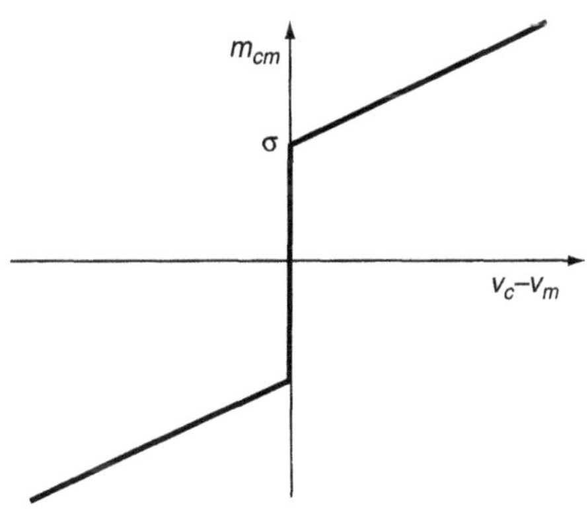

Fig. 7.4 Viscoplastic cell–ECM interaction.

$$\zeta = \frac{b(F_{\text{mic}})}{B(F_{\text{mic}})}, \qquad \text{where} \quad B(F_{\text{mic}}) = \int_{F_{\text{mic}}}^{F_M} b(\phi)\, d\phi \tag{7.60}$$

is the survival function and F_M is the supremum of the support of F_{mic}. (see Preziosi and Vitale (2011) for more details).

7.6 Modeling the cell–ECM interaction force

Coming back to the original problem of modeling the interaction between cells and the ECM and keeping in mind the discussion in the previous section, the simplest assumption of a constant rate of detachment of focal adhesion sites implies eqn (7.53) and therefore a linear relationship between the interaction term and the relative velocity of the cells with respect to the ECM, which can be rephrased in vector form as

$$\mathbf{m}_{cm} = -M_{cm}(\mathbf{v}_c - \mathbf{v}_m), \tag{7.61}$$

similarly to eqn (7.29).

The introduction of a threshold before the breaking of the bonds as in eqn (7.57) is reflected, for instance, in the following simplified constitutive model proposed by Preziosi and Tosin (2009):

$$\begin{aligned} &\mathbf{v}_c = \mathbf{v}_m \qquad \text{if} \quad |\mathbf{m}_{cm}| \le \sigma_{cm}, \\ &(|\mathbf{m}_{cm}| - \sigma_{cm})\frac{\mathbf{m}_{cm}}{|\mathbf{m}_{cm}|} = -M_{cm}(\mathbf{v}_c - \mathbf{v}_m) \qquad \text{if} \quad |\mathbf{m}_{cm}| > \sigma_{cm}. \end{aligned} \tag{7.62}$$

The coefficient σ_{cm} is expected to depend on the adhesion mechanisms and on the volume ratio of the actors, i.e., the cells and the ECM.

Neglecting body forces, from eqn (7.34) we obtain

$$\mathbf{m}_{cm} = -\nabla \cdot (\phi_c \mathbf{T}_c),$$

and therefore eqn (7.62) can be rewritten as

$$\mathbf{v}_c - \mathbf{v}_m = K_{cm}\left(1 - \frac{\sigma_{cm}}{|\nabla \cdot (\phi_c \mathbf{T}_c)|}\right)_+ \nabla \cdot (\phi_c \mathbf{T}_c), \tag{7.63}$$

where $(\cdot)_+$ stands for the positive part of the parenthesis, and $K_{cm} = M_{cm}^{-1}$ is called the motility coefficient of the cell population in this chapter. Equation (7.63) replaces eqn (7.34) and has to be solved jointly with eqn (7.35) or (7.36).

To better understand the meaning of the constitutive equations above, we can make some calculations for $|\mathbf{m}_{cm}| > \sigma_{cm}$. Taking the modulus of eqn (7.62), we have

$$M_{cm}|\mathbf{v}_m - \mathbf{v}_c| = |\mathbf{m}_{cm}| - \sigma_{cm}. \tag{7.64}$$

If we replace $|\mathbf{m}_{cm}|$ in the same equation, it can be rewritten as

$$M_{cm}|\mathbf{v}_c - \mathbf{v}_m|\frac{\mathbf{m}_{cm}}{\sigma_{cm} + M_{cm}|\mathbf{v}_c - \mathbf{v}_m|} = M_{cm}(\mathbf{v}_m - \mathbf{v}_c), \tag{7.65}$$

or

$$\mathbf{m}_{cm} = -\sigma_{cm} \frac{\mathbf{v}_c - \mathbf{v}_m}{|\mathbf{v}_c - \mathbf{v}_m|} - M_{cm}(\mathbf{v}_c - \mathbf{v}_m), \tag{7.66}$$

which allows us to distinguish a static contribution to $\mathbf{m}_{cm}$ in the direction of the relative motion (the first term) from a drag contribution proportional to the velocity difference. Equation (7.66) can also be compared with eqn (7.57). Of course, a viscous drag force is recovered in the limit $\sigma_{cm} = 0$.

As already mentioned, the friction force σ_{cm} depends strongly on the concentration of the ECM. For instance, increasing the concentration of the ECM leads to an increase in activated adhesion sites and therefore in a stronger friction threshold. In addition, it is known that there is an optimal concentration of ECM for motility (Palecek *et al.* 1997), because the content of ECM cannot become too small, otherwise the lack of a substratum would lead to a decrease in cell motility. The observation that cells hardly move when there is too little or too much ECM can then, be translated as an increase in σ_{cm} for small and "large" ϕ_m, thus effectively prohibiting cellular motion.

We can take the interaction between the cell and the ECM to be proportional to the volume ratio of cells and ECM, and assume that there is a threshold above which cells cannot pass through the ECM. A function with the above properties is

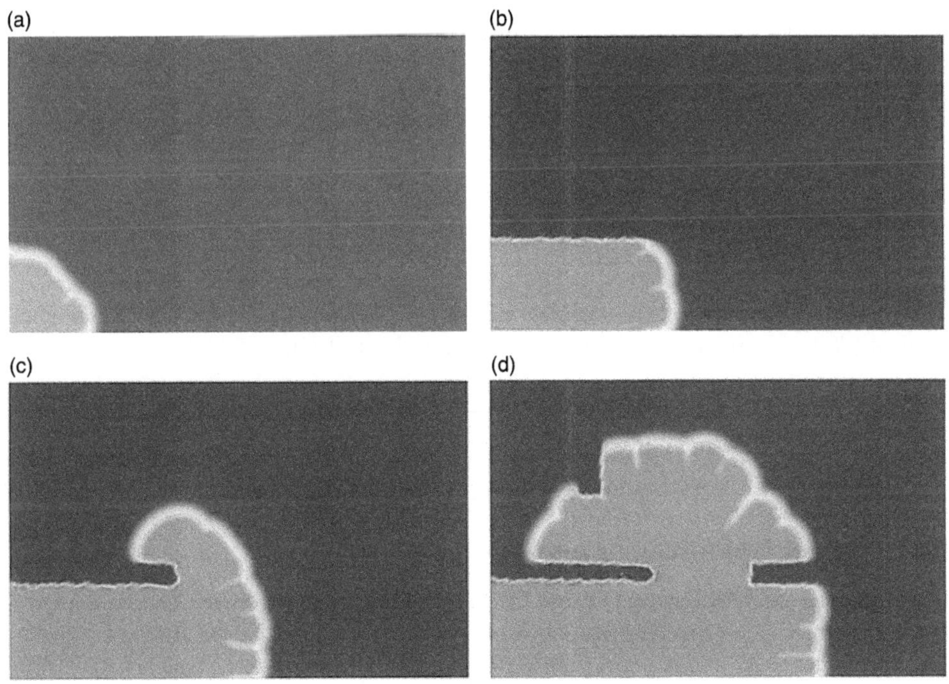

Fig. 7.5 Growth of a tumor mass in the presence of thick basal membranes at $t = 30, 60, 90, 120$.

$$\sigma_{cm} = \frac{\hat{\sigma}_{cm}\phi_c\phi_m}{\phi_m^* - \phi_m}, \tag{7.67}$$

where $\phi_m^* < 1$. This is important to consider in the case of cell compartmentalization by ECM barriers or basal membranes, as shown in Fig. 7.5.

7.7 Tumor cell constituent as a liquid

Reasoning similar to that above can be followed for the deduction of a constitutive equation for the stress tensor of the cellular constituent. Qualitatively speaking, the adhesive interaction force is replaced by the stress within a multicell spheroid, and the relative velocities between the interacting constituents are replaced by a rate-of-strain tensor. For the moment, we will not do that; instead we assume, like most of the papers in the literature, the easiest constitutive equation for the ensemble of cells, i.e., that it behaves like an elastic fluid,

$$\mathbf{T}_c = -\Sigma_c \mathbf{I},$$

where Σ_c is taken as positive in compression. The use of this constitutive equation results in a multicellular spheroid that, in the absence of ECM, is not able to sustain shear.

In this case one can substitute eqn (7.63) into the mass balance equation (7.4) to obtain the following model:

$$\begin{cases} \dfrac{\partial \phi_c}{\partial t} + \nabla \cdot (\phi_c \mathbf{v}_m) = \nabla \cdot \left[\phi_c K_{cm} \left(1 - \dfrac{\sigma_{cm}}{|\nabla(\phi_c \Sigma_c)|} \right)_+ \nabla(\phi_c \Sigma_c) \right] + \Gamma_c, \\ \dfrac{\partial \phi_m}{\partial t} + \nabla \cdot (\phi_m \mathbf{v}_m) = \Gamma_m, \\ \nabla \cdot (\phi_m \mathbf{T}_m) - \nabla(\phi_c \Sigma_c) = \mathbf{0}. \end{cases} \tag{7.68}$$

A possible extension is to consider a viscous behavior, as done by Byrne and Preziosi (2004), Franks *et al.* (2003*a*,*b*), and Franks and King (2003):

$$\hat{\mathbf{T}}_c = (-\Sigma_c + \lambda \nabla \cdot \mathbf{v}_c)\mathbf{I} + 2\mu \mathbf{D}_c, \tag{7.69}$$

where $\mathbf{D} = (\nabla \mathbf{v}_c + \nabla \mathbf{v}_c^{\mathrm{T}})/2$ is the rate-of-strain tensor. This constitutive equation has the advantage that it confers more stability on the growing mass.

7.7.1 Contact inhibition of growth

As an example, and still having in mind the focus of this chapter on mechanical aspects of tumor growth, we describe here the phenomenon of contact inhibition of growth (Dietrich *et al.* 1997; Kato *et al.* 1997; Nelson and Chen 2003; Polyak *et al.* 1994; StCroix *et al.* 1998), which is the property of normal cells that they decrease their proliferation rate when they realize that they are coming into contact with other cells. A quantification of this phenomenon is represented in Fig. 7.6, which reports some

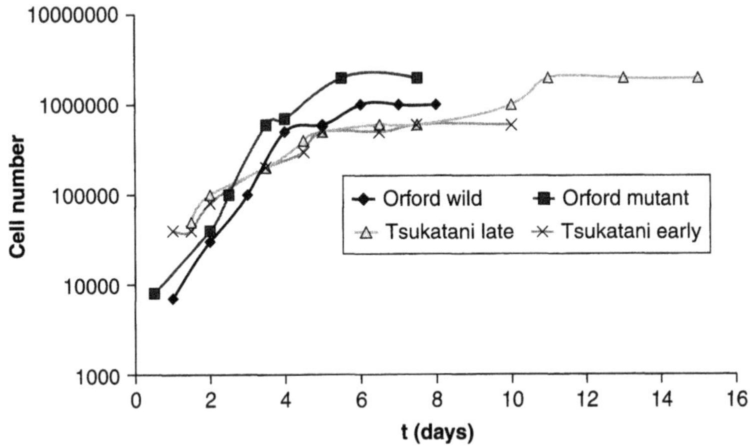

Fig. 7.6 Examples of contact inhibition of growth reported in the experiments of Tzukatani *et al.* (1997) using human breast epithelial cells and of Orford *et al.* (1999) using canine kidney-derived nontransformed epithelial cells.

experimental results by Tzukatani *et al.* (1997) on human breast epithelial cells grown *in vitro* over a suitable substratum and by Orford *et al.* (1999) on canine kidney-derived nontransformed epithelial cells.

It can be seen that after an initial exponential growth, the cell density saturates, forming a monolayer of cells. This phenomenon is called *growth to confluence.* On the other hand, tumor cells continue proliferating, forming a multilayer, leading to the conjecture that they need to feel more contacts or a larger pressure to stop their proliferation program.

A key role in transferring this information is played by mechanotransduction pathways involving cadherin–cadherin junctions. Cadherins are transmembrane receptors involved in homophilic cell–cell interactions. Several experiments (Warchol 2002; Caveda *et al.* 1996; Castilla *et al.* 1999) have shown that if a cell is not sufficiently sensitive to the control mechanisms above, it is subject to deregulated growth, a phenomenon that is considered such an important milestone in the development of tumors that it deserves to be named the *cadherin switch.*

In fact, it is known that loss of contact responsiveness is commonly associated with the formation of hyperplasia and malignant transformation, such as in gastric carcinoma (Becker *et al.* 1994; Oda *et al.* 1994), adenocarcinoma (Tzukita *et al.* 1993), epithelial tumors (Cavallaro *et al.* 2002; Christofori and Semb 1999), colon polyps and carcinoma (Gottardi *et al.* 2001), gynecological cancers (Risinger *et al.* 1994), and intimal thickening (Uglow *et al.* 2000) (see also the review by Harja and Fearon 2002).

However, cadherins represent only the tip of the iceberg. They are more visible than other hidden players because of their transmembrane location, but there are many other candidates that may be responsible for a possible incorrect mechanotransduction. The second family of suspects are the catenins, the proteins cadherins link to for functional cell-to-cell adhesion. In fact, Stockinger *et al.* (2001) showed that epithelial cells exhibited a strong β-catenin activity at low densities ($\leq 40\%$ confluency), which was reduced by a factor of five to seven when the cells reached a confluency $> 80\%$. In fact, it is thought that in physiological conditions, upon reaching confluency, the expressed cadherins sequester catenins, downregulating their activity. Since it is known that the upregulation of catenins, is necessary for cell duplication, the final result is that cell adhesion negatively affects cell proliferation.

Dietrich *et al.* (1997) explained the mechanism of contact inhibition of growth in more detail as follows:

- Tissue compression and overexpression of cadherins cause the underexpression of catenins, which are sequestered by cadherins at the cell membrane.
- The underexpression of catenins determines the accumulation of the cyclin-dependent kinase (cdk) inhibitors p16, p21, and p27.
- Their overexpression inhibits the entry into the S phase, causing cell cycle arrest in the G1 phase. In more detail (Fig. 7.7),
  - p16 blocks the activity of cdk4 by dissociating cyclin D from cdk4 and binding to cdk4;
  - p27 inhibits cdk2-cyclin E activity directly by binding to the complex.

The above biochemical description of the mechanotransduction pathway involved in contact inhibition can be introduced into the model by stating that mitosis stops when the volume ratio (or the compression) exceeds a given threshold.

The behavior of the cells in terms of growth and motion then depends crucially on how they feel the presence of other cells and how they translate the mechanical cues. In Chaplain *et al.* (2005), Graziano and Preziosi (2007), and Galle *et al.* (2009), it was shown that a fault in the mechanotransduction pathway might lead to a misperception of the compression state of the local tissue, and could then confer a clonal advantage on the surrounding cells, leading to the replacement and invasion of healthy tissue with the formation of hyperplasia and therefore tumor lesions.

In the description to follow, we denote by $\Omega_t(t=0)$ the region initially occupied by tumor cells, with a volume ratio ϕ_t, and denote by $\Omega_n(t=0) = \Omega \setminus \Omega_t(t=0)$ the region occupied by normal host cells, with a volume ratio ϕ_n. The interface $\partial\Omega_t(t)$ between the tumor and the environment is a material surface moving with the common velocity of the cells

$$\mathbf{n} \cdot \frac{d\mathbf{x}_t}{dt} = \mathbf{n} \cdot \mathbf{v}_{c_t} = \mathbf{n} \cdot \mathbf{v}_{c_n} \qquad \text{on} \quad \partial\Omega_t(t). \tag{7.70}$$

It can be shown that the two cell populations, initially segregated, stay segregated at all times.

From the mathematical point of view, the phenomenological description of contact inhibition of growth can be formalized by saying that the threshold value for tumor cells to overcome the restriction point and commit themselves to divide is slightly larger than the physiological value. Actually, it may even tend to infinity, meaning that the cells are completely insensitive to compression and continue replicating independently of the compression level.

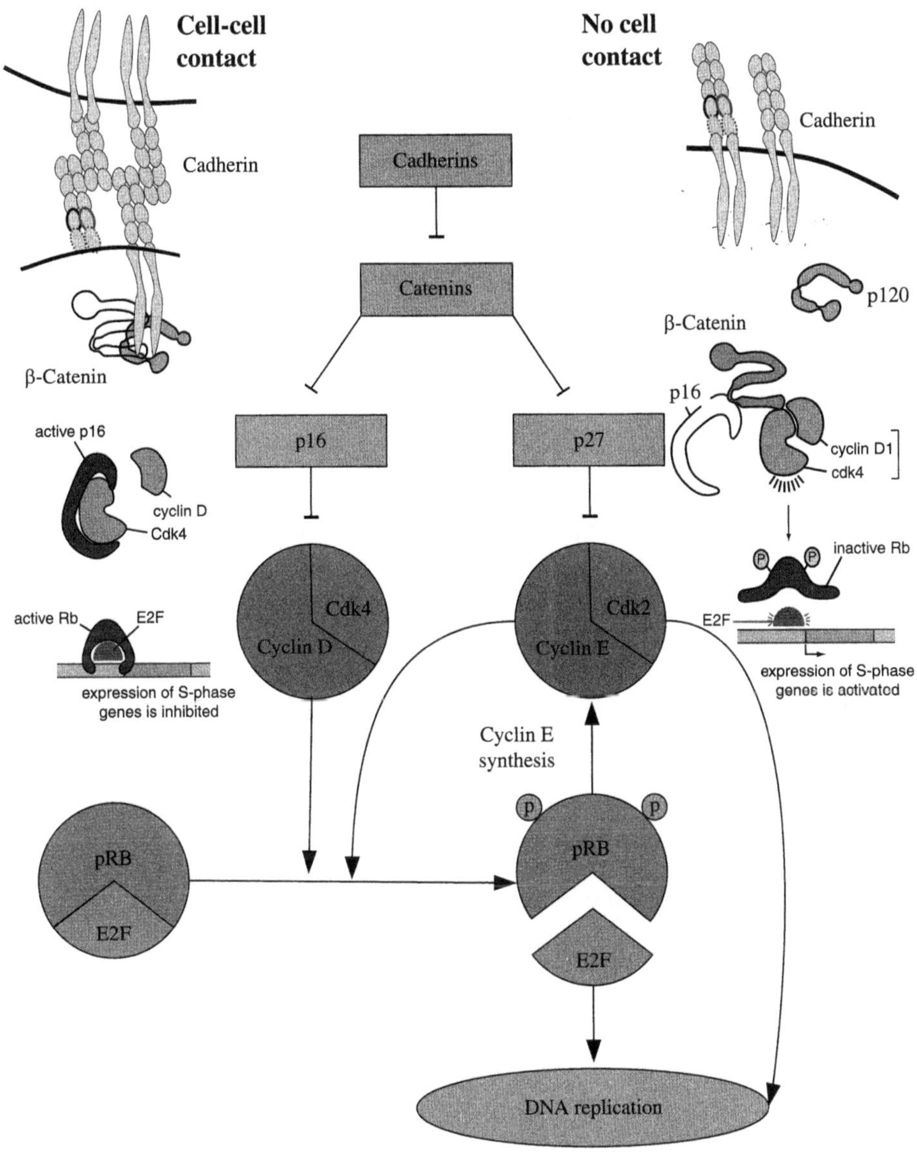

Fig. 7.7 Sketch of the mechanotransduction pathway. The arrows and the lines terminated by bars indicate stimulatory and inhibitory activities respectively; cdk stands for cyclin-dependent kinase and pRB for hypophosphorylated retinoblastoma, and the added p's indicate phosphorylation. On the left, the inhibitory response related to contact inhibition. On the right, the activation of the duplication program owing to lack of cell–cell contact.

We shall then consider the following growth terms:

$$\Gamma_{c_i} = [\gamma_i H_\sigma(\psi_i - \hat{\psi}_i) - \delta_i(\psi_i)]\phi_{c_i}, \qquad i = n, t, \tag{7.71}$$

where $\psi_i = \phi_{c_i} + \phi_m$, and $H_\sigma(\psi_i - \hat{\psi}_i)$ is a mollifier of the step function, which is at least continuous, is equal to 1 for ψ smaller than the threshold value $\hat{\psi}_i$, and vanishes for $\psi > \hat{\psi}_i + \sigma$.

Of course, cellular mechanotransduction is not the only cause of formation of hyperplasia and tumors. In fact, chemical factors operate to regulate the reproduction rates so that the growth terms depend crucially on the presence of growth-promoting factors, growth inhibitory factors, and, of course, nutrients. However, the aim of this chapter is to examine the mechanical aspects of tumor growth, and therefore we assume that all the constituents required to sustain growth and mitosis can be found abundantly in the extracellular liquid.

According to the discussion above, the threshold values $\hat{\psi}_n$ and $\hat{\psi}_t$ are such that $\hat{\psi}_n < \hat{\psi}_t$. For the following discussion it is useful to observe that a balance between cell growth and death occurs when $\gamma_i H_\sigma(\psi_i - \hat{\psi}_i) = \delta_i(\psi_i)$, or, if δ_i is considered constant, as in the following simulations,

$$\psi_i = \hat{\psi}_i + H_\sigma^{-1}\left(\frac{\delta_i}{\gamma_i}\right). \tag{7.72}$$

We assume that what makes the difference between a normal and a tumor cell is contained in the growth term and its dependence on the stress level. Taking into account the fact that the interface conditions are enforced through continuity of stress and velocity, and treating the ensemble of cells as an elastic fluid for the sake of simplicity, we have the following free-boundary problem:

$$\begin{cases} \dfrac{\partial \phi_{c_i}}{\partial t} = \nabla \cdot \left(\phi_{c_i} K_{c_i m}\left(1 - \dfrac{\sigma_{c_i m}}{|\nabla \cdot (\phi_{c_i}\Sigma_{c_i})|}\right)_+ \nabla(\phi_{c_i}\Sigma_{c_i})\right) + \Gamma_{c_i} & \text{in } \Omega_i(t), \\ \mathbf{v}_{c_t} \cdot \mathbf{n} = \mathbf{v}_{c_n} \cdot \mathbf{n} \qquad \text{on } \partial\Omega_t(t), \\ \phi_{c_t}\Sigma_{c_t}(\phi_{c_t}) = \phi_{c_n}\Sigma_{c_n}(\phi_{c_n}) \qquad \text{on } \partial\Omega_t(t), \end{cases} \tag{7.73}$$

where $i = t, n$ and we have assumed that the ECM is rigid ($\mathbf{v}_m = \mathbf{0}$) and nonremodeling ($\phi_m$ constant in time). This makes it unnecessary to consider its stress tensor in detail because the internal stress is indeterminate owing to the rigidity constraint.

In Fig. 7.8, following Graziano and Preziosi (2007) and Galle *et al.* (2009), we show the result of applying the model above to the growth of Widr colonies (a colorectal adenocarcinoma cell line). Two clones of tumor cells were virtually seeded at a dimensional distance of 320 μm apart. They were characterized by different motilities, which was the only difference between the two clones. Specifically, the left one had a motility coefficient ten times larger than the right one. We mention that comparable results would be achieved if the cells of the colony on the left were

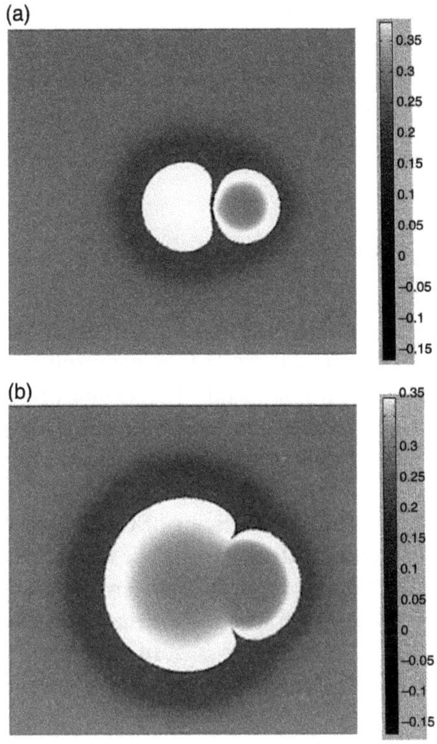

Fig. 7.8 Growth of two colonies which differ in their cell motility. The clone on the left is more motile and therefore able to relax the stress much faster, triggering growth. The grayscale represent the growth rate, which is positive near the boundary. The center of the clones is contact-inhibited, as well as the region between them when they come into contact. The images show the results 6 (top) and 8 (bottom) days after seeding.

assumed to be stiffer. At the very beginning, there was almost no difference in the growth of the two colonies. However, while all cells duplicated in the left colony, in the right one the cells in the center duplicated less. This is due to the fact that the less motile colony on the right did not release its internal stress so fast. The radius of the colony on the right was a little smaller than that of the colony on the left. As time proceeded, the cells in the center became too compressed to activate their duplication program. The difference between the two clones became more and more evident, with cell proliferation confined to the outer proliferating rim (see Fig. 7.8, top clone). This is similar to what has been observed in many papers on tumor growth and described as an effect of nutrient availability inside the tumor. However, we remind readers that nutrient distribution was not important in these experiments, because the colonies were growing on a flat Petri dish with nutrients coming from above. The proliferating rim is simply due to contact inhibition, and for this reason duplication is mainly restricted to cells close to the boundary.

After nearly 6 days the two colonies touched and growth was inhibited on the contact line and also in the center of the clone on the left (see Fig. 7.8 again). However, the cells on the outer border kept duplicating, and those on the left were always more active, so at the end the colony on the right was almost engulfed by the one on the left.

7.7.2 Integrin switch and fibrosis

As a second example, we consider ECM remodeling and the formation of fibrosis. In fact, a variety of stromal cells, mainly fibroblasts, are involved in constantly renewing the ECM through the production of new ECM components and of matrix metalloproteinases (MMP). This is a physiologically functional process because it allows the stroma to be kept young and reactive. In fact, as everybody knows, prolonged rest is detrimental to bones and muscles, while exercise and physical training have the opposite effect.

Under stationary conditions, the remodeling of ECM is a slow process. For instance, in the human lung, the physiological turnover of the ECM is 10–15% per day (Johnson 2001), which leads to an estimated complete turnover in a period of nearly a week. However, when a new tissue has to be formed, for example to repair a wound, then the rate of production is one or two orders of magnitude faster (Caveda *et al.* 1996). In addition, this process is strongly affected by stress, as is well known for bones, teeth, and muscles.

The percentage of ECM content changes considerably from tissue to tissue (see Tables 7.1 and 7.2), from normal to tumor tissues, and also within the same tumor with tumor progression (Zhang *et al.* 2003). For instance, Takeuchi *et al.* (1976) found that breast tumors presented a denser and more fibrous stroma, with several differences in chemical composition. In fact, it is well known to everybody that the first hints of the possible presence of a breast nodule are obtained by palping the breast and feeling stiffer regions. A strong variability in the collagen content has also been found in prostate cancer (Zhang *et al.* 2003), where it can range from 7% to 26% according to the grade of the tumor.

Table 7.1 Examples of elastic moduli of normal and abnormal breast tissue and stroma (data from Paszek *et al.* (2005))

Tissue	Elastic modulus (Pa)
Normal mammary gland	167 ± 31
Average breast tumor	4049 ± 938
Stroma attached to tumor	916 ± 269
Reconstituted basement membrane	175 ± 37
Collagen (2.0 mg/ml)	328 ± 87
Collagen (4.0 mg/ml)	1589 ± 380

Table 7.2 Average fit parameters for the exponential fit $T_c = (C_1/C_2)(e^{C_2\epsilon} - 1)$ determined by Wellman *et al.* (1999) for several normal and cancer breast tissues. No standard deviation was reported for those cases with only one specimen.

Tissue	C_1 (Pa)	C_2
Fat	4460 ± 2345.6	7.4 ± 4
Normal mammary gland	15174.5 ± 6750.7	12.3 ± 7.4
Phyllodes tumor	50312.8	11.9
Papilloma	17765.2 ± 4201.6	21.4 ± 2.8
Lobular carcinoma	28269.6	20.9
Fibroadenoma	37572.4 ± 6047.4	20.0 ± 1.4
Infiltrating ductal carcinoma	37958.7 ± 6146.7	19.9 ± 5.5
Ductal carcinoma *in situ*	55776	24.4

An increased presence of ECM also characterizes other pathologies such as cardiac hyperthrophy, intima hyperplasia, cardiac fibrosis, liver fibrosis, pulmonary fibrosis, asthma, glomerulonephritis, and colon cancer (Johnson 2001).

The alteration in the ECM composition can be due to several probably concurrent reasons:

- increased synthesis of ECM proteins;
- decreased activity of matrix-degrading enzymes (MDEs);
- upregulation of tissue-specific inhibitors of metalloproteinases (TIMPs).

On the other hand, excessive degradation of ECM due to excessive production of MMP-13 characterizes chronic inflammatory diseases such as osteoarthritic cartilage, rheumatoid synovium, chronic ulcers, intestinal ulcerations, periodontitis, and many malignant tumors.

The interaction between cells and the ECM is very important because cells need to properly adhere in order to survive. They duplicate only if they are anchored to the ECM. The mechanotransduction cascade is activated mainly by integrins.

On the other hand, in the process of invasion and formation of metastases, tumor cells detach from the original site, invade the surrounding tissue, intravasate by entering the blood or lymphatic system, and extravasate to reach a secondary site. It is then clear that the formation and diffusion of mestatases require that cells acquire the ability to survive without interacting with the ECM. In fact, as for cadherins, it is found that tumors have altered integrins, which, in turn, alter the downstream signaling pathway, so that one could argue that there is an *integrin switch* in addition to the cadherin and angiogenic switches mentioned earlier.

Actually, Paszek *et al.* (2005) have shown that the stiffness of the ECM promotes malignant behavior through the integrin signaling pathway, consisting in growth enhancement and loss of tissue polarity, which leads, for instance, to the absence of lumen formation in ductal carcinoma and the formation of hyperplasia, the first step toward tumorigenesis.

The remodeling and degradation of ECM is due to the motion of cells within the scaffold and the action of MMPs, whose concentration per unit volume is denoted here by e. One can then argue that

$$\Gamma_m = \mu_t(\phi_{c_t}, \phi_m) H_\epsilon(\psi_m - \psi)\phi_{c_t} + \mu_n(\phi_{c_n}, \phi_m) H_\epsilon(\psi_m - \psi)\phi_{c_n} - \nu e \phi_m, \qquad (7.74)$$

where $\psi = \phi_{c_t} + \phi_{c_n} + \phi_m$. Here, μ_α, where $\alpha = t, n$, is a nonnegative, nonincreasing function representing the net matrix production rate by the cell population c_α tempered by the free-space function H_ϵ, and $\nu > 0$ is the rate of degradation by the MMPs. As usual, the evolution of the concentration of MMPs is governed by a reaction–diffusion equation,

$$\frac{\partial e}{\partial t} = D\,\nabla^2 e + \pi_n \phi_{c_n} + \pi_t \phi_{c_t} - \frac{e}{\tau}, \qquad (7.75)$$

for net production rates $\pi_\alpha > 0$, $\alpha = t, n$, and an enzyme half-life $\tau > 0$. Actually, enzyme dynamics is much faster than the dynamics of cell growth and death, and hence it is possible to work under a quasi-stationary approximation. Furthermore, enzyme action is usually very local, so that diffusion can also be neglected and, finally,

$$e = \tau(\pi_n \phi_{c_n} + \pi_t \phi_{c_t}). \qquad (7.76)$$

Inserting this expression into eqn (7.74) and defining $\nu_\alpha = \nu\tau\pi_\alpha$, $\alpha = t, n$, allows us to write the mass balance equation for the ECM (7.12),

$$\frac{\partial \phi_m}{\partial t} = \sum_{\alpha=t,n} \left[\mu_\alpha(\phi_{c_\alpha}, \phi_m) H_\epsilon(\psi_m - \psi) - \nu_\alpha \phi_m\right] \phi_{c_\alpha}. \qquad (7.77)$$

The pathological cases possibly leading to fibrosis are either $\mu_t(\cdot) > \mu_n(\cdot)$ or $\nu_t < \nu_n$. In eqn (7.77), it is important that the production coefficients of ECM for normal and tumor cells be different in order to describe the formation of the fibrosis that characterizes many tumors. In particular, $\mu_t(\cdot) > \mu_n(\cdot)$. Alternatively, $\nu_t < \nu_n$ implies that tumor cells produce less MMPs than do healthy cells.

One by-product of this model is a description of the formation of fibrotic tissues and of tissues stiffer than normal, such that they may sometimes be felt with a self-test. This is the aim of the simulation shown in Fig. 7.9. The ECM was initially distributed homogeneously with $m = 0.2$. On the other hand, while proliferating, tumor cells produced matrix-degrading enzyme like the host cells but produced more ECM than normal. This led to the formation of a tumor characterized by an amount of ECM with a volume ratio close to $m = 0.3$. From the mechanical point of view, this increase in the percentage of ECM would lead to an increase of almost one order of magnitude in tissue stiffness.

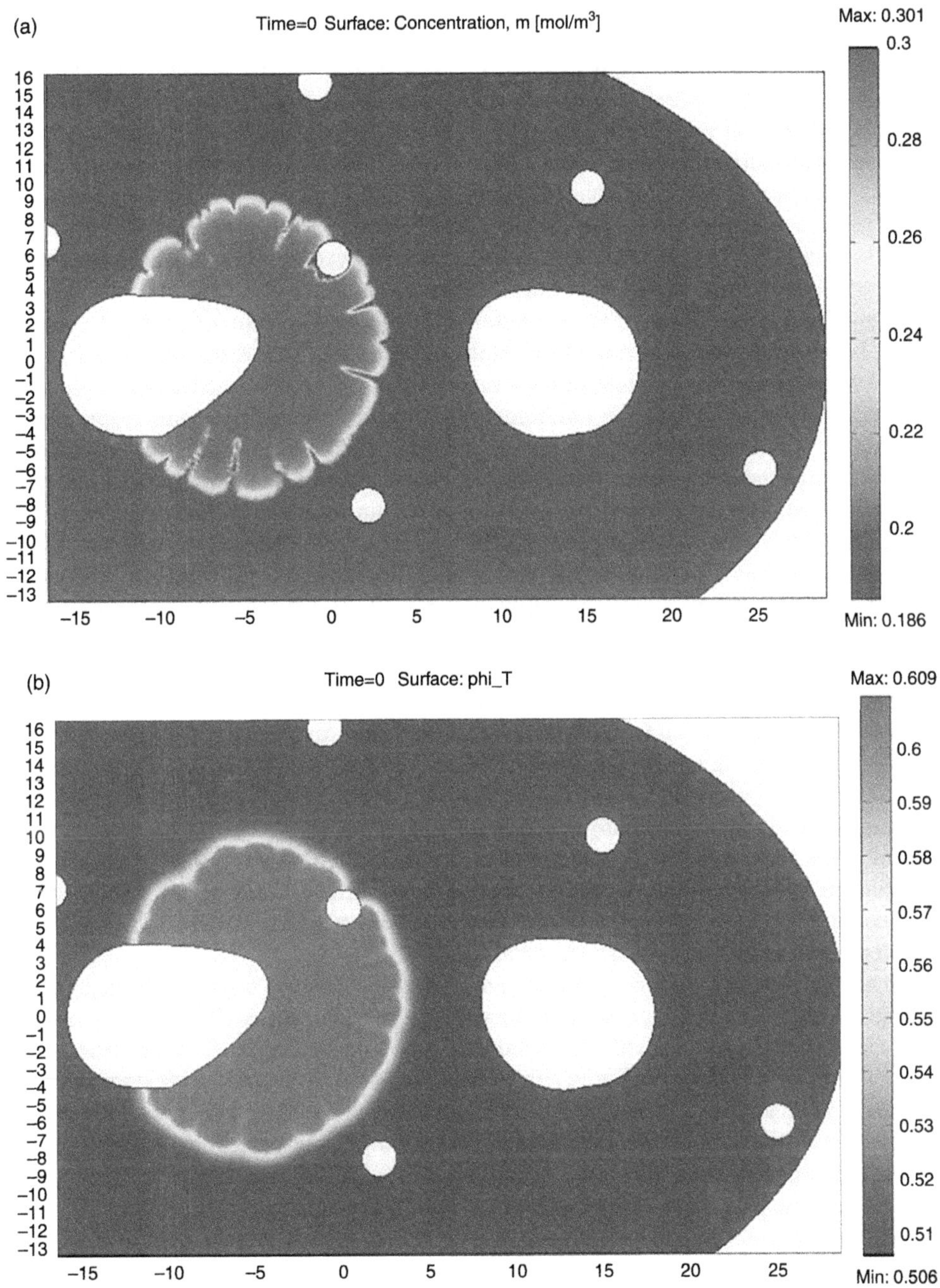

Fig. 7.9 Growth of a fibrotic tumor in a homogeneous tissue surrounding a bone. (a) Cell volume ratio; (b) ECM volume ratio.

7.8 The tumor mass as a solid: Evolving natural configurations

Of course, tumors are not liquids. However, treating cells like a liquid brings many useful simplifications, starting from the fact that it is possible to use an Eulerian approach and to deal with velocities rather than deformations with respect to a not so well-specified natural configuration. Though the models are relatively easy to use, the fact that the cellular liquid is contained in a solid structure formed by the ECM implies that the entire tissue behaves like a viscoelastic solid, or even a rigid solid if the newtork of ECM fibers is assumed to be rigid. However, this is still far from reality because of the presence of adhesion bonds between cells.

A big theoretical difficulty is encountered in describing tumors as solid masses instead, because the cells forming them duplicate and die, the stroma and, in particular, the extracellular matrix continuously remodel, and, even in the absence of growth and death, the ensemble of cells undergoes an internal reorganization in response to deformation. There is then a difficulty in defining a reference configuration and in using a Lagrangian description. In particular, the meaning of deformation loses the immediate meaning that it has in classical continuum mechanics when dealing with inert matter. In fact, when dealing with a living tissue such as a tumor, it is not clear with respect to what we should measure deformations, because the material is always changing. That is why the concept of evolving natural configurations introduced by Rajagopal and coworkers can be of help. Actually, the basic idea of this formalism, also present in plasticity theory, was first applied in a biomechanical context to describe growth by Klisch and Hoger (2003), and Rodriguez *et al.* (1994). In the recent past, Humphrey and Rajagopal have applied this concept to describe the growth and remodeling of several tissues (Humphrey and Rajagopal 2002, 2003; Malik *et al.* 2008; Rao *et al.* 2003). Ambrosi and Mollica (2002, 2004) used a purely elastic one-component model to evaluate residual-stress formation in a growing multicellular spheroid. This approach was developed by Ambrosi and Preziosi (2009), working in a multiphase framework and also taking internal reorganization and ECM deformation into account. This gave rise to an elastoviscoplastic description of the cell population and a compressive elastic description for the ECM.

In this section, we follow Ambrosi and Preziosi (2009), distinguishing in the deformation gradient of the tumor component $\mathbf{F}_c$ the contributions due to pure growth, plastic deformation, and elastic deformation by a multiplicative decomposition, as shown in Fig. 7.10. This splitting is suggested also by the biological observation that growth occurs on a much longer timescale (hours to days) than deformation. Therefore, it is possible to separate the contribution due to growth from that due to deformation (without growth) and to model each of them individually, not only from the theoretical point of view but also from the experimental point of view.

The deformation gradient $\mathbf{F}_c$ is a mapping from a tangent space onto another tangent space, and it indicates how the body is deformed locally on going from the initial (reference) configuration K_0 to the current configuration K_c. An imaginary intermediate configuration can be introduced, assuming that a point of the body can relieve its state of stress if the continuity, i.e. the integrity of the body, is relaxed. The body then relaxes to a stress-free configuration. The atlas of these pointwise

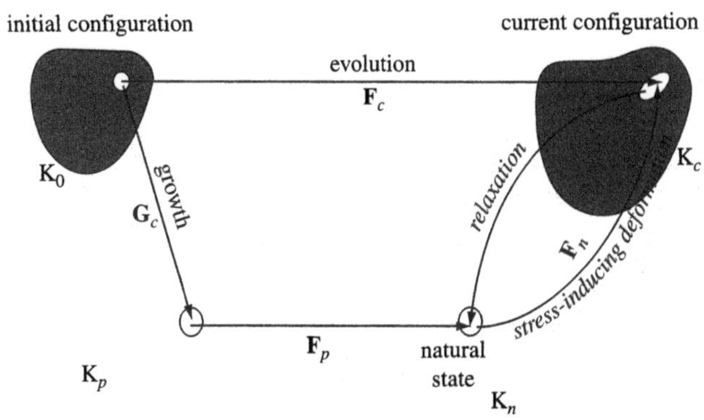

Fig. 7.10 Multiple natural configurations.

configurations forms what we define as a *natural configuration* with respect to K_c and denote by K_n. Referring to Fig. 7.10, we identify this deformation without growth with the tensor $\mathbf{F}_n$, which then describes how the body is deformed locally on going from the natural configuration K_n to K_c.

The particle in the configuration K_n has possibly undergone growth and plastic deformation. One can then again consider the map from K_0 to K_n as composed of two parts: the first one related to growth/death processes (and therefore to mass variations in the volume element), and the second one due to internal reorganization, which implies rearrangement of the adhesion links among the cells, without change of mass in the volume element. Denoting by K_p the "grown configuration", i.e., the intermediate configuration of the body between K_0 and K_n, we assume that for any given point the volume ratio in K_p is the same as in the natural configuration K_n and in the original reference configuration K_0, i.e., $\phi_p = \phi_c(t=0) = \phi_n$.

According to the three-step process outlined above, the deformation gradient is split as

$$\mathbf{F}_c = \mathbf{F}_n \mathbf{F}_p \mathbf{G}_c. \tag{7.78}$$

For the sake of simplicity, we also take growth to be isotropic, i.e., $\mathbf{G}_c = g\mathbf{I}$, so that $J_g = \det \mathbf{G}_c = g^3$.

In Ambrosi and Preziosi (2009), it was shown that

$$\frac{\dot{J}_g}{J_g} = 3\frac{\dot{g}}{g} = \frac{\Gamma_c}{\phi_c}, \tag{7.79}$$

where the dots indicate the time derivative following the cell population. This relation links the evolution of g to the death/growth term Γ_c appearing on the right-hand side of the mass balance equation for the cellular constituent, i.e., eqn (7.4).

Introducing

$$\mathbf{D}_p = sym(\dot{\mathbf{F}}_p \mathbf{F}_p^{-1}), \tag{7.80}$$

we obtain

$$\dot{J}_p = J_p \text{tr}\, \mathbf{D}_p \tag{7.81}$$

From standard tensor calculus, but since it is assumed that the volume ratio does not change during plastic reorganization, $J_p = \det \mathbf{F}_p = 1$, and therefore

$$\text{tr}\, \mathbf{D}_p = \text{tr}\, \mathbf{L}_p = 0. \tag{7.82}$$

Taking the derivative of

$$\mathbf{B}_n = \mathbf{F}_n \mathbf{F}_n^{\mathrm{T}} \tag{7.83}$$

with respect to time, we have

$$\dot{\mathbf{B}}_n = \mathbf{L}_n \mathbf{B}_n + \mathbf{B}_n \mathbf{L}_n^{\mathrm{T}}, \tag{7.84}$$

and, taking the derivative of $\mathbf{F}$ with respect to time, we have

$$\dot{\mathbf{F}} = \dot{g}g^{-1}\mathbf{F} + g\dot{\mathbf{F}}_n \mathbf{F}_p + g\mathbf{F}_n \dot{\mathbf{F}}_p = (\dot{g}g^{-1}\mathbf{I} + \mathbf{L}_n + \mathbf{F}_n \mathbf{L}_p \mathbf{F}_n^{-1})\mathbf{F}. \tag{7.85}$$

Using the definition of $\mathbf{L}$, this can be rewritten as

$$\mathbf{L}_n = \mathbf{L} - \dot{g}g^{-1}\mathbf{I} - \mathbf{F}_n \mathbf{L}_p \mathbf{F}_n^{-1}. \tag{7.86}$$

Substituting this back into eqn (7.84) gives

$$\dot{\mathbf{B}}_n = \mathbf{L}\mathbf{B}_n + \mathbf{B}_n \mathbf{L}^{\mathrm{T}} - 2\dot{g}g^{-1}\mathbf{B}_n - 2\mathbf{F}_n \mathbf{D}_p \mathbf{F}_n^{\mathrm{T}}, \tag{7.87}$$

or, if the definition of the upper convected Maxwell derivative

$$\frac{D\mathbf{M}}{Dt} = \dot{\mathbf{M}} - \mathbf{L}\mathbf{M} - \mathbf{M}\mathbf{L}^{\mathrm{T}} \tag{7.88}$$

is introduced,

$$\frac{D\mathbf{B}_n}{Dt} = -2\dot{g}g^{-1}\mathbf{B}_n - 2\mathbf{F}_n \mathbf{D}_p \mathbf{F}_n^{\mathrm{T}}. \tag{7.89}$$

We want now to include elastoviscoplastic effects in the mechanics of cell aggregates. We do this on the basis of the following observations:

1. When and where the cell population is subject to a moderate amount of stress, the body behaves elastically.
2. When and where the stress exceeds a threshold yield stress, the body undergoes viscoplastic deformations.

Given the resistance of a single bond, the threshold of the onset of plastic deformations is proportional to the area of the cell membranes in contact, which depends on the number of cells per unit volume. We call this threshold value $\tau(\phi)$ the yield stress and compare it with a frame-invariant measure f of the stress of the cellular constituent $\phi_c \mathbf{T}_c$. Note that a very small volume ratio can correspond to highly dispersed single cells or very clustered ensembles. In the former case, the yield stress is expected to be very low. In the second case, borrowing ideas from the dynamics of colloidal particles and flocculated suspensions, the yield stress should increase as the second or third power of the volume ratio (Buscall *et al.* 1988; Snabre and Mills 1996). Iordan *et al.* (2008) measured $\tau(\phi_c)$ to be proportional to $\phi_c^{8.4}$.

On this basis, the following elastic-type constitutive equation can be suggested in the elastic regime:

$$\mathbf{T}_c = \hat{\mathbf{T}}_c(\mathbf{B}_n) \qquad \text{if } f(\phi\mathbf{T}_c) \leq \tau(\phi). \tag{7.90}$$

As a frame-invariant measure of the stress, Basov and Shelukhin (1999) suggested the use of

$$\mathbf{t}(\mathbf{n}) = \phi[\mathbf{T}_c\mathbf{n} - (\mathbf{n}\cdot\mathbf{T}_c\mathbf{n})\mathbf{n}], \tag{7.91}$$

which represents the tangential stress vector relative to the surface identified by the normal $\mathbf{n}$. In particular, we will use

$$f(\phi_c\mathbf{T}_c) = \max_{|\mathbf{n}|=1} |\mathbf{t}(\mathbf{n})|, \tag{7.92}$$

which represents the maximum shear stress magnitude occurring in the plane identified by the eigenvector corresponding to the maximum of $|\mathbf{t}(\mathbf{n})|$. It can be proved that f is given by half of the difference between the maximum and the minimum eigenvalue of $\phi_c\mathbf{T}_c$.

Following Ambrosi and Preziosi (2009), above the yield stress, the excess tension originates from cell unbinding at the microscopic scale and then cell rearrangement at the macroscopic scale. Such a pictorial description is put into formal terms by the following constitutive equation:

$$\left[1 - \left(\frac{\tau(\phi_c)}{f(\phi\mathbf{T}'_c)}\right)^{\alpha}\right](\phi_c\mathbf{F}_m^{\mathrm{T}}\mathbf{T}'_c\mathbf{F}_m^{-\mathrm{T}}) = 2\eta(\phi_c)\mathbf{L}_p \qquad \text{if } f(\phi_c\mathbf{T}'_c) > \tau(\phi_c), \tag{7.93}$$

where $\mathbf{T}'_c = \mathbf{T}_c - \frac{1}{3}(\mathrm{tr}\ \mathbf{T}_c)\mathbf{I}$ operates on the current configuration and, for compatibility, we have mapped $\mathbf{D}_p$ into the same configuration using $\mathbf{F}_n$. We can merge the above equation with the condition that there is no evolution for shear stresses smaller than the yield stress by writing

$$\mathbf{L}_p = \frac{1}{2\tilde{\eta}}\left[1 - \frac{1}{\tilde{f}(\mathbf{T}'_c)}\right]_+ \mathbf{F}_m^{\mathrm{T}}\mathbf{T}'_c\mathbf{F}_m^{-\mathrm{T}} \tag{7.94}$$

where $[\cdot]_+$ stands for the positive part of the argument, $\tilde{\eta} = \eta(\phi_c)/\phi_c$, and $\tilde{f}(\mathbf{T}'_c) = [f(\phi_c\mathbf{T}'_c)/\tau(\phi_c)]^{\alpha}$.

7.9 Growth, cell re-organisation and stress tensor

The equation (7.93) can be justified following the guidelines of Gurtin (2010)[1]. First, Gurtin introduces the Cauchy tensor (for us $\mathbf{T}_c$) which is work–conjugate with the elastic deformation rate $\mathbf{L}^n$ and a plastic tensor, say $\mathbf{T}^p$, work–conjugate with $\mathbf{L}_p$. Then, by a virtual work argument, Gurtin is able to prove:

$$\mathbf{T}^p = J\mathbf{F}_n^T\mathbf{T}_c'\mathbf{F}_n^{-T} \tag{7.95}$$

Eventually, he postulate a dissipation principle, introducing a constitutive free energy ψ and writing the inequality:

$$\dot{\psi} - \mathbf{T}_c\colon \mathbf{L}^n - \mathbf{T}^p\colon \mathbf{L}^p \leq 0 \tag{7.96}$$

If we take $\psi = \hat{\psi}(\mathbf{F}_n)$ for example (see DiCarlo and Quiligotti (2002)), using a standard Coleman Noll argument we find:

$$\mathbf{T}_c = \frac{\partial\hat{\psi}}{\partial\mathbf{F}_n}\mathbf{F}_n^{-T} \tag{7.97}$$

$$\mathbf{T}^p\colon \mathbf{L}^p \geq 0 \tag{7.98}$$

to satisfy the (7.98) we can simply choose $\mathbf{L}^p = D\mathbf{T}^p$ with D a fourth order positive tensor. Substituting (7.96) in (7.94) we can identify consistently:

$$D = \frac{1}{2\tilde{\eta}}\left[1 - \frac{1}{\tilde{f}(\mathbf{T}_c')}\right]_+ \tag{7.99}$$

If we take the following elastic-type constitutive equation,

$$\mathbf{T}_c' = \mu\left(\mathbf{B}_n - \frac{1}{3}\text{tr }\mathbf{B}_n\mathbf{I}\right), \tag{7.100}$$

then $\mathbf{F}_p$ evolves according to

$$\dot{\mathbf{F}}_p = \frac{g^{-1}}{\lambda}\left[1 - \frac{1}{\mu\tilde{f}\left(\mathbf{B}_n - \frac{1}{3}(\text{tr }\mathbf{B}_n)\mathbf{I}\right)}\right]_+\left(\mathbf{C}_n - \frac{1}{3}(t_2\mathbf{C}_n)\mathbf{I}\right)\mathbf{F}_p, \tag{7.101}$$

where $\lambda = \tilde{\eta}/\mu$ is called the *cell reorganization time* and $\mathbf{C}_n = \mathbf{F}_m^T\mathbf{F}_m$.

Equation (7.101) can be explained phenomenologically in the following way. Assuming for the moment that there is no growth, if the body undergoes a deformation corresponding to a stress below the yield stress, then $\mathbf{F}_p$ does not change, i.e., the intermediate configuration does not evolve and all the energy is stored elastically. If the measure of tension f takes a value larger than the yield stress, then the reference configuration changes to release the excess stress, until the yield surface defined by

[1] Gurtin, M.E. Fried, E. and Anand, L. The Mechanics and Thermodynamics of Continua, Cambridge Press 2010

f is reached again. The ratio $\lambda = \tilde{\eta}/\mu$ gives an indication of the characteristic time needed to relax the internal stress through the reorganization of cells and to reach the yield surface.

Taking the derivative of eqn (7.100), one can write

$$\begin{aligned}\dot{\mathbf{T}}'_c = \mu[&\mathbf{L}\mathbf{B}_n + \mathbf{B}_n\mathbf{L}^{\mathrm{T}} - 2\dot{g}g^{-1}\mathbf{B}_n - 2\mathbf{F}_n\mathbf{D}_p\mathbf{F}_n^{\mathrm{T}} \\ &- \frac{1}{3}\mathrm{tr}(\mathbf{L}\mathbf{B}_n + \mathbf{B}_n\mathbf{L}^{\mathrm{T}} - 2\dot{g}g^{-1}\mathbf{B}_n - 2\mathbf{F}_n\mathbf{D}_p\mathbf{F}_n^{\mathrm{T}})\mathbf{I}].\end{aligned}$$

Using eqn (7.94), one then has

$$\begin{aligned}&\dot{\mathbf{T}}'_c + \frac{\mu}{\tilde{\eta}}\left[1 - \frac{1}{\tilde{f}(\mathbf{T}'_c)}\right]_+ \left(\mathbf{T}'_c\mathbf{B}_n + \mathbf{B}_n\mathbf{T}'_c - \frac{1}{3}\mathrm{tr}(\mathbf{T}'_c\mathbf{B}_n + \mathbf{B}_n\mathbf{T}'_c)\mathbf{I}\right) \\ &= \mu[\mathbf{L}\mathbf{B}_n + \mathbf{B}_n\mathbf{L}^{\mathrm{T}} - 2\dot{g}g^{-1}\mathbf{B}_n - \frac{1}{3}\mathrm{tr}(\mathbf{L}\mathbf{B}_n + \mathbf{B}_n\mathbf{L}^{\mathrm{T}} - 2\dot{g}g^{-1}\mathbf{B}_n)\mathbf{I}],\end{aligned}$$

which can be written in terms of the upper convective derivative as

$$\begin{aligned}&\frac{D\mathbf{T}'_c}{Dt} + \frac{1}{\lambda}\left[1 - \frac{1}{\tilde{f}(\mathbf{T}'_c)}\right]_+ \left(\mathbf{T}'_c\mathbf{B}_n + \mathbf{B}_n\mathbf{T}'_c - \frac{1}{3}\mathrm{tr}(\mathbf{T}'_c\mathbf{B}_n + \mathbf{B}_n\mathbf{T}'_c)\mathbf{I}\right) \\ &\quad = \frac{2}{3}\mu[\mathrm{tr}(\mathbf{B}_n)\mathbf{D} - \mathrm{tr}(\mathbf{B}_n\mathbf{D})\mathbf{I} - \dot{g}g^{-1}(3\mathbf{B}_n - (\mathrm{tr}\mathbf{B}_n)\mathbf{I})].\end{aligned} \tag{7.102}$$

Note that the Maxwell derivative of an objective tensor is itself objective.

We now assume that the deformations from the natural configuration are small throughout the evolution of the system and that $g = 1$ (no growth). The small-deformation assumption applies depending on the value of the yield stress: it has to be small enough that the condition $|\mathbf{B}_n \cdot \mathbf{I} - 3| \ll 1$ is always satisfied during the motion. The experiments of Iordan *et al.* (2008) (Fig. 7.11) give an indication of the order of magnitude of the yield stress, which depends on the volume ratio and is below 1 Pa (for $\phi_c = 0.6$, the maximum volume ratio tested). The advantage of this hypothesis is that for small elastic strains one can use linear elasticity. For larger stresses, the natural configuration evolves.

In the limit of small deformations,

$$\mathbf{L}\mathbf{B}_n + \mathbf{B}_n\mathbf{L}^{\mathrm{T}} \approx 2\mathbf{D} \qquad \text{and} \qquad \mathbf{B}_n\mathbf{T}'_c\mathbf{T}'_c\mathbf{B}_n \approx \mathbf{T}'_c, \tag{7.103}$$

and therefore eqn (7.102) simplifies to

$$\dot{\mathbf{T}}'_c + \frac{1}{\lambda}\left[1 - \frac{1}{\tilde{f}(\mathbf{T}'_c)}\right]_+ \mathbf{T}'_c = 2\mu\left(\mathbf{D} - \frac{1}{3}\mathrm{tr}\,\mathbf{D}\,\mathbf{I}\right), \tag{7.104}$$

where the trace of $\mathbf{D}$ is related to the growth term through the mass balance equation.

We observe that in eqn (7.104), the term containing the yield stress plays the role of a stress relaxation term that switches on as soon as the stress is above the yield value. Otherwise, for $\tilde{f}(\mathbf{T}'_c) < 1$, eqn (7.104) can be integrated to give back an elastic-like constitutive equation.

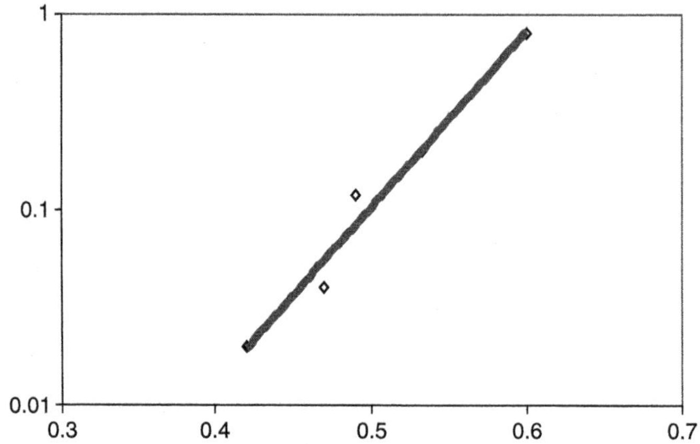

Fig. 7.11 Yield stress in Pascals as a function of the cell volume ratio as measured by Iordan *et al.* (2008).

In the limit λ much larger than the characteristic time of the process of interest, this leads in principle to the models used by Araujo and McElwain (2004, 2005*a*,b). However, in this case the procedure is incompatible with the small-deformation assumption because the stress relaxes very slowly, and so large stresses and deformations can build up. On the other hand, rewriting eqn (7.104) as

$$\lambda\dot{\mathbf{T}}'_c + \left[1 - \frac{1}{\tilde{f}\,(\mathbf{T}'_c)}\right]_+ \mathbf{T}'_c = 2\eta\left(\mathbf{D} - \frac{1}{3}\mathrm{tr}\ \mathbf{D}\,\mathbf{I}\right), \tag{7.105}$$

it is easy to realize that for processes with characteristic times larger than λ and stresses much larger than τ (i.e., $\tilde{f} \gg 1$), the model behaves like the viscous models used for instance by Franks *et al.* (2003*a*,*b*).

Following the same argument Preziosi and Joseph (1987), one can state that in the case of transient phenomena, for times much larger than the cell reorganization time, the natural configuration evolves and relaxes the stress, leaving the material in a state of stress that is mostly on the yield surface.

7.10 Response to shear tests

In these final two sections, we present some descriptions of the response of materials satisfying eqn (7.101) to several tests. More comparisons can be found in Preziosi *et al.* (2010).

For pure shear tests, eqn (7.104) can be rewritten as

$$\dot{T}_c + \frac{1}{\lambda}\left[1 - \left(\frac{\hat{\tau}}{|T_c|}\right)^{\alpha}\right]_+ T_c = \mu\dot{\gamma}, \tag{7.106}$$

where $\dot{\gamma}$ is the shear rate, $f(\mathbf{T}_c) = |T_{xy}| = |T_c|$, and $\hat{\tau} = \tau(\phi)/\phi$. The case $\alpha = 1$ corresponds to a Bingham fluid, and the limiting case $\alpha = 0$ corresponds to an elastic fluid. In the following, sometimes the particular case $\alpha = 1$ will be considered in more detail, because it allows us to write analytical solutions in an easy way and to easily understand the main features of the constitutive model.

When a cell aggregate obeying the constitutive model above is sheared with a smoothly increasing load, initially the body deforms elastically and the stress grows until the yield value $\hat{\tau}$ is reached. In this initial regime, the cell bonds are simply stretched, no internal reorganization occurs, the natural configuration does not evolve, and the body would be able to return to the initial stress-free configuration if allowed to do so, for instance if the stress were released (see upper plots in Fig. 7.12). When the stress locally exceeds the yield value, some adhesion bonds break, new ones form, and the natural configuration evolves. The body is then not able to return to the original configuration any longer, as shown in the lower plots in Fig. 7.12.

7.10.1 Steady shear

As an example, consider the case of a shear increasing linearly in time, i.e., $\gamma = \dot{\gamma}_0 t$ and $\alpha = 1$. Integration of eqn (7.106) with respect to time gives

$$T_c(t) = \begin{cases} \mu\dot{\gamma}_0 t & \text{for} \quad t \le t_\tau = \hat{\tau}/\mu\dot{\gamma}_0, \\ \hat{\tau} + \mu\dot{\gamma}_0\lambda\left[1 - e^{(-t+t_\tau)/\lambda}\right] & \text{for} \quad t > t_\tau. \end{cases} \tag{7.107}$$

If $t \to \infty$, then $T_c(t) \to \hat{\tau} + \tilde{\eta}\dot{\gamma}_0$. This means that if the apparent viscosity $\eta(\dot{\gamma}_0) = T_c/\dot{\gamma}_0$ is plotted as a function of the shear rate $\dot{\gamma}_0$, the plot should go like $\dot{\gamma}_0^{-1}$ for small $\dot{\gamma}_0$. This is the case in the experiments reported by Iordan *et al.* (2008), shown in Fig. 7.13.

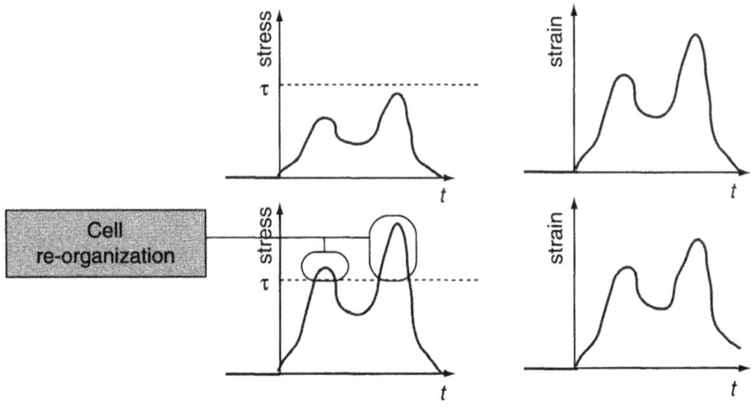

Fig. 7.12 Unrecovered deformation due to stress history above the yield stress.

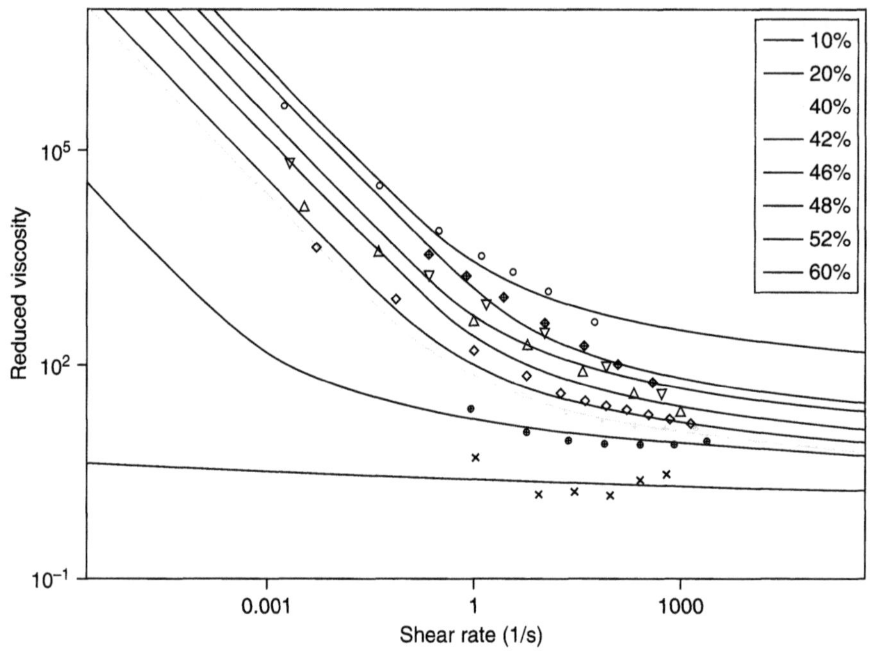

Fig. 7.13 Comparison of the model with the experiments of Iordan *et al.* (2008). Viscosity versus shear rate at different volume ratios.

In general, for $0 \leq \alpha \leq 1$, the relation between $\dot{\gamma}_0$ and the apparent viscosity $\eta(\dot{\gamma}_0)$ is given by

$$\dot{\gamma}_0 = \frac{\hat{\tau}}{\eta(\dot{\gamma}_0)} \left(1 - \frac{\tilde{\eta}}{\eta(\dot{\gamma}_0)}\right)^{-1/\alpha}. \tag{7.108}$$

If $\eta \to \infty$, then $\dot{\gamma}_0$, goes to zero as $\hat{\tau}/\eta(\dot{\gamma}_0)$, which implies that for small $\dot{\gamma}_0$, the apparent viscosity $\eta(\dot{\gamma}_0)$ behaves like $\dot{\gamma}_0^{-1}$. On the other hand, if $\eta \to \tilde{\eta}$, then $\dot{\gamma}_0$ goes to infinity, which means that at high shear rates, the apparent viscosity $\eta(\dot{\gamma}_0) = T/\dot{\gamma}_0$ goes towards its limiting value $\tilde{\eta}$. This limit is concentration-dependent and has been obtained experimentally (Chien *et al.* 1967; Iordan *et al.* 2008). A plot of the viscosity against shear rate is presented in Fig. 7.14. Note the limiting behavior at small shear rates (slope -1 on this log–log figure). At high $\dot{\gamma}_0$, the viscosity reaches its limit $\tilde{\eta}$.

In order to validate this model with a true biological aggregate, Preziosi *et al.* (2010) compared the results of the model with the experiments of Iordan *et al.* (2008), and the results are reported here in Fig. 7.13. The only adjustable parameters were found to be $\hat{\tau}/\tilde{\eta}$ and α, by simple scaling arguments. As $\tilde{\eta}$ was fixed to be the culture medium viscosity $\tilde{\eta} = 0.0013\,\text{Pa}\cdot\text{s}$, only $\hat{\tau}$ and α were allowed to vary. For instance, for a concentration of 42%, $\hat{\tau} = 0.05$ Pa and $\alpha = 0.01$. It was found that the values

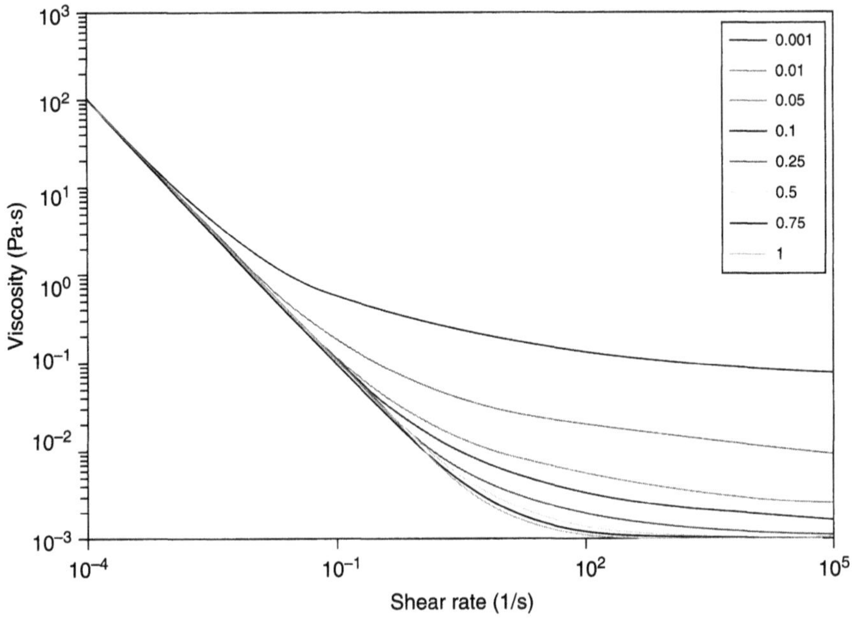

Fig. 7.14 Apparent viscosity $\eta(\dot{\gamma}_0) = T/\dot{\gamma}_0$ as a function of $\dot{\gamma}_0$ for different values of the parameter α.

of the yield stress were close to those obtained by Iordan *et al.* (2008), and showed a typical dependence of the type $\hat{\tau} \sim 962.2\,\phi^{11}$ at large concentrations $0.4 \le \phi \le 0.6$.

7.10.2 Stress relaxation tests

A standard relaxation test is performed by applying a sudden constant deformation γ_0. If $\gamma_0 \le \hat{\tau}/\mu$, then the stress in the body is $T_c = \mu\gamma_0 (< \hat{\tau})$ at any time corresponding to the straight line in Fig. 7.15(a), an elastic response. If, on the other hand, $\gamma_0 > \hat{\tau}/\mu$, the solution of eqn (7.106) is

$$T(t) = \hat{\tau}\left\{1 + \left[\left(\frac{\mu\gamma_0}{\hat{\tau}}\right)^{\alpha} - 1\right] e^{-\alpha t/\lambda}\right\}^{1/\alpha}. \tag{7.109}$$

In particular, for $\alpha = 1$,

$$T(t) = \hat{\tau} + (\mu\gamma_0 - \hat{\tau})e^{-t/\lambda}, \tag{7.110}$$

which is plotted in Fig. 7.15 for increasing values of γ_0.

Hence for small strains the body behaves elastically, while for large strains part of the stress relaxes to the yield value $\hat{\tau}$, regardless of the magnitude of the applied strain. The decrease toward the asymptotic state is faster for larger exponents.

Forgacs *et al.* (1998) obtained similar plots when compressing a multicellular spheroid in a uniaxial compression test that will be analyzed more deeply in the next section. They observed that when a fixed deformation was applied to the cell aggregate,

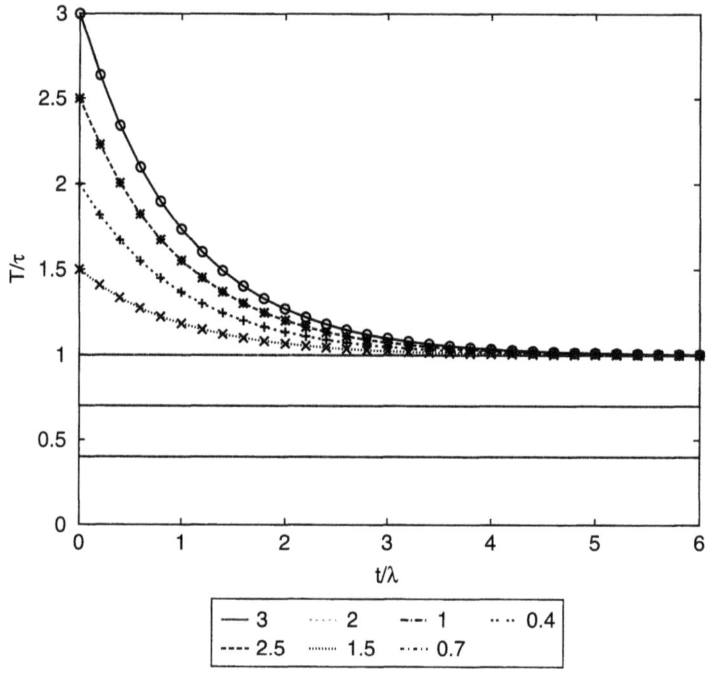

Fig. 7.15 Stress relaxation response $T/\hat{\tau}$ for different values of $\mu\gamma_0/\hat{\tau}$ and $\alpha = 1$. For $\gamma_0 \leq \hat{\tau}/\mu$, the response is elastic. Time is normalized with respect to λ.

the internal stress relaxed until it reached an asymptotic value. They interpreted this long-time behavior as an effect of surface tension. This idea is supported by the observation that doubling the deformation would give rise to the measurement of a similar surface tension coefficient, whereas a constitutive model for a linear elastic solid would give rise to a doubling of the stress. They also observed that the coefficient depended linearly on the density of cadherins on the cell surface, which is closely related to the adhesion properties of the cell. This information and the above results suggest that the same results can be interpreted using the constitutive model presented here, relating the asymptotic behavior to a measure of the yield stress. In fact, similarly to the experiments of Forgacs *et al.* (1998), doubling the deformation in the present virtual experiment would give rise to the same asymptotic value when the yield stress was measured. In addition, the present model would also be in agreement with the results of Iordan *et al.* (2008), which cannot be explained by the concept of surface tension. Hence, using the values given by Forgacs *et al.* (1998), the yield stress for the various tissues tested there ranges from 1 to 100 Pa. The results obtained by Iordan *et al.* (2008) for a cell suspension with a volume ratio of 0.6 (the maximum value they used) give a yield stress around 1Pa, and higher concentrations would probably lead to data in the same range, especially since the yield stress has a concentration dependence like $\hat{\tau} \sim \phi^{11}$, as observed previously.

7.11 Uniaxial compression tests

7.11.1 Uniaxial compression

To compare the results that can be obtained from the model with the experimental results described by Forgacs *et al.* (1998), Foty *et al.* (1996), and Winters *et al.* (2005), without restricting to small deformations we assume here that the deformation generated by a compressive stress P_{appl} generates a homogeneous deformation along the z-axis is homogeneous, taking the following form:

$$x = \frac{X}{\sqrt{\psi(t)}}, \qquad y = \frac{Y}{\sqrt{\psi(t)}}, \qquad z = \psi(t)Z, \tag{7.111}$$

where $\mathbf{F}_p$ is given by

$$\mathbf{F}_p = \text{diag}\left\{\frac{1}{\sqrt{\Psi(t)}}, \frac{1}{\sqrt{\Psi(t)}}, \Psi(t)\right\}. \tag{7.112}$$

How $\Psi(t)$ differs from 1 is then a measure of how much the aggregate has reorganized and the natural configuration has evolved. Hence $\mathbf{F}_c$ and $\mathbf{F}_n$ are given by

$$\mathbf{F}_c = \text{diag}\left\{\frac{1}{\sqrt{\psi(t)}}, \frac{1}{\sqrt{\psi(t)}}, \psi(t)\right\} \tag{7.113}$$

and

$$\mathbf{F}_n = \text{diag}\left\{\sqrt{\frac{\Psi(t)}{\psi(t)}}, \sqrt{\frac{\Psi(t)}{\psi(t)}}, \frac{\psi(t)}{\Psi(t)}\right\}, \tag{7.114}$$

respectively. Therefore, dropping t for the sake of simplicity,

$$\mathbf{B}_n = \text{diag}\left\{\frac{\Psi}{\psi}, \frac{\Psi}{\psi}, \frac{\psi^2}{\Psi^2}\right\} \tag{7.115}$$

and

$$\mathbf{T}_c = \mu\,\text{diag}\left\{-\Sigma + \frac{\Psi}{\psi}, -\Sigma + \frac{\Psi}{\psi}, -\Sigma + \frac{\psi^2}{\Psi^2}\right\} = \text{diag}\{0, 0, P_{\text{appl}}\}, \tag{7.116}$$

where P_{appl} is the time-dependent applied stress in the z-direction. In conclusion,

$$\Sigma = \frac{\Psi}{\psi} \quad \text{and} \quad P_{\text{appl}} = \mu\frac{\psi^3 - \Psi^3}{\psi\Psi^2}. \tag{7.117}$$

On the other hand,

$$\mathbf{L}_p = \dot{\mathbf{F}}_p\mathbf{F}_p^{-1} = \text{diag}\left\{-\frac{1}{2}, -\frac{1}{2}, 1\right\}\frac{\dot{\Psi}}{\Psi}, \tag{7.118}$$

and therefore, from eqn (7.94),

$$\frac{\dot{\Psi}}{\Psi} = \frac{1}{3\tilde{\eta}}\left[1 - \frac{1}{\tilde{f}(\mathbf{T}')}\right]_+ P_{\rm appl} = \frac{1}{3\tilde{\eta}}\left[1 - \left(\frac{2\hat{\tau}}{|P_{\rm appl}|}\right)^{\alpha}\right]_+ P_{\rm appl}. \tag{7.119}$$

Substituting eqn (7.117) in the equation above, we obtain the evolution equation for Ψ, i.e., for the natural configuration:

$$\frac{\dot{\Psi}}{\Psi} = \frac{\mu}{3\tilde{\eta}}\left[1 - \left(\frac{2\hat{\tau}}{\mu}\frac{\psi\Psi^2}{|\psi^3 - \Psi^3|}\right)^{\alpha}\right]_+ \frac{\psi^3 - \Psi^3}{\psi\Psi^2}. \tag{7.120}$$

In order to mimic Forgacs *et al.'s* experiment, assume that a given strain $\psi_0 < 1$ is imposed on the cell aggregate. Then, if the compression is sufficiently strong that the quantity in the square brackets above is positive, i.e.,

$$\frac{1}{\psi_0} - \psi_0^2 > \frac{2\hat{\tau}}{\mu}, \tag{7.121}$$

Ψ will decrease from $\Psi = 1$ according to

$$\dot{\Psi} = -\frac{\mu}{3\tilde{\eta}}\left[1 - \left(\frac{2\hat{\tau}}{\mu}\frac{\psi_0\Psi^2}{\Psi^3 - \psi_0^3}\right)^{\alpha}\right]\frac{\Psi^3 - \psi_0^3}{\psi_0\Psi}. \tag{7.122}$$

The result of integrating this equation is shown in Fig. 7.16.

If the compression is applied for a long enough time, then Ψ will tend towards the value $\Psi_\infty > \psi_0$ corresponding to the vanishing of the quantity in the square brackets in eqn (7.122), i.e.,

$$\frac{\Psi_\infty^3 - \psi_0^3}{\psi_0\Psi_\infty^2} = \frac{2\hat{\tau}}{\mu}, \tag{7.123}$$

which is independent of α. In particular, $P_{\rm appl}$ will tend to

$$P_{\rm appl,\infty} = \mu\frac{\psi_0^3 - \Psi_\infty^3}{\psi_0\Psi_\infty^2} = -2\hat{\tau}, \tag{7.124}$$

(see bottom of Fig. 7.16), which in addition to being independent of α is also independent of ψ_0, as in the experiments of Forgacs *et al.* (1998).

In Fig. 7.16, the results of applying several different compressions are shown. If $\psi_0 > 0.8351$, the compression is not strong enough to trigger cell reorganization, and therefore $\Psi(t) = 1$ does not evolve and $|P_{\rm appl}|$ is constantly below the yield value, which in this case is $2\hat{\tau} = 0.5$. For larger ψ_0, the natural configuration evolves and $P_{\rm appl}$ again tends to the constant value given in eqn (7.124).

The behavior is then similar to that described for stress relaxation tests in Section 9.2, although eqn (7.122) cannot be easily solved. Finally, if at an instant t_2 (when $\Psi = \Psi_2 > \Psi_\infty$) the compression is suddenly released, then ψ will readily adjust to the value $\psi = \Psi_2$.

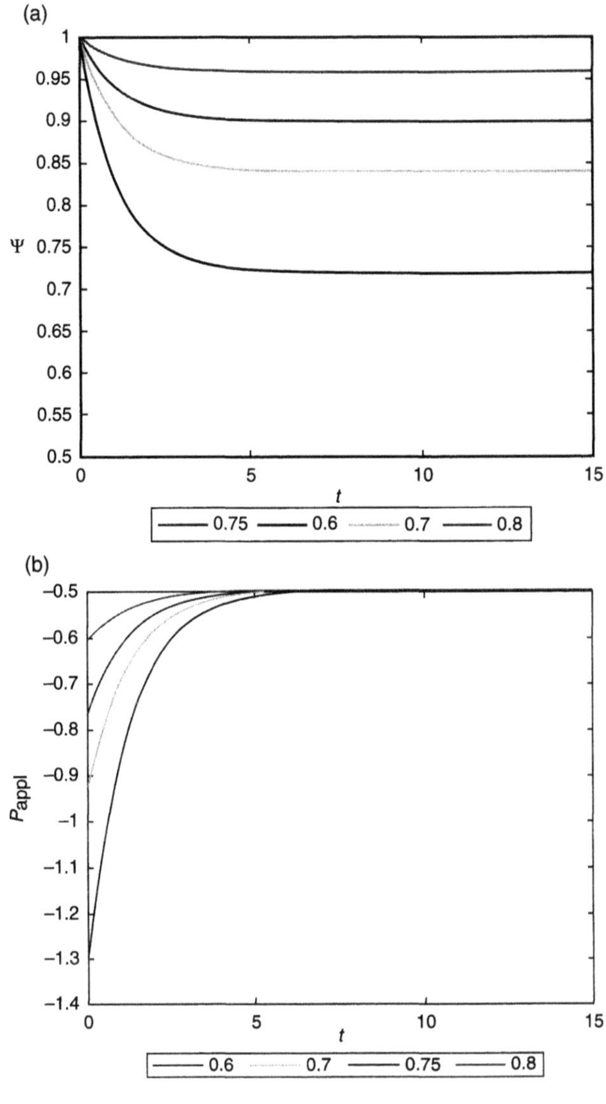

Fig. 7.16 Uniaxial compression for $\hat{\tau}/\mu = 0.25$. Time is normalized with respect to $\tilde{\eta}/\mu$ and the applied stress is normalized with respect to μ. If $\psi_0 > 0.8351$, the compression is not strong enough to trigger cell reorganization, and therefore $\Psi(t) = 1$ does not evolve and $P_{\rm appl}/\mu$ is constantly equal to $\psi_0^2 - 1/\psi_0 > -0.5$. The curves represent the behavior for $\psi_0 < 0.8351$, when a stronger deformation triggers cell reorganization and stress relaxation. The asymptotic value of the volume ratio is in all cases 0.8351, that of Ψ is given by eqn (7.123), and that of $P_{\rm appl}/\mu$ is always -0.5, corresponding to the yield value. α is fixed at 1.

As already discussed in the previous section, this result can be compared with the stress relaxation tests performed by Forgacs *et al.* (1998). These authors observed that the measured stress decreased asymptotically to a value that depended on the tissue but was independent of the deformation imposed. They interpreted this long-time behavior as an effect of surface tension, but, as just shown, the same result can be interpreted by using the constitutive model above and relating the asymptotic behavior to a measure of the yield stress.

Furthermore, Foty *et al.* (1996) correlated the surface tension coefficient obtained from the asymptotic value with the density of surface cadherin per cell. The same proportionality obviously holds for the yield stress τ, because this coefficient is clearly related to the number of adhesion bonds. In our opinion, the relation between yield stress and surface tension also explains the observation by Winters *et al.* (2005) that cell invasiveness is related to the inverse of the surface tension: the smaller the adhesion between cells, the smaller the yield stress, and the more invasive the cell clone.

We also mention that Forgacs *et al.* (1998) pointed out the existence of two relaxation times in the cellular matter (one of the order of few seconds and the other of the order of tens of seconds), which could be obtained in the present approach by considering two reorganization mechanisms or the detachment of different types of adhesion proteins. From the modeling point of view, this means generalizing the model above to allow two relaxation times.

7.11.2 Elastic recovery

A second part of the experiment done by Foty *et al.* (1996) consisted in releasing the imposed stress after some time from the beginning of the experiment. As already stated, if the imposed deformation is small compared with the yield condition, then the body will go back to the original configuration. Otherwise, an internal reorganization of the cells is induced and the body does not recover its initial configuration, because in the meantime the natural configuration has changed. In fact, denoting the value of the stress before the sudden stress release (which is much faster than the cell reorganization time λ) by T_2, the yield stress is reached for

$$\gamma = \gamma_2 - \frac{T_2 - \hat{\tau}}{\mu}, \tag{7.125}$$

independently of α. After that, the body relaxes the stress, reaching the stress-free configuration when

$$\gamma_{\text{fin}} = \gamma_2 - \frac{T_2}{\mu}. \tag{7.126}$$

To be more specific, if in the stress relaxation experiment described by eqn (7.109) the compressing plate is removed at time t_{compr} so that the specimen is stress-free,

$$\gamma_{\text{fin}} = \gamma_0 - \frac{\hat{\tau}}{\mu}\left\{1 + \left[\left(\frac{\mu\gamma_0}{\hat{\tau}}\right)^{\alpha} - 1\right] e^{-\alpha t_{\text{compr}}/\lambda}\right\}^{1/\alpha}. \tag{7.127}$$

In particular, if $t_{\text{compr}} \ll \lambda$, the exponential and the power in eqn (7.127) can be approximated to give

$$\gamma_{\text{fin}} \approx \gamma_0 \left[1 - \left(\frac{\hat{\tau}}{\mu\gamma_0}\right)^{\alpha}\right] \frac{t_{\text{compr}}}{\lambda}. \tag{7.128}$$

This means that if the compression is applied for such a small time that the body does not have enough time to reorganize, then the body will recover almost everything and return to close to the original configuration, showing mainly an elastic-like behavior. If, on the other hand, $t_{\text{compr}} \gg \lambda$, then

$$\gamma_{\text{fin}} = \gamma_0 - \frac{\hat{\tau}}{\mu}, \tag{7.129}$$

meaning that the body will still recover by an amount $\hat{\tau}/\mu$, corresponding to the elastic component. In other words, it will not maintain the value $\gamma = \gamma_0$ that was imposed, even if this is done for a very long time.

The above description is consistent with the observations of Foty *et al.* (1996) (Figs. 3 and 5 in that paper) that if the spheroid was compressed for few seconds then it would bounce back almost to the original configuration (there was only a small flattening at the poles). If it was compressed for a longer time (a few hours in the experiments), this did not occur, though a minor shape recovery was observed. This cannot be explained using the concept of surface tension, but is compatible with the model presented here.

On the other hand, we have to point out that in this model the process is instantaneous, whereas it seems that in the experiments of Foty *et al.* (1996) it took some times for the spheroid to return to the natural configuration, as in standard indentation tests. This might be due to the fact that here we have completely neglected the fact that the spheroid was a porous material filled with the liquid in which the experiment was done. Such an effect can be included by, for instance, taking into account the fact that T_c is only one of the components of the stress and a further viscous component needs to be added to obtain the stress for the mixture as a whole, similarly to what was done by Saramito (2007, 2008). This would give rise to a constitutive model that can more properly describe creep tests.

References

Ambrosi, D. and Mollica, F. (2002). On the mechanics of a growing tumor. *Int. J. Engng.*, **40**, 1297–1316.

Ambrosi, D. and Mollica, F. (2004). The role of stress in the growth of a multicell spheroid. *J. Math. Biol.*, **48**, 477–499.

Ambrosi, D. and Preziosi, L. (2002). On the closure of mass balance models for tumor growth. *Math. Mod. Methods. Appl. Sci.*, **12**, 737–754.

Ambrosi, D. and Preziosi, L. (2009). Cell adhesion mechanisms and stress relaxation in the mechanics of tumors. *Biomech. Model. Mechanobiol.*, **8**, 397–413.

Araujo, R. P. and McElwain, D. L. S. (2004). A linear-elastic model of anisotropic tumour growth. *Eur. J. Appl. Math.*, **15**, 365–384.

Araujo, R. P. and McElwain, D. L. S. (2005*a*). A mixture theory for the genesis of residual stresses in growing tissues, I: A general formulation. *SIAM J. Appl. Math.*, **65**, 1261–1284.

Araujo, R. P. and McElwain, D. L. S. (2005*b*). A mixture theory for the genesis of residual stresses in growing tissues, II: Solutions to the biphasic equations for a multicell spheroid. *SIAM J. Appl. Math.*, **65**, 1285–1299.

Astanin, S. and Preziosi, L. (2008). Multiphase models of tumour growth. In *Selected Topics on Cancer Modelling: Genesis – Evolution – Immune Competition – Therapy* (ed. M. C. N. Bellomo and E. D. Angelis) pp. 223–253. Birkhäuser.

Basov, I. V. and Shelukhin, V. V. (1999). Generalized solutions to the equations of compressible Bingham flows. *ZAMM*, **79**, 185–192.

Baumgartner, W., Hinterdorfer, P., Ness, W., Raab, A., Vestweber, D., and Schindler, H. D. D. (2000). Cadherin interaction probed by atomic force microscopy. *Proc. Natl. Acad. Sci. USA*, **97**, 4005–4010.

Bear, J. and Bachmat, Y. (1990). *Introduction to Modeling of Transport Phenomena in Porous Media.* Kluwer Academic.

Becker, K. F., Atkinson, M. J., Reich, U., Becker, I., Nekarda, H., Siewert, J. R., and Hofler, H. (1994). E-cadherin gene mutation provide clues to diffuse gastric carcinoma. *Cancer Res.*, **54**, 3845–3852.

Bowen, R. M. (1976). *The Theory of Mixtures*, Volume 3, Chapter 1. Academic Press.

Buscall, R., Mills, P. D. A., Goodwin, J. W., and Lawson, D. W. (1988). Scaling behaviour of the rheology of aggregate networks formed from colloidal particles. *J. Chem. Soc. Faraday Trans.*, **84**, 4249–4260.

Byrne, H. and Preziosi, L. (2004). Modeling solid tumor growth using the theory of mixtures. *Math. Med. Biol.*, **20**, 341–360.

Canetta, E., Duperray, A., Leyrat, A., and Verdier, C. (2005). Measuring cell viscoelastic properties using a force-spectrometer: Influence of the protein–cytoplasm interactions. *Biorheology*, **42**, 298–303.

Castilla, M. A., Arroyo, M. V. A., Aceituno, E., Aragoncillo, P., Gonzalez-Pacheco, F. R., Texeiro, E., Bragado, R., and Caramelo, C. (1999). Disruption of cadherin-related junctions triggers autocrine expression of vascular endothelial growth factor in bovine aortic endothelial cells. Effect on cell proliferation and death resistance. *Circ. Res.*, **85**, 1132–1138.

Cavallaro, U., Schaffhauser, B., and Christofori, B. (2002). Cadherin and the tumour progression: Is it all in a switch? *Cancer Lett.*, **176**, 123–128.

Caveda, L., Martin-Padura, I., Navarro, P., Breviario, F., Corada, M., Gulino, D., Lampugnani, M. G., and Dejana, E. (1996). Inhibition of cultured cell growth by vascular endothelial cadherin (cadherin-5/ve-cadherin). *J. Clin. Invest.*, **98**, 886–893.

Chaplain, M., Graziano, L., and Preziosi, L. (2005). Mathematical modelling of the loss of tissue compression responsiveness and its role in solid tumour development. *Math. Med. Biol.*, **23**, 197–229.

Chauviere, A. and Preziosi, L. (2010). A mathematical framework to model migration of a cell population in the extracellular matrix. In *Cell Mechanics: From Single*

Scale-Based Models to Multiscale Modeling (ed. L. P. A. Chauviere and C. Verdier), pp. 279–312. Chapman and Hall/CRC Press.

Chien, S., Usami, S., Dellenback, R. J., Gregersen, M. I., Nanninga, L. B., and Mason-Guest, M. (1967). Blood viscosity: Influence of erythrocyte aggregation. *Science*, **157**, 827–829.

Christofori, G. and Semb, H. (1999). The role of cell-adhesion molecule e-cadherin as a tumors-suppressor gene. *Trends Biochem. Sci.*, **24**, 73–76.

DiCarlo, A. and Quiligotti, S. (2002). Growth and balance. *Mech. Res. Commun.*, **29**, 449–456.

Dietrich, C., Wallenfrang, K., Oesch, F., and Wieser, R. (1997). Differences in the mechanisms of growth control in contact-inhibited and serum-deprived human fibroblasts. *Oncogene*, **15**, 2743–2747.

Forgacs, G., Foty, R., Shafrir, Y., and Steinberg, M. (1998). Viscoelastic properties of living embryonic tissues: A quantitative study. *Biophys. J.*, **74**, 2227–2234.

Foty, R. A., Forgacs, G., Peger, M., and Steinberg, M. S. (1996). Surface tensions of embryonic tissues predict their mutual envelopment behavior. *Development*, **122**, 1611–1620.

Franks, S. and King, J. (2003). Interactions between a uniformly proliferating tumor and its surrounding uniform material properties. *Math. Med. Biol.*, **20**, 47–89.

Franks, S., Byrne, H., King, J., Underwood, J., and Lewis, C. (2003*a*). Modelling the early growth of ductal carcinoma in situ of the breast. *J. Math. Biol.*, **47**, 424–452.

Franks, S., Byrne, H., Mudhar, H., Underwood, J., and Lewis, C. (2003*b*). Mathematical modelling of comedo ductal carcinoma in situ of the breast. *Math. Med. Biol.*, **20**, 277–308.

Galle, J., Preziosi, L., and Tosin, A. (2009). Contact inhibition of growth described by a multiphase and an individual cell-based models. *Appl. Math. Lett.*, **22**, 1483–1490.

Gottardi, G. J., Wong, E., and Gumbiner, B. M. (2001). E-cadherin suppresses cellular transformation by inhibiting β-catenin signaling in an adhesion-independent manner. *J. Cell. Biol.*, **153**, 1049–1060.

Graziano, L. and Preziosi, L. (2007). Mechanics in tumour growth. In *Modeling of Biological Materials* (ed. K. R. F. Mollica, L. Preziosi), pp. 267–328. Birkhäuser.

Harja, K. M. and Fearon, E. R. (2002). Cadherin and catenin alterations in human cancer. *Genes Chromosomes Cancer*, **34**, 255–268.

Humphrey, J. and Rajagopal, K. (2002). A constrained mixture model for growth and remodeling of soft tissues. *Math. Mod. Methods. Appl. Sci.*, **22**, 407–430.

Humphrey, J. and Rajagopal, K. (2003). A constrained mixture model for arterial adaptations to a sustained step change in blood flow. *Biomech. Model. Mechanobiol.*, **2**, 109.

Iannelli, M., Martcheva, M., and Milner, F. A. (2005). Gender-structured population modeling: Mathematical Methods, Numerics, and Simulations, SIAM Frontiers in Applied Mathematics. SIAM. *SIAM Frontiers in Applied Mathematics*. SIAM.

Iordan, A., Duperray, A., and Verdier, C. (2008). Fractal approach to the rheology of concentrated cell suspensions. *Phys. Rev. E*, **77,011911**.

Johnson, P. R. A. (2001). Role of human airway smooth muscle in altered extracellular matrix production in asthma. *Clin. Exp. Pharm. Physiol.*, **28**, 233–236.

Kato, A., Takahashi, H., Takahashi, Y., and Matsushime, H. (1997). Inactivation of the cyclin d-dependent kinase in the rat fibroblast cell line, 3y1, induced by contact inhibition. *J. Biol. Chem.*, **272**, 8065–8070.

Klisch, S. M. and Hoger, A. (2003). Volumetric growth of thermoelastic materials and mixtures. *Math. Mech. Solids*, **8**, 377–402.

Lowengrub, J. S., Frieboes, H. B., Jin, F., Chuang, Y. L., Li, X., Macklin, P., Wise, S. M., and Cristini, V. (2010). Nonlinear modelling of cancer: Bridging the gap between cells and tumours. *Nonlinearity*, **23**, in press.

Malik, W. A., Prasad, S. C., Rajagopal, K. R., and Preziosi, L. (2008). On the modeling of the viscoelastic response of embryonic tissues. *Math. Mech. Solids*, **13**, 81–91.

Nelson, C. M. and Chen, C. S. (2003). Ve-cadherin simultaneously stimulates and inhibits cell proliferation by altering cytoskeletal structure and tension. *J. Cell Sci.*, **116**, 3571–3581.

Oda, T., Kanai, Y., Okama, T., Yoshiura, K., Shimoyama, Y., Birchmeier, W., Sugimura, T., and Hirohashi, S. (1994). E-cadherin gene mutation in human gastric carcinoma cell lines. *Proc. Natl. Acad. Sci. USA*, **91**, 1858–1862.

Olz, D. and Schmeiser, C. (2010). *How Do Cells Move? Mathematical Modelling of Cytoskeleton Dynamics and Cell Migration.* Chapman and Hall/CRC Press.

Olz, D., Schmeiser, C., and Small, V. (2008). Modelling of the actin-cytoskeleton in symmetric lamellipodial fragments. *Cell Adhes. Migrat.*, **2**, 117–126.

Orford, C., Orford, C. C., and Byers, S. W. (1999). Exogenous expression of β-catenin regulates contact inhibition, anchorage-independent growth, anoikis, and radiation-induced cell cycle arrest. *J. Cell. Biol.*, **146**, 855–867.

Palecek, S. P., Loftus, J. C., Ginzberg, M. H., Lauffenburger, D. A., and Horwitz, A. (1997). Integrin-ligand binding properties govern cell migration speed through cell-substratum adhesiveness. *Nature*, **385**, 537–540.

Paszek, M. J., Zahir, N., Johnson, K. R., Lakins, J. N., Rozenberg, G. I., Gefen, A., Reinhart-King, C. A., Margulies, S. S., Dembo, M., Boettiger, D., Hammer, D. A., and Weaver, V. M. (2005). Tensional homeostasis and the malignant phenotype. *Cancer Cell*, **8**, 241–254.

Polyak, K., Kato, J., Solomon, M. J., Sherr, C. J., Massague, J., Roberts, J. M., and Koff, A. (1994). p27kip1, a cyclin-cdk inhibitor, links transforming growth factor-β and contact inhibition to cell cycle arrest. *Genes Devel.*, **8**, 9–22.

Preziosi, L. and Farina, A. (2001). On Darcy's law for growing porous media. *Int. J. Nonlinear Mech.*, **37**, 485–491.

Preziosi, L. and Joseph, D. D. (1987). Stokes' first problem for viscoelastic fluids. *J. Non-Newtonian Fluid Mech.*, **25**, 239–259.

Preziosi, L. and Tosin, A. (2009). Multiphase modeling of tumor growth and extracellular matrix interaction: Mathematical tools and applications. *J. Math. Biol.*, **58**, 625–656.

Preziosi, L. and Vitale, G. (2011). A multiphase model of tumour and tissue growth including cell adhesion and plastic re-organisation, *Math. Models Methods. Appl. Sci.*, doi:10.1142/50218202511005593

Preziosi, L., Ambrosi, D., and Verdier, C. (2010). An elasto-visco-plastic model of cell aggregates. *J. Theor. Biol.*, **262**, 35–47.

Rajagopal, K. and Tao, L. (1995). *Mechanics of Mixtures.* World Scientific.

Rao, I., Humphrey, J., and Rajagopal, K. (2003). Biological growth and remodeling: A uniaxial example with possible application to tendons and ligaments. *Comp. Mod. Eng. Sci.*, **4**, 439–495.

Risinger, J. I., Berchuck, A., Kohler, M. F., and Boyd, J. (1994). Mutation of e-cadherin gene in human gynecological cancers. *Nature Genetics*, **7**, 98–102.

Rodriguez, E. K., Hoger, A., and McCulloch, A. (1994). Stress dependent finite growth in soft elastic tissues. *J. Biomech.*, **27**, 455–467.

Saramito, P. (2007). A new constitutive equation for elastoviscoplastic fluid flows. *J. Non-Newtonian Fluid Mech.*, **145**, 1–14.

Saramito, P. (2008). A new elastoviscoplastic model based on the Herschel–Bulkley viscoplasticity. *J. Non-Newtonian Fluid Mech.*, submitted.

Snabre, P. and Mills, P. (1996). Rheology of weakly occulated suspensions of rigid particles. *J. Phys.*, **6**, 1811–1834.

StCroix, B., Sheehan, C., Rak, J. W., Florenes, V. A., Slingerland, J. M., and Kerbel, R. S. (1998). E-cadherin-dependent growth suppression is mediated by the cyclin-dependent kinase inhibitor p27(kip1). *J. Cell. Biol.*, **142**, 557–571.

Stockinger, A., Eger, A., Wolf, J., Beug, H., and Foisner, R. (2001). E-cadherin regulates cell growth by modulating proliferation-dependent β-catenin transcriptional activity. *J. Cell. Biol.*, **152**, 1185–1196.

Sun, M., Graham, J., Hegedus, B., Marga, F., Zhang, Y., Forgacs, G., and Grandbois, M. (2005). Multiple membrane tethers probed by atomic force microscopy. *Biophys. J.*, **89**, 4320–4329.

Takeuchi, J., Sobue, M., Sato, E., Shamoto, M., and Miura, K. (1976). Variation in glycosaminoglycan components of breast tumours. *Cancer Res.*, **36**, 2133–2139.

Tracqui, P. (2009). Biophysical models of tumour growth. *Rep. Prog. Phys.*, **72**(056701).

Tzukatani, Y., Suzuki, K., and Takahashi, K. (1997). Loss of density-dependent growth inhibition and dissociation of α-catenin from e-cadherin. *J. Cell. Physiol.*, **173**, 54–63.

Tzukita, S., Itoh, M., Nagafuchi, A., and Yonemura, S. (1993). Submembrane junctional plaque proteins include potential tumor suppressor molecules. *J. Cell. Biol.*, **123**, 1049–1053.

Uglow, E. B., Angelini, G. D., and George, S. J. (2000). Cadherin expression is altered during intimal thickening in humal sapphenous vein. *J. Submicrosc. Cytol. Pathol.*, **32**, C113–C119.

Warchol, M. E. (2002). Cell proliferation and n-cadherin interactions regulate cell proliferation in the sensory epithelia of the inner ear. *J. Neurosci.*, **22**, 2607–2616.

Wellman, P. S., Howe, R. D., Dalton, E., and Kern, K. A. (1999). Breast tissue stiffness to compression is correlated to histological diagnosis. Harvard Biorobotics Laboratory Technical Report,.

Winters, B., Shepard, S., and Foty, R. (2005). Biophysical measurement of brain tumor cohesion. *Int. J. Cancer*, **114**, 371–379.

Zhang, Y., Nojima, S., Nakayama, H., Yulan, J., and Enza, H. (2003). Characteristics of normal stromal components and their correlation with cancer occurrence in human prostate. *Oncol. Rep.*, **10**, 207–211.

Part IV

Biomembranes and bioshells

8
Microbial mechanics: The growth and form of filamentary microorganisms

Michael TABOR

Program in Applied Mathematics, University of Arizona

8.1 Introduction

The growth of filamentous cells such as fungi, actinomycetes, root hairs, and pollen tubes has been a long-standing topic of interest to microbiologists for several centuries. Although there are many fundamental differences in the structure, life cycles, size, and function of these diverse cells, their overall pattern of growth appears to share common morphological features that reflect an interplay between the anisotropy of the cell wall and the mechanical stresses experienced by the growing cell—all to be discussed in these lectures. To model this fascinating and fundamental process, one can call on two very different approaches: the geometric/mechanical approach, as espoused in D'Arcy Thompson's famous book *On Growth and Form* (Thompson 1992), and the molecular biological approach, as now taught in the outstanding textbook by Alberts *et al.* (2002), *The Molecular Biology of the Cell.* While these two works are philosophically a century apart, today, I believe, we have now the chance to marry these two views of the natural world; namely, we can use the tools of geometry and modern continuum mechanics reinforced by an understanding of the underlying molecular biology to describe many biological processes involving growth and form.

8.2 The size and nature of things

A wise man once said, "The noblest order in science is the order of magnitude". So, let us remind ourselves of some important characteristic scales. Organisms span an enormous range of sizes (Table 8.1), as do their genomes (Table 8.2). Materials come in many different strengths. And even a small scoop of dirt contains enormous numbers of different microorganisms.

Before we can develop mathematical models for the growth of microorganisms, we need to have a summary understanding of the differences in the underlying cell biology and cell wall structure of bacteria and fungi. Bacteria are prokaryotes and fungi are eukaryotes. From the evolutionary point of view, prokaryotes are much older than eukaryotes. They are also much simpler organisms: they do not have a membrane-bound nucleus; instead, their genetic material is located in a nucleoid, with additional satellite DNA found in the form of plasmids in the surrounding cytoplasm. By contrast,

Table 8.1 The size of things

Thing	Size (meters)
Atoms	10^{-10}
Small molecules	10^{-9}
Proteins	$10^{-9}-10^{-8}$
Viruses	$10^{-8}-10^{-7}$
Bacteria (prokaryotes)	$10^{-6}-10^{-5}$
Fungi (eukaryotes)	$10^{-5}-10^{-3}$
Human hair	100μm

Table 8.2 The genomes of things

Organism	Base pairs	Genes
Bacteria: *Bacillus subtilis*	4M	4000
Bacteria: *E. coli*	5M	4800
Fungi: *A. nidulans*	30M	9500
Fungi: *Neurospora crassa*	40M	10 000
Plants: *Arabidosis thaliana*	115M	37 000
Plants: rice	4×10^8	10 000
Mammals: dogs	2.9×10^9	19 000
Mammals: humans	3.3×10^9	20 000–25 000

eukaryotes have a membrane-bound nucleus surrounded by a very complex structure (the endoplasmic reticulum), and there are large numbers of additional membrane-bound organelles in the cytoplasm. Eukaroytes also have a cytoskeleton, consisting of actin filaments that provide a cellular "scaffold", and microtubules that perform many important functions and act as the "highways" to transport building materials to the cell wall. It is generally believed that prokaryotes do not have a cytoskeleton. Although in both fungi and bacteria the cell wall is separated from the cytoplasm by a plasma membrane, the cell wall structures are very different. Historically, bacteria have often been characterized as being either Gram-positive or Gram-negative (based on a dye test developed in the late 19th century). Gram-positive bacteria have a thick outer layer structure of peptidoglycan and teichoic acid. By contrast, Gram-negative cells have an outer membrane with a thin peptidoglycan layer between it and the plasma membrane. The outer membrane of Gram-negative bacteria tends to be resistant to antibiotics such as penicillin, and these bacteria tend to be the ones responsible for undesirable bacterial infections. The cell wall structure of fungi is quite different. A key component is chitin, a long-chain polymer $(C_8H_{13}O_5N)_n$. The chitin molecules, which are rod-like, are layered outside the plasma membrane and, in turn

Table 8.3 The strength of things

Material	Elastic modulus (GPa)
Rubber (small strain)	0.01–0.1
Nylon	2–4
Oak (along the grain)	11
Brass	100–125
Steel	200
Diamond	1220
Fungal cell wall	100 MPa

Table 8.4 Getting dirty: what you can find in 1 cm^3 of dirt

Organism	Amount
Nematodes	50
Algae	6×10^4
Fungi	10^5
Actinomycetes	3×10^6
Bacteria	2.5×10^7

are surrounded by a network of β(1-3)-glucans and β(1-6)-glucans, as well as other molecules (mannoproteins).

8.3 Hyphal growth

There are many websites displaying the growth of filamentary fungi and bacteria, including

http://www.fungalcell.org/
http://129.215.156.68/Movies/hypha.htm
http://www.gerosteinberg.com/movies.php

As can be seen at these websites, growing filamentary fungi display (as do filamentary bacteria) common morphological features with respect to their growth dynamics, (Fig. 8.1). The growth is concentrated at the tip of the filament—this is often referred to as "apical growth"—and in the case of fungi, the filaments are usually referred to as hyphae. For fungi, the hyphae might vary in diameter from 5 to 10 μm and can grow to up to millimeters in length; bacterial filaments are typically an order of magnitude smaller, typically 0.5–1 μm in diameter and growing to 100 μm in length. Hyphal growth rates of fungi have been recorded in the range 14–34 μm/min, while bacterial filaments grow at a rate of the order of 1 μm/min. We first briefly review the characteristics of the various filamentous organisms whose hyphal growth we aim to model.

8.3.1 Actinomycetes

Despite the relative simplicity of their cellular structure, these prokaryotic organisms are genomically quite sophisticated, and the chromosome of a typical streptomycete, such as *Streptomyces coelicolor* A3(2), is approximately 8 Mbase pairs long. Their cell wall structure is dominated by a characteristic network of peptidoglycan polymers (Chater and Hopwood 1993). The streptomycetes are of particular interest, since they are a natural source of pharmaceutically important antibiotics.

In a typical growth cycle, the streptomycete spores bud into long filamentary hyphae, which grow in and on a nutrient source. The growing hyphae eventually

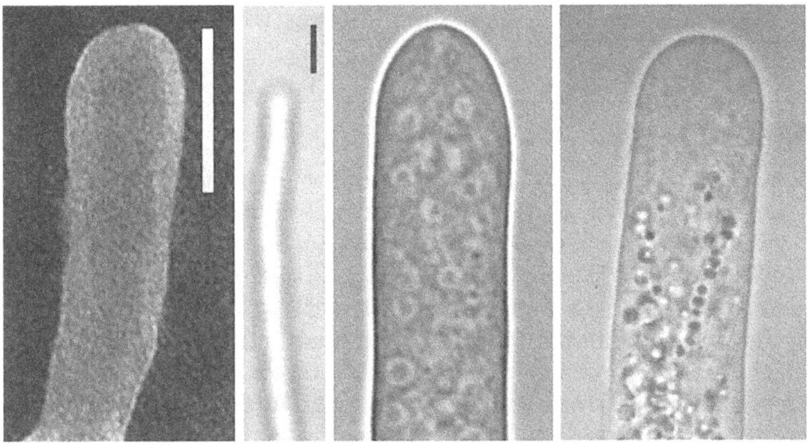

Fig. 8.1 Pictures of growing tips. From left to right: actinomycetes from a spore (bar is 1 μm) (source: Society for Actinomycetes, Japan); streptomyces A3(2) (bar is 1 μm); allomyces; lily root hair (typical diameter 15–20 μm) (images courtesy of J. Dumais). Actinomyces and streptomyces are bacteria, and allomyces is a fungus.

undergo branching, leading to the formation of a dense mycelium. In this "vegetative" phase, apical growth is energetically more efficient than growth along the entire filament, since the latter would result in more friction with the nutrient medium. A typical filament of *Streptomyces coelicolor* in this phase is less than 1 μm in diameter and can grow to lengths of 50–100 μm. The vegetative phase, is usually followed by an aerial growth phase, in which the hyphae grow out of the mycelium, sometimes deforming into helical structures, and ultimately break up into spores—which then begin the growth cycle again. We note that in the aerial phase, hyphal growth is no longer apical.

The apical growth of streptomycetes in the vegetative phase has been demonstrated in a variety of studies tracking the presence of tritiated GlcNAc, a precursor of peptidoglycan. For example, Gray *et al.* (1990) used this technique in their study of *Streptomyces coelicolor* A3(2) to demonstrate that the bulk of the labeled material was incorporated close to the hyphal apex. The interplay between turgor pressure and cell wall rigidity has also been demonstrated by the action of lysing agents which degrade the cell wall. In their studies of *S. coelicolor*, Gray *et al.* (1990) showed that *b*-lactam antibiotics caused apical swelling of the hyphae leading, in some cases, to lysis. These and other experimental studies of hyphal extension in *S. coelicolor* show that the main factors determining growth in the vegetative phase are (i) turgor pressure, (ii) addition of newly synthesized wall-building material in the apical extension zone, and (iii) the rigidification of the distal cell wall. Thus the overall picture of the growth mechanism consistent with these observations is a continuous complex process in which wall-building materials are transported to and incorporated into the tip, which is stretched by the turgor pressure, and as the tip is continually stretched and rebuilt, the more remote portions of the hyphal wall become (relatively) rigid (Prosser and Tough 1991).

8.3.2 Fungi

Filamentous fungi are typically at least an order of magnitude bigger than actinomycetes and have a much more complex internal cellular structure. Their highly complex and multilayered cell wall typically consists of a network of chitin with cross-linking *b*-glucan chains (Wessels 1990). Despite all these differences, there are still striking similarities with the overall growth dynamics of other filamentous cells—especially that of apical hyphal growth. For example, studies of *Neurospora crassa* showed that the incorporation of tritiated wall precursors decreased rapidly from the hyphal tip (Gooday and Trinci 1980). For most fungi, turgor pressure also appears to be a factor in the growth process, but there is some experimental evidence indicating that oomycetes continue to grow even when they do not maintain turgor (Money 1997). But, for all fungi, it seems clear that the cytoskeleton plays a central role in the growth process and in transporting wall-building materials to the tip, although a complete understanding of the different roles played by actin filaments and microtubules in the tip growth process is still evolving. It has also been proposed that a vesicle supply center, or spitzenkorper, organizes and/or directs the traffic of vesicles containing the wall-building materials to the apical region (Grove and Bracker 1970). The website of Steinberg's laboratory cited above has some remarkable experimental results displaying these ideas, which are discussed in (Steinberg 2007).

A striking result concerns normal cell wall expansion in filamentous fungi. By tracking carbon particle markers on the tip of *Rhizoctonia solani*, Bartnicki-Garcia *et al.* (2000) were able to show that the markers moved in a direction perpendicular to the surface of the hyphal apex, thereby confirming an old prediction of Reinhardt (Reinhardt 1892). Their results also support the view that turgor pressure is (still) a major driving force in the hyphal growth process.

However, for all these complexities and subtleties, the overall picture parallels that of hyphal growth in filamentous bacteria: namely, apical growth driven by a complex process involving transport of wall-building materials and a tip-straining mechanism.

8.3.3 Root hairs

Root hairs have been one of the most enduring topics of study in microbiology, with scientific publications dating back almost two hundred and fifty years; for example, a study of root hairs with excellent illustrations can be found in Duhamel du Monceau's dissertation published in 1758 (Duhamel du Monceau 1758), in which he describes the process of tip growth. As with filamentary fungi, it was established long ago that the cell wall expansion is concentrated at the tip. More refined microscopic studies have also shown that the expansion is both inhomogeneous (the rate of expansion being maximal around the tip) and anisotropic (the meridional expansion being greater than the circumferential expansion), the latter observation suggesting a varying stiffness of the cell wall, perhaps due to a corresponding variation in microfibril cross-linking.

Experimental studies of the growth of *M. truncatula* root hairs (Shaw *et al.* 2000) support the notion of normal motion of material points on the growing tip.

8.3.4 Modeling hyphal growth

Hyphal-growth modeling involves several different possible levels of description:

- the *biomechanical level*, which aims to describe the growth and morphology based on mechanical principles;
- the *biophysical level*, which aims to translate the structural details of the cell wall into macroscopic material properties that must be incorporated into the mechanical models;
- the *biomolecular level*, which aims to understand the molecular basis for the detailed structure and expression of the cell.

An alternative approach is to describe the growth at the *geometrical level*, in which the increase in wall area of an advancing tip is balanced with the rate of incorporation of wall-building material (da Riva Ricci and Kendrick 1972; Trinci and Saunders 1977; Saunders and Trinci 1970). While this approach can sometimes yield formulae for idealized tip shapes (Bartnicki-Garcia *et al.* 1989; Goriely *et al.* 2005), it cannot capture the fundamental interplay between the mechanical stresses and the material response of the cell wall. By contrast, the mechanical models can capture the tip growth process through a proper description of the physical stresses and strains experienced by the growing hypha.

At the mechanical level, several different approaches have been proposed. The use of (linear) plasticity theory yields irreversible extension of the cell wall—which is taken as a manifestation of growth. This approach has been used to model the growth of plant cells (Lockhart 1965) and root hairs (Dumais *et al.* 2006). An alternative approach, discussed here, is to use the techniques of nonlinear elasticity theory, which enables one to follow large deformations of the cell, and to model the actual growth mechanism by an incremental process in which mass is added to the elastically strained cell wall. Overall, we will show that it is possible to formulate quite general mechanical models based on nonlinear elasticity theory that, when combined with an appropriate growth mechanism, can describe the morphology and mechanics of a wide variety of growing filamentous cells. Our results suggest that this "morphoelastic" description is capable of describing many of the observed features of hyphal growth, and that the addition of biophysical information about the cell wall structure can give further insights into the growth process and, given sufficient information about the cell wall structure, help distinguish details of the growth in different filamentous cells. However, in modeling any biological process, especially a complex one, it is worth remembering the old adage, "All models are wrong, but some may be useful".

8.4 Geometrical models

Although geometrical models lack mechanical or biological content, they provide an intuitively appealing view of idealized tip growth that to some extent might be thought of as "universal", since the same ideas can be used to model the the propagation of two-dimensional interfaces, such as those seen in fluid dynamics and solidification—the essential common feature being that of curvature-driven growth. Here we summarize the key ideas (the details of this approach have been given by Goriely *et al.* 2005). The hyphal tip (or the interface of interest) is modeled geometrically as parameterized space curves $\mathbf{X}(\sigma, t)$, whose kinematics is governed by dynamical rules reflecting the underlying process (physical or biological) driving the growth. The motion of such a curve is studied by decomposing its velocity into tangential and normal components $\mathbf{t}$ and $\mathbf{n}$ namely

$$\frac{\partial}{\partial t}\mathbf{X}(\sigma, t) = W\mathbf{t} + U\mathbf{n}, \tag{8.1}$$

where W and U denote, respectively, the tangential and normal velocities at each material point σ along the curve. These velocities are typically taken to be functions of the curve geometry, namely its curvature κ and curvature derivatives, and plausible biological/physical arguments can be used to determine specific forms of U and W. By considering the space curve as a representation of a (growing) cell wall, this formalism is quite easily adapted to biological problems such as hyphal growth.

The hyphal filament is modeled as an axisymmetric surface of revolution centered about the y-axis, thereby reducing the description to that of a plane curve lying in the (x, y) plane pointing in the y-direction. In keeping with Reinhardt's century-old hypothesis of normal growth—now supported by experimental observations—we set $W = 0$ and, furthermore, the curve representing the cell profile is assumed to propagate with a steady velocity U_0 in the y-direction. The curve, parameterized in terms of the arc length s, is taken to be of the form

$$\mathbf{X}(s, t) = (x(s, t), y(s, t) + U_0 t) \equiv (f(s, t), g(s, t) + U_0 t), \tag{8.2}$$

with $s = s(\sigma, t)$, where σ denotes a material coordinate. In what follows, we will assume that there is no explicit time dependence in the as yet unspecified functions f and g. The unit tangent and outward normal vectors are simply

$$\mathbf{t} = \frac{\partial \mathbf{X}}{\partial s} = (f_s, g_s), \quad \mathbf{n} = (-g_s, f_s), \tag{8.3}$$

and the curve can be reconstructed by integrating the differential equations

$$\frac{\partial x}{\partial s} = f_s = \cos\theta, \quad \frac{\partial y}{\partial s} = g_s = -\sin\theta, \tag{8.4}$$

where θ is the "orientation" angle (see Fig. 8.2). There are two principal curvatures: the longitudinal (or meridional) curvature κ, and the latitudinal (or azimuthal) curvature k, where

$$\kappa = \theta_s, \qquad k = \frac{\sin\theta}{f}.$$

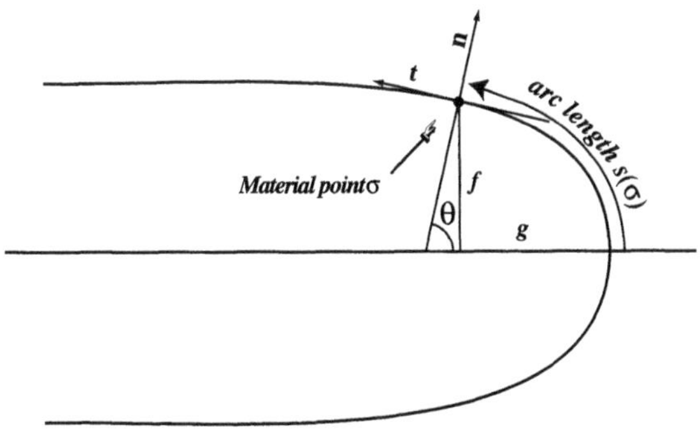

Fig. 8.2 Geometry of axisymmetric finger.

Of central importance is the quantity

$$\lambda(s,t) = \frac{ds}{d\sigma}, \tag{8.5}$$

which measures the stretching of material points along the curve. Its time derivative, $\dot{\lambda}$, gives the rate of stretching of the curve. It is straightforward to show that the velocity of the curve at any material point is

$$\frac{\partial \mathbf{X}(\sigma,t)}{\partial t} = (\dot{s} f_s, \dot{s} g_s + U_0), \tag{8.6}$$

where $\dot{}$ denotes the time derivative. The normal and tangential velocity components are thus

$$U = \mathbf{n}.\frac{\partial \mathbf{X}}{\partial t} = U_0 f, \quad W = \mathbf{t}.\frac{\partial \mathbf{X}}{\partial t} = \dot{s} + U_0 g_s. \tag{8.7}$$

We note that the normal velocity at a material point is the same as that for a fixed value of s, whereas the tangential velocity at a material point is the sum of the tangential velocity at a fixed value of s, namely $U_0 g_s$, and the rate at which s changes at that point. The *normal-growth* model $W = 0$ leads to

$$\dot{s} = -U_0 g_s, \tag{8.8}$$

and a little analysis (Goriely *et al.* 2005) shows that

$$\frac{\dot{\lambda}}{\lambda} = -U_0 g_s = U_0 \kappa f_s = \kappa U_0 \cos\theta \equiv \kappa U, \tag{8.9}$$

where $U = U_0 \cos\theta$ is the normal velocity at each point along the curve. Despite its apparent simplicity, this is a useful formula: given experimental images of a propagating tip profile that are well enough resolved to estimate its curvature and tip velocity U_0, it can be used directly to estimate the rate of wall stretching at each point along the tip.

To proceed further with the modeling, one needs a plausible characterization of the wall-building process itself. Here, we propose that it is determined by the rate at which material is incorporated into the tip and, as demonstrated by Goriely *et al.* (2005), one can show that the rate of growth of a local area element ΔA around a three-dimensional axisymmetric tip is given by

$$\frac{1}{\Delta A}\frac{d}{dt}\Delta A = \left(\dot{s}\frac{f_s}{f} + \frac{\dot{\lambda}}{\lambda}\right), \tag{8.10}$$

where the first term on the right represents circumferential expansion and the second term represents longitudinal expansion. We make the ansatz

$$\frac{1}{\Delta A}\frac{d}{dt}\Delta A = N_0 n(s,t), \tag{8.11}$$

where $n(s,t)$ is some dimensionless geometry-dependent function that reflects the details of the wall-building process, and N_0, which has dimensions T^{-1}, sets the fundamental timescale of that process.

It is now possible to determine different tip shapes by choosing different rate functions $n(s,t)$. Given the widely accepted "soft-spot" hypothesis of Reinhardt that growth is concentrated at the tip, where the cell membrane is "softest", it is natural to make n proportional to the tip curvature. In fact, one can opt for either a "two-dimensional" or a "three-dimensional" model. In the case of the former, one considers only longitudinal stretching and examines a simplified version of eqn (8.10), namely

$$\frac{\dot{\lambda}}{\lambda} = N_0 n(s,t), \tag{8.12}$$

whereas in the three-dimensional model one uses the full equation. For the two-dimensional model, the choice

$$n(s,t) = \kappa^2(s)/\kappa_0^2,$$

where $1/\kappa_0$ is a characteristic length scale associated with the tip, results in the tip shape

$$y(x) = \frac{1}{\alpha\kappa_0^2}\log(\cos(\alpha\kappa_0^2 x)), \tag{8.13}$$

where $\alpha = N_0/U_0$, which has dimensions L^{-1}, identifies a natural length scale associated with the wall-building process (albeit at the level of the curve dynamical model). In fact, this particular form of tip shape is a reasonable match to observed hyphal tips. It has also been derived in the literature several times as an idealized form of a single dendrite used in studies of solidification (Pelcé 1988). Furthermore, if we set $\kappa_0 = \kappa(0)$, namely the curvature at the hyphal apex, one can show that $\alpha = 1/\kappa(0)$. Thus the tip curvature is set by the fundamental parameter $\alpha = N_0/U_0$.

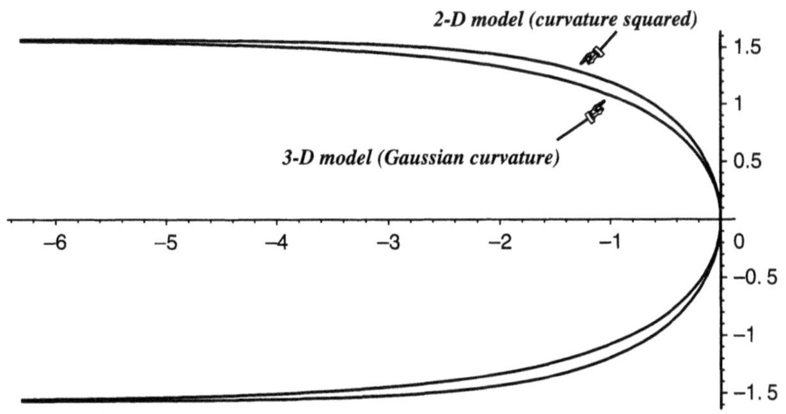

Fig. 8.3 The hyphal profiles obtained with the two-dimensional and three-dimensional geometrical growth models.

In the case of the three-dimensional model, one can propose that

$$n(s,t) = K_{\mathrm{g}}/K_0, \tag{8.14}$$

where K_{g} is the Gaussian curvature of the tip. Recalling that the Gaussian curvature is the product of the principal curvatures, it is easy to see that for our problem, $K_{\mathrm{g}} = \kappa k$. Here K_0 represents a local area scale to ensure that $n(s,t)$ is dimensionless. A little analysis results in the remarkably simple (and possibly novel) tip shape

$$y(x) = 4\alpha \log\left(1 - \frac{x^2}{16\alpha^2}\right), \tag{8.15}$$

where, on setting $K_0 = K_{\mathrm{g}}(0)$, it is easy to show that

$$K_0 = \frac{1}{4\alpha^2}. \tag{8.16}$$

In fact, there is not a lot of difference between the two-and three-dimensional tip profiles. A comparison of the two- and three-dimensional solutions (Fig. 8.3) shows that the three-dimensional model has a slightly sharper tip than the two-dimensional model.

8.5 Nonlinear elastic models of hyphal growth

Here, I will describe a general model in which the cell wall is represented as a growing nonlinear elastic shell that can also incorporate the effects of tangential stresses on the cell wall. By doing so, we are able to represent the effects of friction between a growing hypha and the external environment. As will be described, the new formulation and calculations based on it have enabled us to study the changes in tip shape resulting from frictional effects and to give a rather striking demonstration of the normal-growth hypothesis. The model represents a generalization of our earlier work on this problem (Goriely and Tabor 2003*a*). However, before proceeding, a few comments about elastic sheets are in order.

8.5.1 Two-dimensional elasticity

Modeling cell walls or other biological membranes as some form of two-dimensional elastic sheet is a staple of contemporary biophysics. At first glance, one might think that two-dimensional elasticity would be somehow "easier" than three-dimensional elasticity—but in fact it is more difficult. The problem is that there is not really such a thing as a purely two-dimensional elastic body, since even a thin sheet has a finite, albeit small, thickness. It is also worth pausing to recall the uses and abuses of terminology. In classical elasticity, a thin *plate* is, well, a thin plate: a flat body whose thickness is much smaller than its dimensions in the other two directions. Here, as you will find elsewhere, the term elastic *sheet* is used loosely to mean a thin plate. A *shell* is an elastic sheet whose undeformed configuration is curved. A general deformation of a sheet or a shell supports both stretching and bending, but in the case of a deformation in which bending stresses can be neglected, the sheet (or shell) is referred to as an elastic *membrane*. The fundamental problem in formulating the mechanics of sheets and shells is this: how well can we represent a real elastic sheet, which is physically a three-dimensional body, as a two-dimensional body? In the end, certain approximations have to be made. There are two important concepts in two-dimensional elasticity: *plane strain* and *plane stress*.

Plane strain is a concept that can be applied to three-dimensional bodies such as long cylinders or pipes whose lateral loads do not vary appreciably in the longitudinal direction, for example a pipe carrying a fluid under pressure. Although there will clearly be lateral displacements in the pipe, it is reasonable to assume that there will not be any significant strain in the longitudinal direction (taken here to be along the z-axis). In the standard notation of linear elasticity, a "state of plane strain" corresponds to the following conditions on the strain tensor ϵ_{ij}:

$$\epsilon_{zz} = \epsilon_{zx} = \epsilon_{zy} = 0.$$

However, a state of plane strain does not mean that there is no longitudinal stress, and, at the level of linear elasticity theory, the components of the stress tensor σ_{ij} satisfy

$$\sigma_{zz} = \nu(\sigma_{xx} + \sigma_{yy}), \quad \sigma_{zx} = \sigma_{zy} = 0,$$

where ν is the Poisson ratio. Plane stress is a concept that applies to thin sheets and shells. Here, the idea is that the loading is such that all the stress is in the plane of the body, but that there is negligible stress through the body, i.e. in the normal direction. A standard example of this situation is a horizontal plate spinning about the vertical axis. In the framework of linear elasticity, a "state of plane stress" corresponds to

$$\sigma_{zz} = \sigma_{zx} = \sigma_{zy} = 0.$$

However, a state of plane stress does not mean that there is no strain through the body, and, at the level of linear elasticity theory, the components of the strain tensor satisfy

$$\epsilon_{zz} = \nu(\epsilon_{xx} + \epsilon_{yy}), \quad \epsilon_{zx} = \epsilon_{zy} = 0.$$

Thus as a sheet stretches laterally, it becomes thinner. However, the plane stress assumption is an approximation and there is a large body of literature (and much debate and argument) about how to derive a rigorous two-dimensional theory as the limit of three-dimensional elasticity. The central idea is to invoke the concept of a "middle surface" through the body across which the stresses are averaged. Thus, for a sheet of thickness h, one considers the averaged stresses given by

$$\bar{\sigma}_{ij} = \frac{1}{h}\int_{-h/2}^{h/2} \sigma_{ij}\, dz.$$

At the end of the day, it is possible to show that for thin sheets the plane stress assumption is a good approximation except in regions whose distances from the sheet boundaries are of the order of the plate thickness.

The plane stress approximation can be carried over to nonlinear elasticity in the following way. Consider an incompressible elastic sheet that deforms along the principal directions, say a rectangular sheet lying in the x, y plane being pulled along the x-and y-axes. Following Professor Ogden's lecture (not published in this volume), you will recognize the basic relationship

$$\sigma_{ii} = \lambda_i \frac{\partial W}{\partial \lambda_i} - p, \tag{8.17}$$

where the λ_i are the principal stretches, $W = W(\lambda_1, \lambda_2, \lambda_3)$ is the elastic free energy (i.e. we are assuming a hyperelastic material), and p is the hydrostatic pressure. Here we will take the subscripts $(1, 2, 3)$ to correspond to the directions (x, y, z). For an incompressible material, the stretches satisfy the condition

$$\lambda_1 \lambda_2 \lambda_3 = 1.$$

The plane stress approximation corresponds to $\sigma_{33} = 0$, and it follows from eqn (8.17) for $i = 3$ that

$$p = \lambda_3 \frac{\partial W}{\partial \lambda_3},$$

and hence

$$\sigma_{11} = \lambda_1 \frac{\partial W}{\partial \lambda_1} - \lambda_3 \frac{\partial W}{\partial \lambda_3},$$

$$\sigma_{22} = \lambda_2 \frac{\partial W}{\partial \lambda_2} - \lambda_3 \frac{\partial W}{\partial \lambda_3}.$$

If the original thickness of the sheet is h, then the thickness after deformation is $\lambda_3 h$. For sheets it is often convenient to work with the *tensions* which have the dimensions of force per unit length (as opposed to the stresses, which have units of force per unit area), and one may show that the lateral tensions in the principal directions are given by

$$t_i = \lambda_3 h_0 \sigma_{ii}, \qquad i = 1, 2.$$

With these preliminary ideas in mind, it is now possible to set up a nonlinear elastic formulation for hyphal extension and growth.

8.5.2 Hyphal geometry

Elastic-shell theory has long been used to model red blood cells (Evans and Skalak 1980; Secomb and Gross 1983) and arteries and, more recently, to describe hyphal growth (Goriely and Tabor 2003*a*,*b*) and the mechanical action of the fungus *Magnaporthe grisea* (Goriely and Tabor 2006). In all these models, a basic but reasonable assumption of an axisymmetric cell shape is made. This enables one to give a concise formulation of the shell geometry and deformation.

In our models, the hyphal wall is represented as an elastic shell, whose geometry is shown in Fig. 8.4. The shape, which is assumed to be axisymmetric about the z-axis, is described in terms of $s(\sigma)$, which represents the arc-length distance of a material point σ from the tip, and $r(\sigma)$, which represents the distance of the point from the z-axis. For a given initial shape $(\sigma, \rho(\sigma))$, the deformation is described in terms of of the variables

$$\alpha_s = \frac{ds}{d\sigma}, \qquad \alpha_\varphi = \frac{r(\sigma)}{\rho(\sigma)}. \tag{8.18}$$

The variables α_s and α_φ measure the meridional and azimuthal stretches, respectively, of the shell. A third deformation variable, α_3, measures changes in the normal thickness

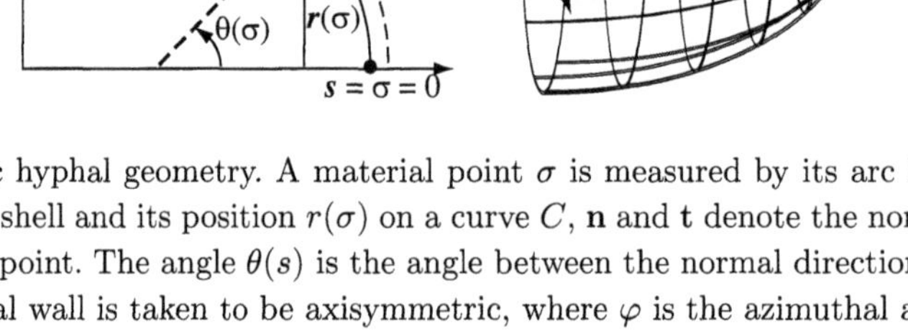

Fig. 8.4 Basic hyphal geometry. A material point σ is measured by its arc length $s(\sigma)$ from the apex of the shell and its position $r(\sigma)$ on a curve C, **n** and **t** denote the normal and tangent vectors at that point. The angle $\theta(s)$ is the angle between the normal direction and the hyphal axis. The hyphal wall is taken to be axisymmetric, where φ is the azimuthal angle.

of the shell, and for an incompressible shell it is simply related to α_s and α_φ through the incompressibility condition $\alpha_s \alpha_\varphi \alpha_3 = 1$. The geometric variables satisfy the equations

$$\frac{dr}{ds} = \cos(\theta), \quad \frac{dz}{ds} = \sin(\theta), \tag{8.19}$$

and the principal curvatures of the shell are given by

$$\kappa_s = \frac{d\theta}{ds}, \quad \kappa_\varphi = \frac{\sin\theta}{r}. \tag{8.20}$$

Apart from some minor notational differences, this is the exactly the same setup as is used in the geometrical model described in Section 8.4.

8.5.3 Mechanics

The equations of mechanical equilibrium, which here includes bending moments, are (Evans and Skalak 1980)

$$\frac{1}{r}\frac{d(rq_s)}{ds} = q_n - (\kappa_s t_s + \kappa_\varphi t_\varphi), \tag{8.21}$$

$$\frac{1}{r}\frac{d(rt_s)}{ds} = \frac{\cos\theta}{r} t_\varphi + \kappa_s q_s - \tau_s, \tag{8.22}$$

$$\frac{1}{r}\frac{d(rm_s)}{ds} = \frac{\cos\theta}{r} m_\varphi + q_s, \tag{8.23}$$

where t_s and t_φ are, respectively, the meridional and azimuthal stresses; m_s and m_φ are the bending moments; and q_s is the shear stress normal to the surface. In eqn (8.21), which represents the balance of normal stresses, q_n represents the total *normal* stress exerted on the shell. For a pressurized shell, $q_n = \Delta P$, i.e. the pressure difference across the shell—this is referred to as the turgor pressure in the biological context. If there are additional normal forces, represented by some function τ_n, then $q_n = \Delta P + \tau_n$.

It is important to note that at the mechanical level all that the shell "feels" is the total normal force acting on it, not the origins of that force.This simple observation explains why it is possible to have a common mechanical model for hyphal growth for both fungi and bacteria. In eqn (8.22), which represents the balance of tangential stresses, τ_s is the external tangential shear stress acting on the shell. Here it will be used to represent the friction between the growing hypha and its environment.

8.5.4 Constitutive relationship

Constitutive relations provide the connection between the mechanics and the geometry. These relations are developed through the introduction of an elastic free-energy function, specified as an energy per unit volume,

$$W = W(I_1, I_2, I_3), \tag{8.24}$$

where I_1, I_2, I_3 are the strain invariants

$$I_1 = \alpha_s^2 + \alpha_\varphi^2 + \alpha_3^2, \tag{8.25}$$

$$I_2 = \alpha_s^2\alpha_\varphi^2 + \alpha_s^2\alpha_3^2 + \alpha_\varphi^2\alpha_3^2, \tag{8.26}$$

$$I_3 = \alpha_s\alpha_\varphi\alpha_3. \tag{8.27}$$

For incompressible materials, $\alpha_3 = 1/\alpha_s\alpha_\varphi$. Thus $I_3 = 1$, and W depends only on I_1 and I_2. The stresses are given by the relations

$$t_s = 2h\alpha_3(\alpha_s^2 - \alpha_3^2)\left(\frac{\partial W}{\partial I_1} + \alpha_\varphi^2\frac{\partial W}{\partial I_2}\right), \tag{8.28}$$

$$t_\varphi = 2h\alpha_3(\alpha_\varphi^2 - \alpha_3^2)\left(\frac{\partial W}{\partial I_1} + \alpha_s^2\frac{\partial W}{\partial I_2}\right), \tag{8.29}$$

where h is the shell thickness, and the factor $h\alpha_3$ represents the change in wall thickness in the current (i.e. stressed) configuration (Rivlin and Saunders 1951). We comment that the thickness prefactor means that t_s and t_ϕ have units of force per unit length, i.e. they have the dimensions of tensions rather than stresses. The choice of W depends on the problem being considered. For elastomers, a common choice is the Mooney–Rivlin model

$$W = C_1(I_1 - 3) + C_2(I_2 - 3), \tag{8.30}$$

where C_1 and C_2 are certain elastic parameters. When $C_2 = 0$, eqn(8.30) reduces to the so-called neo-Hookean model, and for small deformations, C_1 is related to Young's modulus E by $E = 6C_1$. Although the Mooney–Rivlin and neo-Hookean free energies are linear in the invariants, they are still capable of describing large deformations. A variety of nonlinear material responses can be captured by considering more general functional forms of W; for example, for soft tissue, the Fung energy (Fung 1993)

$$W = \frac{C_1}{\mu}\left(e^{\mu(I_1-3)} - 1\right) \tag{8.31}$$

is often used. Given any model, we write the constitutive relationships (8.28) and (8.29) in the form

$$t_s = Af_s(\alpha_s, \alpha_\varphi), \tag{8.32}$$

$$t_\varphi = Af_\varphi(\alpha_s, \alpha_\varphi), \tag{8.33}$$

where f_s, f_ϕ are dimensionless functions and A provides the dimensional factor appropriate to the scaling of the equilibrium equations. For the neo-Hookean and Mooney–Rivlin energies, $A = 2C_1h$.

Finally, we need to specify a constitutive relationship for the bending moments. The bending moments are assumed to be isotropic and proportional to the change in the surface's mean curvature, i.e.

$$m_\varphi = m_s = B(\kappa_s + \kappa_\varphi - K_0), \tag{8.34}$$

where K_0 is the initial mean curvature and B is the bending modulus (Evans and Skalak 1980).

8.5.5 The complete shell equations and the membrane limit

The geometric and mechanical equations can be combined to give a closed system. It is convenient to express all the derivatives in terms of the material coordinate σ, leading to

$$\frac{dz}{d\sigma} = \alpha_s \sin(\theta), \tag{8.35}$$

$$\frac{dr}{d\sigma} = \alpha_s \cos(\theta), \tag{8.36}$$

$$\frac{d\theta}{d\sigma} = \alpha_s \kappa_s, \tag{8.37}$$

$$\frac{d\kappa_s}{d\sigma} = \alpha_s \left[\frac{\cos\theta}{r}\left(\frac{\sin\theta}{r} - \kappa_s\right) + \frac{q_s}{B}\right], \tag{8.38}$$

$$\frac{dt_s}{ds} = \alpha_s A \left[\frac{\cos\theta}{r}(f_\varphi - f_s) + \kappa_s \frac{q_s}{A} - \frac{\tau_s}{A}\right], \tag{8.39}$$

$$\frac{dq_s}{d\sigma} = \alpha_s A \left[\frac{q_n}{A} - \kappa_s f_s - \frac{\sin\theta}{r} f_\varphi - \frac{q_s}{A}\frac{\cos\theta}{r}\right], \tag{8.40}$$

where eqn(8.38) is obtained from eqn(8.23) using the constitutive relation (8.34), and eqn (8.20) is used to express κ_φ in terms of r and θ. In eqns (8.39) and (8.40), t_s and t_φ are expressed in terms of α_s and α_φ through the scaled constitutive relations (8.32) and (8.33), and eqn (8.39) is converted into a differential equation for α_s by eliminating α_φ through the relation $\alpha_\varphi = r/\rho$.

The six ordinary differential equations (8.35)–(8.40), together with the relationships (8.32) and (8.33) and $\alpha_\varphi = r/\rho$, form a closed system for the variables $(z, r, \theta, \kappa_s, \alpha_s, q_s)$ that can be solved for a given initial profile $\rho(\sigma)$, elastic parameters A, B, prescribed normal and tangential stresses q_n and τ_n, and appropriate boundary conditions.

When bending moments can be neglected, the shell no longer supports an out-of-plane shear force, i.e. $q_s = 0$, and eqn (8.40) reduces to

$$\frac{q_n}{A} = \kappa_s f_s + \kappa_\varphi f_\varphi, \tag{8.41}$$

which is just a generalized form of the Young–Laplace law. The system of shell equations then simplifies to

$$\frac{dz}{d\sigma} = -\alpha_s \sin(\theta), \tag{8.42}$$

$$\frac{dr}{d\sigma} = \alpha_s \cos(\theta), \tag{8.43}$$

$$\frac{d\theta}{d\sigma} = \alpha_s \kappa_s, \tag{8.44}$$

$$\frac{d\kappa_s}{d\sigma} = \alpha_s \left[\frac{\cos\theta}{r}\left(\frac{\sin\theta}{r} - \kappa_s\right)\right], \tag{8.45}$$

$$\frac{dt_s}{ds} = \alpha_s A \left[\frac{\cos\theta}{r}(f_\varphi - f_s) - \frac{\tau_s}{A}\right]. \tag{8.46}$$

As before, using the constitutive relationship $t_s = Af_s(\alpha_s, r/\rho)$ results in a closed system of five equations for $(z, r, \theta, \kappa_s, \alpha_s)$ with an initial profile $\rho(s)$, a tangential shear τ_s, and boundary values.

It is useful to introduce the following dimensionless number for the problem:

$$\xi = \frac{P_{\text{eff}} w}{3A}, \tag{8.47}$$

where w is a characteristic length (typically, the width of the tip), P_{eff} is a measure of the normal force per unit area acting on the walls, and A characterizes the elastic properties of the wall. The number ξ gives a measure of the magnitude of the deformation, i.e. tips under different normal loads and with different elastic properties will experience the same magnitude of deformation from their initial unstressed state as measured in terms of their tip widths. That is, high values of ξ, representing either a soft material or a high turgor pressure, result in large elastic deformations, whereas low values of ξ, corresponding to either a rigid material or a low turgor pressure, result in small elastic deformations. For a neo-Hookean material, we can relate A to the Young's modulus $E = 3A/h$, where h is the thickness of the shell or membrane, so that

$$\xi = \frac{P_{\text{eff}} w}{hE}. \tag{8.48}$$

Detailed studies of the mechanical behavior of microorganisms are difficult, and only a few estimates have been proposed (see Table 8.5).

Table 8.5 Experimental estimates of hyphal parameters. w, width (μm;, h, thickness (mm); P, pressure (MPa); G, growth rate (μm/min); E, Young's modulus (MPa)

Organism	w	h	P	E	ξ	G	Reference
A. nidulans, mature	3	46	1.4	115	0.8	0.2–0.5	Ma *et al.* (2005)
A. nidulans, tip	3	46	1.4	75	1.2	0.2–0.5	Ma *et al.* (2005)
M. gryphiswaldenese	0.5	1	0.1	30	0.003		Arnoldi *et al.* (2000)

8.5.6 Modeling cell wall properties and the soft-spot hypothesis

A fundamental component of any model of apical growth is the representation of the cell wall anisotropy. This will depend on the details of the cell wall architecture: for example, the degree of peptidoglycan cross-linking in actinomycetes or the orientation of the cellulose microfibrils in root hairs (Dumais *et al.* 2006). Such biophysical detail is difficult to obtain experimentally, and here we take a simpler phenomenological approach that captures the essence of the soft-spot hypothesis by prescribing that the elastic response of the cell wall varies along the meridional direction such that the distal regions of the hypha are practically rigid, while the regions near the tip have a much smaller elastic modulus, corresponding to a softer material. At this level of modeling, a simple scaling argument shows that a softer material response can be interpreted as an increase in turgor pressure, and vice versa.

This is the basis for the "effective-pressure" model used in our earlier study of actinomycete growth (Goriely and Tabor 2003*a*), and we continue to use it here. Thus we set

$$q_n^{(\text{eff})} = \frac{Q}{2}\left[1 - \tanh\left(\frac{\sigma - \sigma_1}{a}\right)\right] + b, \tag{8.49}$$

where the parameter Q sets the scale of normal stress and the parameters σ_1 and a describe the length of the apical extension zone. The parameter b describes the effective normal force far from the tip, since $\lim_{\sigma\to\infty} q_n^{(\text{eff})} = b$. In this region, the cell walls are relatively rigid. A typical plot of $q_n^{(\text{eff})}(\sigma)$ is shown in Fig. 8.5.

8.5.7 Modeling growth of elastic materials

At the biochemical level, hyphal growth comes about through a complex process in which the wall-building materials are continuously transported (by different

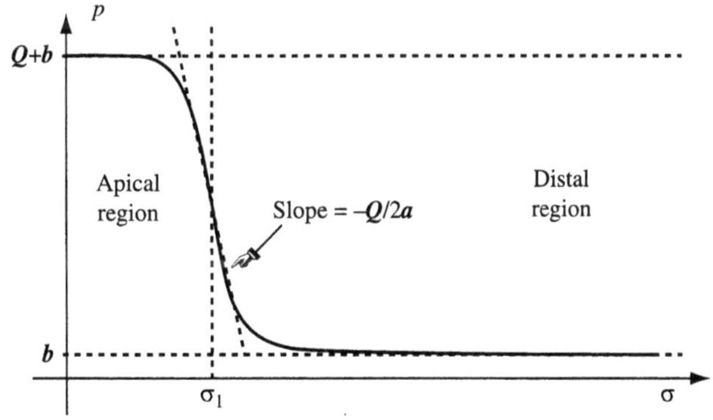

Fig. 8.5 A typical plot of the effective normal stress $q_n^{(\text{eff})}$. Close to the tip ($\sigma = 0$), the walls are soft and the elastic coefficient becomes very small. In the distal regions, the wall are relatively rigid and the elastic coefficient A is, comparatively, very large so that the effective normal stress is small (equal to b).

mechanisms for different cell types) to, and incorporated into, the apical region. At the mechanical level, when an elastic shell is pressurized (or, more precisely, subjected to an internal normal force) it will expand until a new stress equilibrium is achieved. During the course of this deformation, the initially uniformly distributed material points will become redistributed (typically, separated under expansion), as a manifestation of the new strained configuration. Furthermore, the shell expansion leads to a reduction in the shell wall thickness—although this may not be uniform around the shell. Once equilibrium has been achieved, no further expansion will occur unless there is a further increase in pressure or the stresses in the shell are somehow relieved.

At the mathematical level, the process of growth is achieved by considering an evolving reference configuration parameterized by a time t. At any given time, the current configuration is obtained by loading the reference configuration with the turgor pressure P and computing the resulting shape for the given boundary conditions. The problem is then to describe the evolution of the reference configuration. The basic idea is that remodeling is achieved by assuming that it takes place so as to remove stresses present in the current configuration. Explicitly, we consider an *initial configuration* defined by the pair $(\sigma_0, \rho_0(\sigma_0))$ and a *reference configuration* defined by $(\sigma(\sigma_0, t), \rho(\sigma_0, t))$, with initial values $\sigma(\sigma_0, 0) = \sigma_0$ and $\rho(\sigma_0, 0) = \rho_0$. The deformation from the initial to the reference configuration is characterized by the *growth strains*

$$\gamma_s = \frac{d\sigma}{d\sigma_0}, \quad \gamma_\varphi = \frac{\rho}{\rho_0}. \tag{8.50}$$

The *current configuration* at a given time is defined by $(s(\sigma, t), r(\sigma, t))$ and is obtained by solving the mechanical problem of the shell under load, that is, by computing the *elastic strains* from the reference configuration at time t to the current configuration at the same time t,

$$\alpha_s = \frac{ds}{d\sigma}, \quad \alpha_\varphi = \frac{r}{\rho}. \tag{8.51}$$

Note that the mapping from the initial configuration to the current configuration is characterized by the product of the growth and elastic strains:

$$\frac{ds}{d\sigma_0} = \alpha_s \gamma_s = \frac{ds}{d\sigma}\frac{d\sigma}{d\sigma_0}, \quad \frac{r}{\rho_0} = \alpha_\varphi \gamma_\varphi = \frac{r}{\rho}\frac{\rho}{\rho_0}. \tag{8.52}$$

That is, the growth and elastic processes are modeled through a multiplicative decomposition, similar to the one used in the theory of elastic growth (Rodriguez *et al.* 1994), where the deformation tensor is decomposed into a product of two tensors, namely the growth tensor, mapping the initial configuration to an unstressed configuration, and an elastic deformation tensor, mapping the evolving reference configuration to the current configuration.

The introduction of the growth strains require new constitutive relationships. Since they describe the evolution of the material in time, these relationships are usually refered to as *evolution laws* and take the form of differential equations for the growth rate that relate the reference and current configurations at a time t to a new reference configuration at a subsequent time $t + dt$. If we assume that the remodeling process depends on the stress, these evolution laws take the general form

$$\frac{d\gamma_s}{dt} = \eta \mathcal{G}_s(\gamma_s, t_s, t_\varphi), \quad \frac{d\gamma_\varphi}{dt} = \eta \gamma_\varphi \mathcal{G}_\varphi(\gamma_s, t_s, t_\varphi), \tag{8.53}$$

where η^{-1} is a characteristic relaxation time, and $(\mathcal{G}_s, \mathcal{G}_\varphi)$ is a function of the stresses that vanishes at $(\gamma_s, t_s, t_\varphi) = (1, 0, 0)$.

In practice, the evolution laws are not known for microbial growth, and we adopt a much simpler evolution rule by assuming that growth takes place so as to fully remove the stresses. That is, we discretize the evolution law and assume that either the remodeling time is sufficiently fast or the elastic deformation is sufficiently small that at each time step the elastic stresses are fully relieved. We start with an initial shape defined by the function $\rho_0(\sigma_0)$, $0 \leq \sigma_0 \leq L_0$, and compute the new shape $r(\sigma_0)$ by solving the mechanical equations subject to the boundary condition $r(L_0) = R_0, r(0) = 0$. The new shape $r(\sigma_0)$ represents the new mechanical equilibrium of the membrane. Once the stresses are fully relieved, the new reference configuration is characterized by the pair $(\sigma_1 = s(\sigma_0), \rho_1(\sigma_1))$, which defines the new reference shape of length $L_1 = s(L_0)$ computed from eqn (8.18). The new configuration is then used to compute a new mechanical equilibrium with boundary conditions $r(L_1) = R_0, r(0) = 0$ and so on.

8.6 Results

Here we focus on the new results provided by our current formulation, namely, the effect on tip shapes resulting from external friction due to the medium, and an examination of the normal-growth hypothesis. We also revisit the self-similar propagation of hyphal tips. These three topics are covered in the following subsections.

8.6.1 Tip shapes and the effects of external surface stresses

We begin by considering tip growth in the absence of surface friction, i.e. $\tau_s = 0$. Using the membrane equations, a neo-Hookean energy, and an initially spherical shell with a soft spot, we can simulate the initial phases of hyphal growth a typical growth pattern over N time steps is shown in Fig. 8.6. In Fig. 8.7, we plot the corresponding values of the stresses (τ_s, τ_ϕ), the deformation parameters $(\alpha_s, \alpha_\phi, \alpha_3)$, and the curvatures (κ_s, κ_ϕ). All these plots support the hypothesis of maximal stress and strain in the vicinity of the apex. It is interesting to note that the curvature of the tip is not maximal at the tip but instead is maximal close to the tip, as observed in experiments (Dumais *et al.* 2004).

We now consider the effect of external surface friction by increasing the value of τ_s. In Fig. 8.8, we compare tip shapes after $N = 70$ iterations for a series of values of τ_s and the same material parameters and initial conditions. The results clearly show a progressive "flattening" of the tip shape as τ_s is increased. A variety of numerical studies testing the effect of using different elastic energies, such as the Mooney–Rivlin

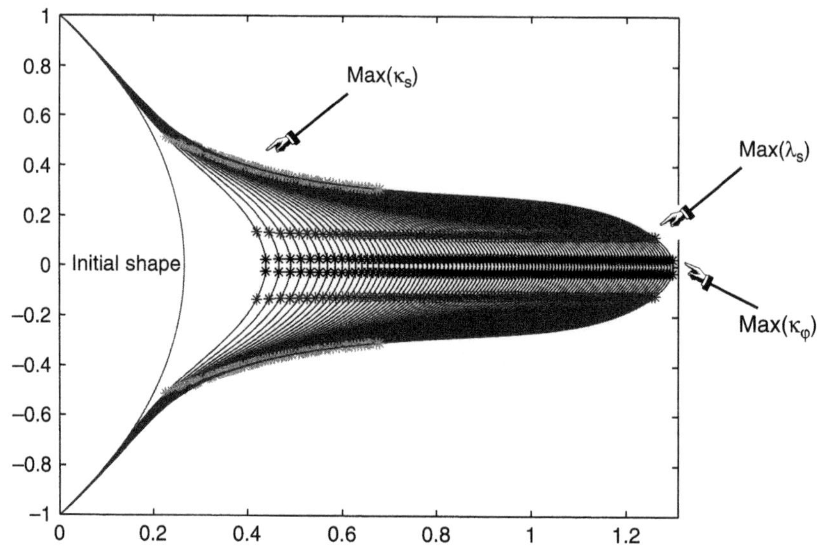

Fig. 8.6 Evolution of a tip in the absence of external surface stress ($\tau_s = 0$, $Q = 1$, $\sigma_1 = \pi/12$, $a = 1/8$, $b = 10^{-4}$).

energy, and the effect of including bending moments did not reveal any significant changes to the above results. Accordingly, we have used the membrane equations and the neo-Hookean energy for the rest of our computations.

8.6.2 The normal-growth hypothesis

We now test the normal-growth hypothesis by tracking the paths of material points on the tip surface and comparing them with what would be the equivalent normal paths. Using the same parameters above, we considered tip growth cases with a series of τ_s values, starting with $\tau_s = 0$. As we see in Fig. 8.9, the particle trajectories do *not* follow normal paths for $\tau_s = 0$, but as τ_s is increased, the particle trajectories become ever more "normal". This is a striking result and, to the best of our knowledge, has not been discussed before. Yet the reason for this may be be quite straightforward: without having to work against external forces, there is no energetic reason for the tip expansion to follow a normal path, but once such forces are included, such paths are the most work-efficient ones to follow.

8.6.3 Self-similar tip propagation

In an earlier paper (Goriely and Tabor 2003*b*), we observed that in long-time simulations the tip propagation appeared to be self-similar, namely that at each time step the newly grown tip looks like a translation of the tip shape at the previous instant. This suggests that in the course of apical growth dynamics, it is this (temporal) self-similarity, which depends delicately on the local environmental conditions and material parameters, that characterizes the propagation rather than a specific mathematical

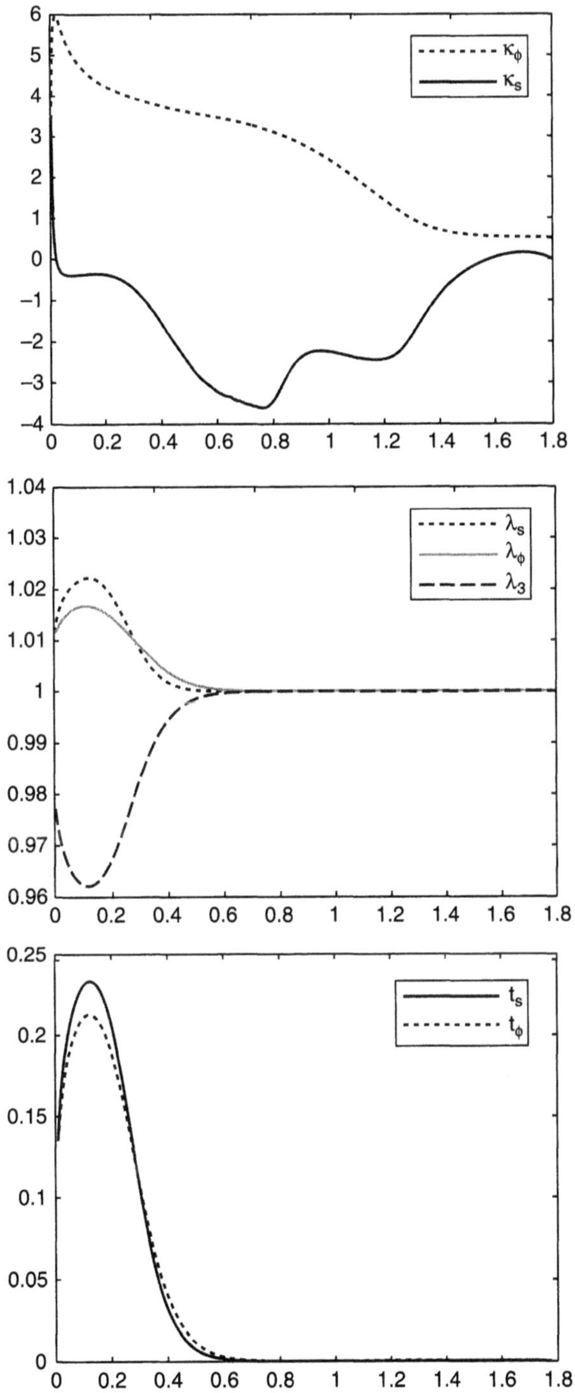

Fig. 8.7 Profiles of curvature, strains, and stresses at $n = 70$ as a function of s (same data as in previous figure).

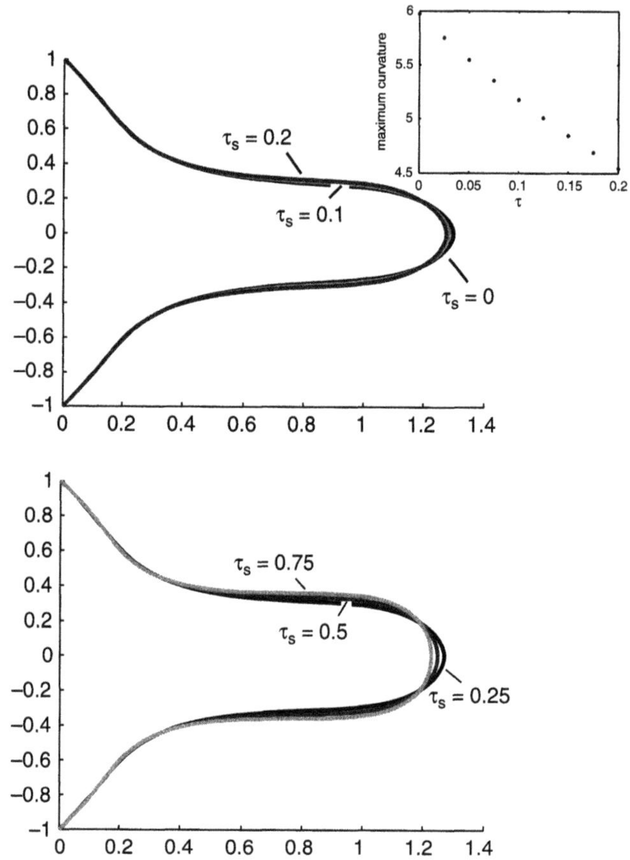

Fig. 8.8 Comparison of the final shape of the tip for increasing values of external surface stress τ_s. As the external stress increases, the curvature at the tip decreases and the tip flattens. Inset: evolution of the maximum of the curvature as τ_s increases.

form of the tip shape. Using our more complete formalism, we have computed, over longer periods, hyphal propagation for a number of different conditions. In each case we see the long-time self-similarity of the pattern (see Fig. 8.10).

8.7 Conclusions

The use of an exact elasticity formulation coupled with a simple representation of growth provides a mechanically self-consistent model that appears capable of exploring and capturing many features of the apical hyphal growth exhibited by a variety of different organisms. The formulation enables us to test the contributions of bending moments and the effects of various elastic energy functions on the hyphal expansion and indicates that the observed phenomena can be effectively captured with relatively simple elastic energies (such as the neo-Hookean or Money–Rivlin energy), and that

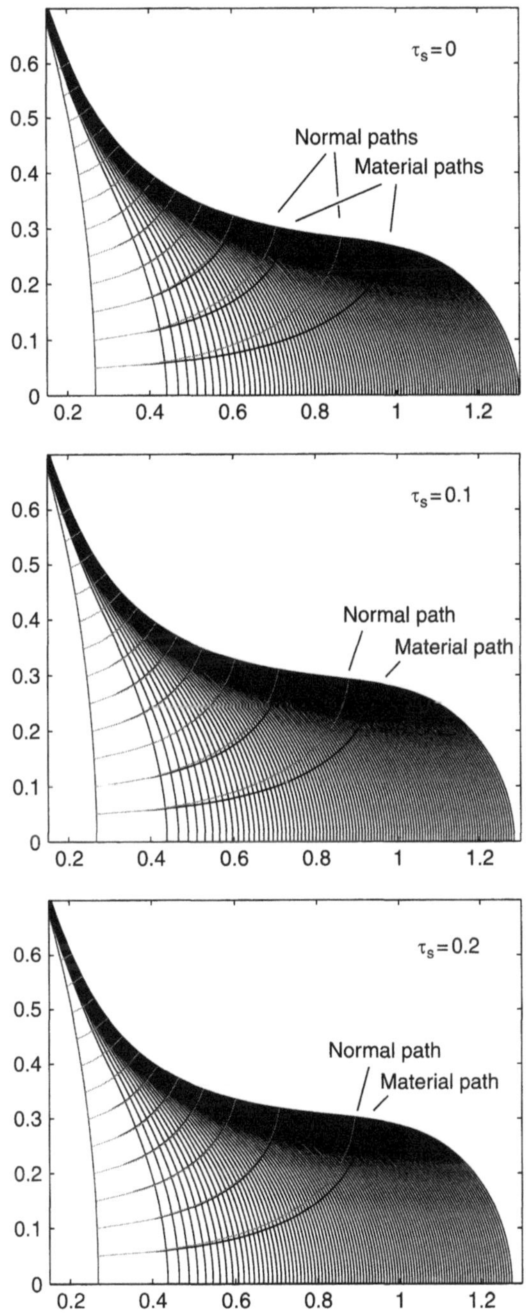

Fig. 8.9 Testing the normal-growth hypothesis. We compare the path of a material point on the tip with the path starting from the same material point but following a trajectory normal to the curve for increasing values of τ_s. As can be seen, the material trajectories tend to align with the normal trajectories.

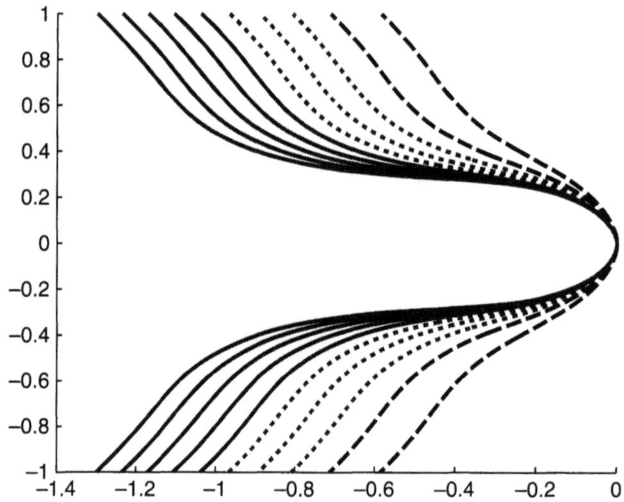

Fig. 8.10 Evolution of a tip shape seen in a reference frame moving with the tip. As growth proceeds, the shape converges towards a self-similar shape. (Growth proceeds from early growth (dashed curves) and intermediate stages (dotted curves) to the final state (solid curves). Curves are shown for every 10 iterations (right to left). ($\tau_s = 0$ and same parameters as in Fig. 8.6).

bending moments do not play a significant role in determining the tip shape under typical growth conditions. In particular, the model makes possible the inclusion of external tangential stresses corresponding to friction between the propagating tip and the surrounding medium. This study reveals a striking result, namely that the long-predicted normal-growth hypothesis is driven by presence of tangential stresses. As the tangential stresses increase on the cell wall, the trajectories of the material points align themselves with the surface-normal trajectories. The model also captures a regime of self-similar tip propagation conditioned by the material properties of the cell wall and the local environment.

The model of growth used here is based on both a reparameterization of the cell wall to take into account local wall buildup and a rigidification of the side wall due to remodeling. It is, implicitly, a stress-induced, position-dependent evolution law. That is, growth occurs in regions of higher stress but is also localized in an active region close to the tip. These two simple basic assumptions, together with the possibility of elastic deformations, determine the most important properties of tip growth observed experimentally, namely, a self-similar growth mechanism in homogeneous conditions, normal trajectories when the tip is subject to tangential stresses, and a variety of tip shapes depending on the elastic parameters, applied load, and remodeling behavior.

Acknowledgments

The ideas and results discussed in these lectures are the result of a long-running research collaboration with Alain Goriely that has been generously supported by the National Science Foundation's Division of Mathematical Sciences.

References

Alberts, B., Johnson, A., Lewis, J., Raff, M., Roberts, K., and Walter, P. (2002). *Molecular Biology of the Cell.* Garland Science, New York.

Arnoldi, M., Fritz, M., Bauerlein, E., Radmacher, M., Sackmann, E., and Boulbitch, A. (2000). Bacterial turgor pressure can be measured by atomic force microscopy. *Phys. Rev. E.*, **62**(1 Pt B), 1034–1044.

Bartnicki-Garcia, S., Hergert, F., and Gierz, G. (1989). Computer simulation of fungal morphogenesis and the mathematical basis for hyphal tip growth. *Protoplasma*, **153**, 46–57.

Bartnicki-Garcia, S., Bracker, C. E., Gierz, G., Lopez-Franco, R., and Lu, H. S. (2000). Mapping the growth of fungal hyphae: Orthogonal cell wall expansion during tip growth and the role of turgor. *Biophys. J.*, **79**(5), 2382–2390.

Chater, K. F. and Hopwood, D. A. (1993). Streptomyces. In *Bacillus subtilis and Other Gram-Positive Bacteria: Biochemistry, Physiology and Molecular Genetics* (ed. J. P. Sonenshein), pp. 83–89. ASM, Washington, DC.

da Riva Ricci, D. and Kendrick, B. (1972). Computer modelling of hyphal tip growth in fungi. *Can. J. Bot.*, **20**, 2455–2462.

Duhamel du Monceau, H. R. (1758). *La physique des arbres* [The physics of trees]. L. Guérin et L.-F. Delatour, Paris.

Dumais, J., Long, S. R., and Shaw, S. L. (2004). The mechanics of surface expansion anisotropy in *Medicago truncatula* root hairs. *Plant Physiol.*, **136**, 3266–3275.

Dumais, J., Shaw, S. L., Steele, C. R., Long, S. R., and Ray, P. M. (2006). An anisotropic-viscoplastic model of plant cell morphogenesis by tip growth. *Int. J. Develop. Biol.*, **50**(2–3), 209–222.

Evans, E. A. and Skalak, R. (1980). *Mechanics and Thermodynamics of Biomembranes.* CRC Press, Boca Raton, FL.

Fung, Y. C. (1993). *Biomechanics: Mechanical Properties of Living Tissues.* Springer, New York.

Gooday, G. W. and Trinci, A. P. J. (1980). The eukaryotic microbial cell. In *Wall Structure and Biosynthesis in Fungi* (ed. G. W. Gooday, D. Lloyd, and A. P. J. Trinci), Vol. 30, pp.207–251. Cambridge University Press, London.

Goriely, A. and Tabor, M. (2003*a*). Biomechanical models of hyphal growth in actinomycetes. *J. Theor. Biol.*, **222**, 211–218.

Goriely, A. and Tabor, M. (2003*b*). Self-similar tip growth in filamentary organisms. *Phys. Rev. Lett.*, **90**(10), 108101.

Goriely, A. and Tabor, M. (2006). Estimates of biomechanical forces in *Magnaporthe grisea. Mycol. Res.*, **110** (Part 7), 755–759.

Goriely, A., Karolyi, G., and Tabor, M. (2005). Growth induced curve dynamics for filamentary micro-organisms. *J. Math. Biol.*, **51**(3), 355–366.

Gray, D. I., Gooday, G. W., and Prosser, J. I. (1990). Apical hyphal extension in *Streptomyces coelicolor* A3(2). *J. Gen. Microbiol.*, **136**, 1077.

Grove, S. N. and Bracker, C. E. (1970). Protoplasmic organization of hyphal tips among fungi: Vesicles and spitzenkörper. *J. Bacteriol.*, **104**, 989–1009.

Lockhart, J. A. (1965). An analysis of irreversible plant cell elongation. *J. Theor. Biol.*, **8**, 264–275.

Ma, H., Snook, L., Kaminskyj, S., and Dahms, T. (2005). Surface ultrastructure and elasticity in growing tips and mature regions of aspergillus hyphae describe wall maturation. *Microbiology*, **151**(Pt 11), 3679–3688.

Money, N. P. (1997). Wishful thinking of turgor revisited: The mechanics of fungal growth. *Fungal Genet. Biol.*, **21**, 173–187.

Prosser, J. I. and Tough, A. J. (1991). Growth mechanisms and growth kinetics of filamentous microorganisms. *Critical Rev. Biotechnol.*, **10**, 253–274.

Pelcé P. (ed.) (1988). *Perspective in Physics: Dynamics of Curved Fronts.* Academic Press, San Diego.

Reinhardt, M. O. (1892). Das Wachsthum der Pilzhyphen. Ein Betrag Zur Kenntniss des Flachenwachstums vegetabilischer Zellmembranen. *Jahrbuch für wissenschaftliche Botanik*, **23**, 479–566.

Rivlin, R. S. and Saunders, D. W. (1951). Large elastic deformations of isotropic materials. VII. Experiments on the deformation of rubber. *Philos. Trans. R. Soc. Lond., Ser. A, Math. Phys. Sci.*, **243**(865), 251–288.

Rodriguez, E. K., Hoger, A., and McCulloch, A. (1994). Stress-dependent finite growth in soft elastic tissue. *J. Biomech.*, **27**, 455–467.

Saunders, P. T. and Trinci, A. P. J. (1970). Determination of tip shape in fungal hyphae. *J. Gen. Microbiol.*, **110**, 469–473.

Secomb, T. W. and Gross, J. F. (1983). Flow of red blood cells in narrow capillaries: Role of membrane tension. *Int. J. Microcirc: Clin. Exp.*, **2**, 229–240.

Shaw, S. L., Dumais, J., and Long, S. R. (2000). Cell surface expansion in polarly growing root hairs of *Medicago truncatula. Plant Physiol.*, **124**, 959–969.

Sonenshein, A. L., Hoch, J. A., and Losick, R. (eds.) (1993). *Streptomyces*, Volume *Bacillus subtilis and Other Gram-Positive Bacteria: Biochemistry and Molecular Genetics.* American Society for Microbiology.

Steinberg, G. (2007). Hyphal growth: A tale of motors, lipids and the spitzenkorper. *Eukaryotic Cell*, **6**, 351–360.

Thompson, D.W. (1992). *On Growth and Form: The Complete Revised Edition.* Dover, New York.

Trinci, A. P. J. and Saunders, P. T. (1977). Tip growth of fungal hyphae. *J. Gen. Microbiol.*, **103**, 243–248.

Wessels, J. G. H. (1990). Role of cell wall architecture in fungal tip growth generation. In *Tip Growth in Plant and Fungal Cells* (ed. I. B. Heath), pp.1–23. Academic Press, San Diego.

9
The physics of the cell membrane

Martin Michael MÜLLER[1] and Martine BEN AMAR[2]

[1]Equipe BioPhysStat, Université Paul Verlaine-Metz, France
[2]Laboratoire de Physique Statistique de l'Ecole Normale Supérieure, Paris, France

9.1 Biological membranes

Biological membranes consist of a bilayer of lipid molecules in which proteins are embedded. Under physiological conditions, this bilayer behaves like a two-dimensional fluid, with individual molecules able to diffuse rapidly in it.

Every cell is surrounded by a closed membrane, the plasma membrane, which physically separates the extracellular environment from the cell's interior. Bacteria and archae possess just this single membrane, whereas animal and plant cells also contain internal membrane-limited subcompartments, the organelles. The nucleus, for example, stores the genetic information, the mitochondria are the power plants of the cell, and the endoplasmic reticulum is the factory where protein molecules are synthesized (Lodish *et al.* 2004; Alberts *et al.* 2002).[1]

To cope with the various tasks in which membranes are involved as a crucial part, a membrane's molecular composition is highly inhomogeneous. Despite that, they all have a common basic structure: the lipid bilayer.

9.1.1 The lipid bilayer

Lipid molecules are the basic building blocks of the membrane. They are *amphiphiles*, that is, they consist of two different parts: a *hydrophilic* (i.e., "water-loving") head and a *hydrophobic* (i.e., "water-fearing") tail. The hydrophobic tail is typically formed by two hydrocarbon chains originating from various fatty acids. One of them is often unsaturated. This means that at least one double bond exists between two of the carbon atoms (see Table 9.1). The fatty acids are attached via ester or amide covalent bonds to a linker group such as glycerol, which also binds to the hydrophilic head group. In phospholipids and glycolipids, the most abundant lipids in the cell, this head group consists of a simple phosphate or sugar group, respectively, and another group such as choline (see Table 9.2).

In an aqueous environment, the hydrophobic carbon tails of the lipid disturb the hydrogen bonds that exist between the water molecules close to it. These can only be maintained if the water sacrifices a part of its entropy. This phenomenon is called the *hydrophobic effect* (Tanford 1991) and is the main reason why the hydrophobic part of an amphiphile tries to avoid water. The hydrophilic part, however, can form hydrogen bonds with the water and therefore likes to stay in it.

The "schizophrenic" nature of amphiphiles causes them to aggregate spontaneously above a certain critical concentration. At that concentration, which is called the *critical micelle concentration* (cmc), it becomes energetically more favorable for them to sacrifice some of their entropy and self-assemble into a condensed structure, thereby shielding their hydrophobic parts from the water. The morphology of the aggregate depends on the nature of the amphiphile (i.e., the size of the head group and the hydrophobic tail), and also on the solution conditions (i.e., salt concentration, pH, or temperature) (Israelachvili 1992). The typical structures that result are spherical or cylindrical micelles, bilayers, and vesicles (Fig. 9.1).

[1]Cells that have a nucleus and other organelles are also called *eukaryotic cells.*

Table 9.1 Common biological fatty acid anions (Gompper *et al.* 2004, A3.8). The number of carbon atoms (#C) is even because biological fatty acids are synthesized by concatenation of C_2 units. Note also that the fatty acid anions can be further divided into saturated (no double bond (db)) and unsaturated (one or more double bonds) ions.

Substance	#C	#db	Chemical structure
Laurate	12	0	$CH_3{-}(CH_2)_{10} - COO^-$
Myristate	14	0	$CH_3{-}(CH_2)_{12} - COO^-$
Palmitate	16	0	$CH_3{-}(CH_2)_{14} - COO^-$
Stearate	18	0	$CH_3{-}(CH_2)_{16} - COO^-$
Arachidate	20	0	$CH_3{-}(CH_2)_{18} - COO^-$
Oleate	18	1	$CH_3{-}(CH_2)_7{-}CH{=}CH{-}(CH_2)_7 - COO^-$
Linoleate	18	2	$CH_3{-}(CH_2)_4 - (CH{=}CH{-}CH_2)_2{-}(CH_2)_6 - COO^-$
γ-Linolenate	18	3	$CH_3{-}(CH_2)_4 - (CH{=}CH{-}CH_2)_3{-}(CH_2)_3 - COO^-$
Arachidonate	20	4	$CH_3{-}(CH_2)_4 - (CH{=}CH{-}CH_2)_4{-}(CH_2)_2 - COO^-$

Most of the lipid molecules in cells form bilayers, a fact already discovered in 1925 (Gorter and Grendel 1925). The bilayer of a biological membrane is approximately 5 nm thick and contains typically on the order of one hundred different lipids. The lipid fractions vary between organisms, cells, different organelles within the cell, and even between the two sides of the bilayer. Still, some common features can be identified that are essential for biological function (Lodish *et al.* 2004; Alberts *et al.* 2002; Gompper *et al.* 2004):

- The hydrophobic core of the bilayer acts as a *barrier* to the passage of most water-soluble molecules. In consequence, a small number of molecular channels and pumps can strictly regulate the exchange of chemicals between the two sides of the membrane.[2]
- The cmc is very low (10^{-6}–10^{-10} M for bilayer-forming lipids, as opposed to 10^{-2}–10^{-5} M for micelle-forming amphiphiles). The bilayer is therefore very *stable*: even though the exterior aqueous environment can vary widely in ionic strength and pH, the bilayer has the strength to retain its characteristic architecture.
- The biological bilayer behaves as a *two-dimensional fluid* at physiological temperatures. The lipids can diffuse freely within each leaflet of the bilayer.[3]

Below a critical transition temperature T_m, the bilayer loses its fluidity and becomes a gel-like solid. Mismatches between the cross-sectional areas of the lipid head and

[2]An alternative method of material transport through the membrane will be discussed in the next section.

[3]A membrane lipid can diffuse the length of a typical bacterial cell (1 μm) in only one second, and the length of a typical animal cell in about 20 s.

Table 9.2 Examples of membrane lipids. R1 and R2 correspond to hydrocarbon chains of fatty acids (see Table 9.1) (Gompper *et al.* 2004, A3.8f).

Substance	Chemical structure	Type
Lysophosphatidylcholine (lysolecithin)	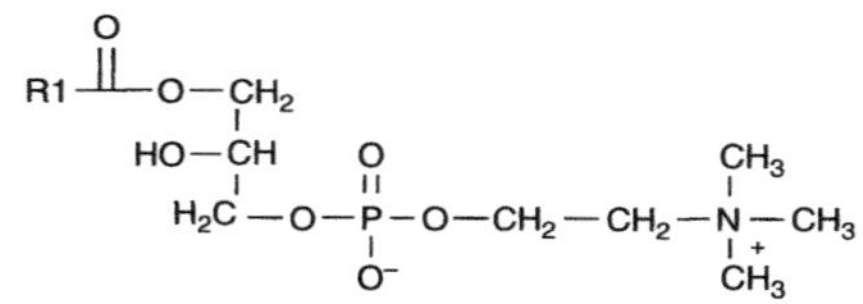	Lysophospholipid
Phosphatidylethanolamine (PE)	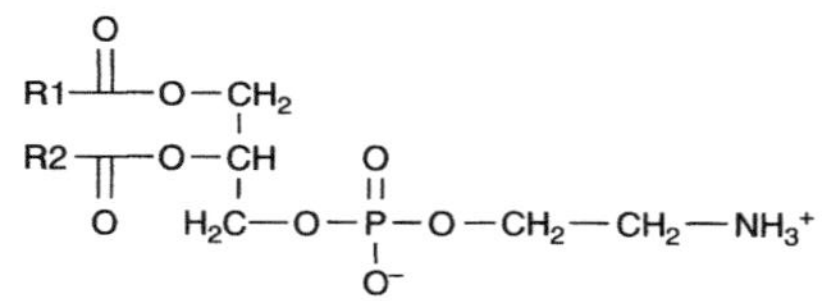	Glycerophospholipid
Phosphatidylcholine (PC)	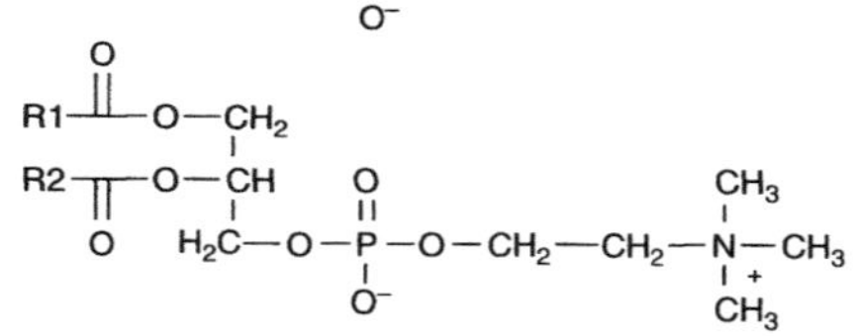	
Digalactosyldiglyceride (DGDG)	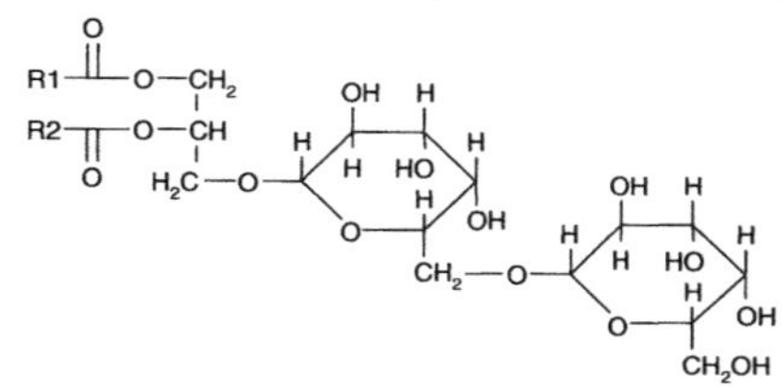	Glyceroglycolipid
Sphingomyelin	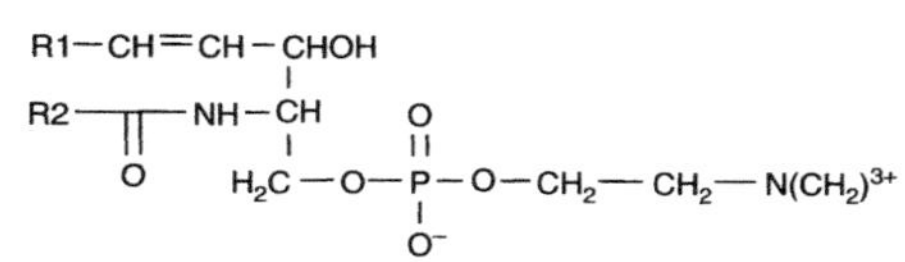	Sphingophospholipid

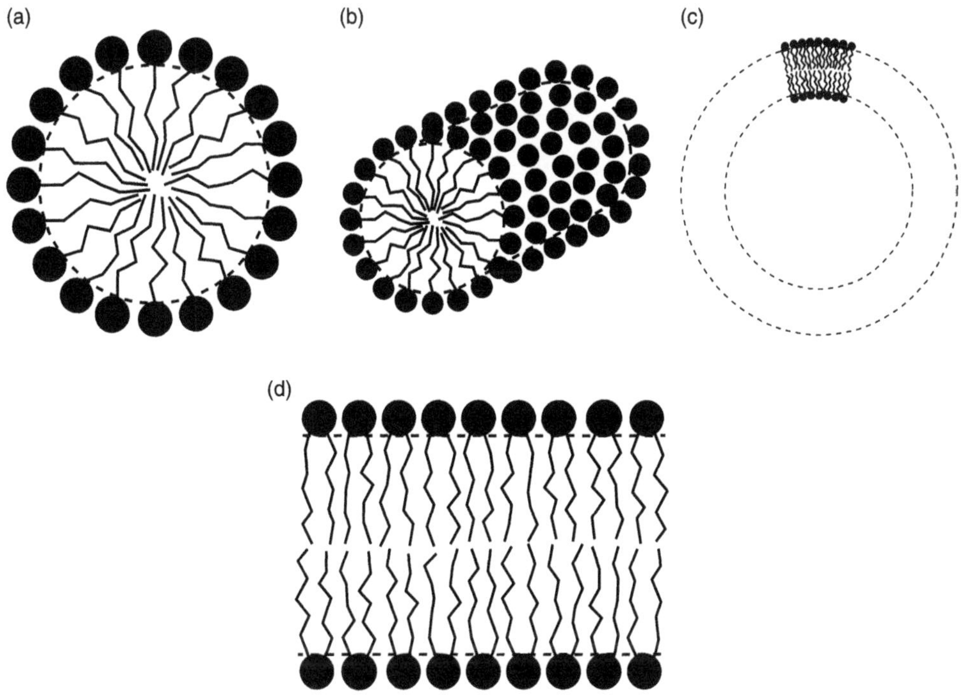

Fig. 9.1 Typical structures that result from self-assembly: (a) spherical micelle, (b) cylindrical micelle, (c) flexible bilayer/vesicle, and (d) planar bilayer. The black circles are the hydrophilic heads, and the chains are the hydrophobic tails of the amphiphiles.

tail groups lead to tilted or interdigitated chains and rippled surfaces (Lipowsky and Sackmann 1995, Chapter 5).

The value of T_m depends, for instance, on the degree of saturation of the fatty acid chains in the bilayer. In unsaturated fatty acids, the chains are kinked and cannot pack closely. Lipids with unsaturated fatty acids thus form bilayers with a lower transition temperature than lipids with saturated fatty acids. For example, a pure DPPC (1,2-dipalmitoylphosphatidylcholine) bilayer has a transition temperature of $(41.3 \pm 1.8)\,°\mathrm{C}$, whereas DOPC, with one *cis* double bond in every chain (Fig. 9.2(a)), is fluid above $(-18.3 \pm 3.6)\,°\mathrm{C}$ (Koynova and Caffrey 1998).[4] The same rule holds for synthetic membrane lipids. Under ambient conditions, a DOTAP bilayer is already fluid, whereas DODAB is still in the gel phase (Figs. 9.2(b) and (c)).

The fluidity of the bilayer also depends on the length of the hydrocarbon tails it contains. A shorter chain length reduces the tendency of the tails to interact with each other, so that the bilayer remains fluid at lower temperatures.

[4]The double bond cannot rotate. If the groups that are connected to it are oriented in the same direction, the molecule is referred to as *cis*, and when they point in opposing directions, it is referred to as *trans*.

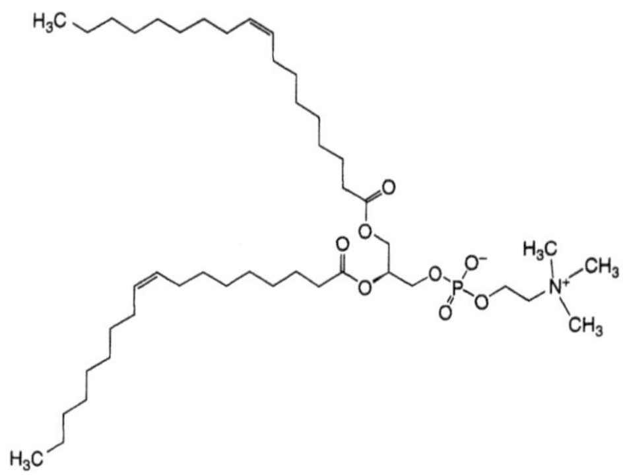

(a) cis-(DOPC) (1,2-dioleoylphosphatidylcholine)

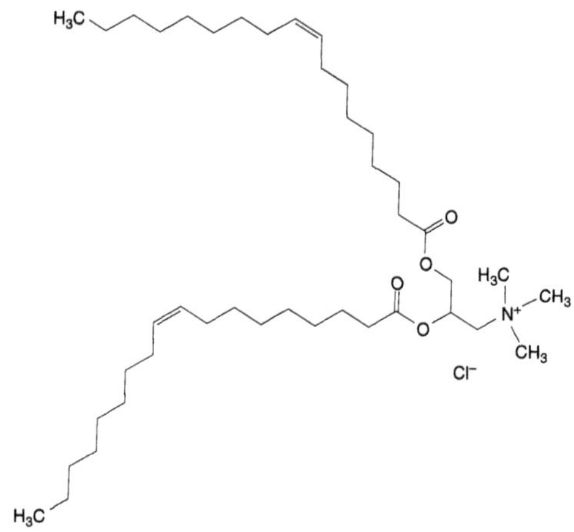

(b) DOTAP (1,2-dioleoyl-3-trimethylammoniumpropane chloride)

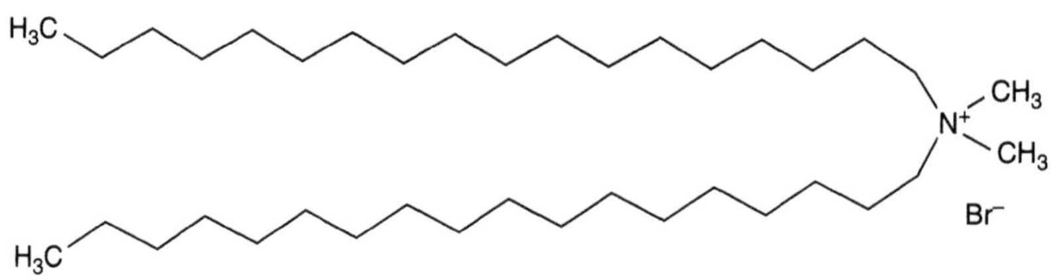

(c) DODAB (*N,N*-dimethyl-*N,N*-dioctadecylammonium bromide)

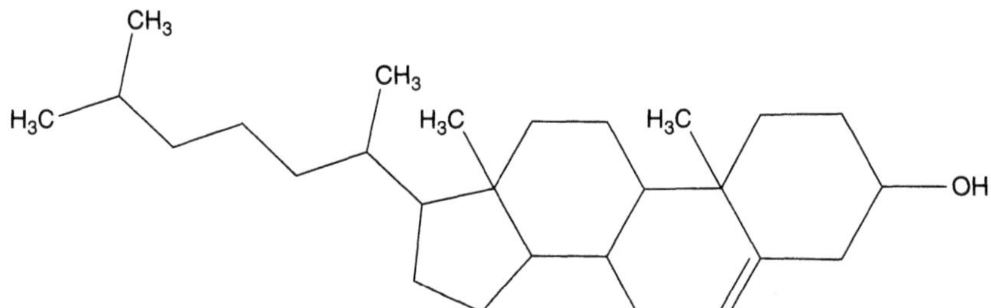

(d) cholesterol

Fig. 9.2 Examples of lipids with different effects on bilayer fluidity: DOPC (a) and DOTAP (b) bilayers are fluid under ambient conditions; DODAB (c) forms a gel-like bilayer. Cholesterol (d) tends to make fluid lipid bilayers less fluid.

Both effects, the degree of saturation and the chain length of the fatty acid tails, are exploited in nature. Under thermal stress (arising from, say, an abnormal decrease in temperature), many organisms, such as bacteria and yeasts, actively adjust the fatty acid composition of their membrane to maintain its fluidity (Thieringer *et al.* 1998).

Eukaryotic cell membranes additionally contain cholesterol for these purposes. Cholesterol consists of a rigid, relatively planar steroid skeleton with a single hydroxyl group as the hydrophilic part (Fig. 9.2(d)). It cannot form a bilayer unless it is mixed with other lipids. By decreasing the mobility of the hydrocarbon chains, it tends to make fluid lipid bilayers less fluid. At the high concentrations found in most eukaryotic plasma membranes, it also prevents the hydrocarbon chains from coming together and crystallizing, thereby inhibiting the fluid–gel phase transition (Lodish *et al.* 2004; Alberts *et al.* 2002). About ten years ago, it was proposed that both cholesterol and sphingolipids accumulate in the plasma membrane to form tiny domains with diameter of a few tens of nanometers (Simons and Ikonen 1997; Brown and London 1997, 1998). These "lipid rafts" are supposed to help to organize certain proteins which are important for biological functions such as cell signaling.

Experiments with synthetic vesicles made from cholesterol, sphingomyelin, and a phospholipid such as POPC have shown a variety of phases, phase transitions, and coexistence regimes (Dietrich *et al.* 2001; Veatch and Keller 2005). Typically, such a mixture can phase-separate into *micrometer-sized* liquid ordered (L_o) domains, which are rich in sphingomyelin and cholesterol, and liquid disordered (L_d) domains of similar size. Whether rafts exist in living cells or not, however, remains the subject of lively debate (Munro 2003; Nichols 2005; Hancock 2006), as it is much harder to search for domains of *nanometer size*, especially in vivo. We will come back to this issue in Section 9.4, where the corresponding model systems are discussed in more detail.

9.1.2 Membrane proteins

Although the lipid bilayer provides the basic structure of the membrane, proteins perform most of its biological functions. They serve as ion channels and specific receptors, for example, or catalyze membrane-associated reactions, such as the synthesis of adenosine 5′-triphosphate (ATP), the principal carrier of chemical energy in cells.

Proteins are macromolecules that consist of amino acids joined together by peptide bonds (Fig. 9.3). One can distinguish four levels of organization in their structure. The linear sequence of amino acids is called the primary structure. It is composed of 20 different types of amino acids and is encoded in the genome of the cell. Individual amino acid residues on a protein may attract or repel one another. Localized parts of the peptide chain can thus fold into three-dimensional structures such as α helices and β sheets (see Fig. 9.3 again). These arrangements are referred to as secondary structures. Their degree of hydrophobicity depends on the character of the side chains of the amino acids involved. The tertiary structure describes the full three-dimensional organization of the chain. A protein can, however, be formed as a complex of more than one polypeptide chain. For instance, hemoglobin, the oxygen carrier of the blood,

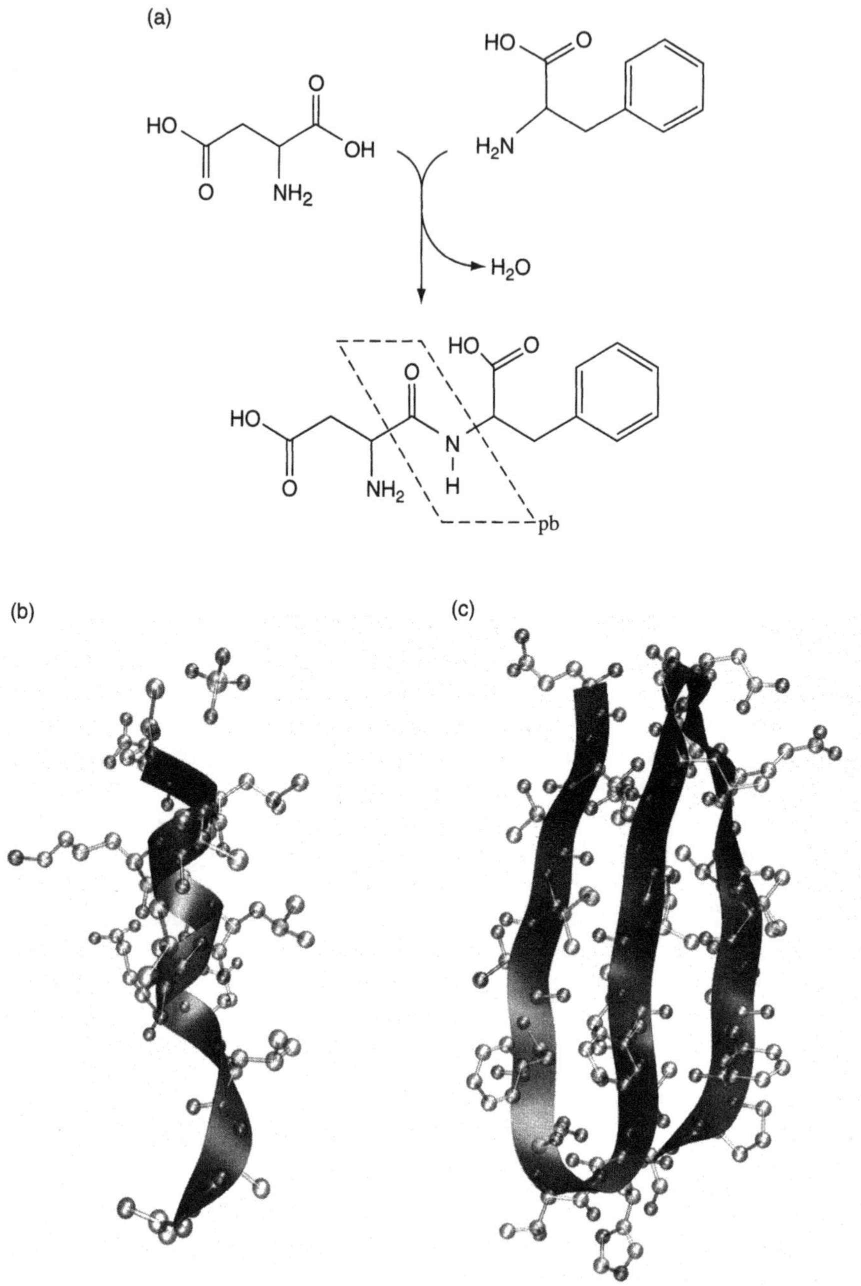

Fig. 9.3 (a) Amino acids can connect to each other via a peptide bond. Two important secondary structures are (b) the α helix and (c) the β sheet.

consists of four subunits. The complete structure is then designated as the quaternary structure (Lodish *et al.* 2004; Alberts *et al.* 2002).

Proteins can be bound to membranes in different ways. (i) *Integral membrane proteins*, also called transmembrane proteins, span the lipid bilayer. They contain one or more hydrophobic domains that are inserted into the core of the bilayer, and hydrophilic domains that extend into the aqueous medium on both sides. (ii) *Lipid-anchored membrane proteins* are bound covalently to one or more lipid molecules. The hydrophobic tail of the attached lipid is embedded in one leaflet of the membrane. The polypeptide chain itself does not enter the bilayer. (iii) *Peripheral membrane proteins* do not interact with the fatty acid chains of the bilayer. Instead, they are bound only indirectly by interactions with integral membrane proteins or directly by interactions with the lipid head groups. How a protein associates with the membrane depends on its biological function. Only transmembrane proteins can function on both sides of the bilayer or transport molecules across it. By contrast, many proteins that function on only one side of the bilayer are often lipid-anchored or peripheral proteins.

Because the bilayer is fluid, all types of proteins can move laterally in the membrane and interact with one another, leading to membrane remodeling above the molecular scale. Such processes are important, for instance, in endocytosis, in which material is ingested by invagination of the plasma membrane and subsequent internalization in a membrane-bound vesicle.

However, living cells have found ways to immobilize specific membrane proteins. In the previous section, we have already discussed lipid rafts as one example. Besides that, the interaction with the *cytoskeleton* restricts the mobility of proteins as well. The cytoskeleton provides structural support for the eukaryotic cell. It consists of a network of filaments built up from proteins such as actin and tubulin and permits directed movements of organelles and the cell itself.[5] Some proteins in the plasma membrane can link permanently to the cytoskeleton, leaving them completely immobile in the membrane. Other proteins, though still mobile, are slowed down in the presence of the cytoskeleton.

We will not go further into the biological details here. The interested reader is referred once more to Lodish *et al.* (2004) and Alberts *et al.* (2002). We will instead focus on a physical description of the biological membrane, which will ultimately enable us to consider interesting problems of fluid membrane mechanics.

9.2 Membrane energy and the shape equation

The energy associated with a fluid membrane can be described by a reparameterization-invariant Hamiltonian that depends only on the surface geometry. The equilibrium shape of the membrane is governed by a nonlinear partial differential equation, the "shape equation", which can be linearized for nearly flat membranes.

[5]For active transport, you of course additionally need some kind of engine and fuel. In the cell, these are provided by special motor proteins and ATP, respectively.

In order to choose the correct physical description of the membrane, it first has to be clear how much detail is needed to answer the questions of interest. To model the exact conformation of a protein embedded in the membrane, for example, computer simulations are the right tool to use (Leach 2001).

If the lateral extension and the size of all deformations of interest are much larger than the membrane thickness, the membrane can be considered from a *mesoscopic* point of view instead. The lipid bilayer can then be modeled as a two-dimensional surface embedded in three-dimensional Euclidean space $\mathbb{R}^3$. Proteins or other entities that can bind to the membrane (e.g., viruses) are represented by solid particles that impose deformations in the surface.

To predict the behavior of the membrane under deformations such as these, one has to know how its energy changes. Let us therefore study possible contributions to the energy of a fluid surface first and discuss their relevance to the case of the membrane.

9.2.1 Surface tension and curvature energy

9.2.1.1 *Surface tension*

When surfaces occur in physical systems, the energy associated with them is often completely described by a Hamiltonian that depends only on the surface geometry. The easiest and best-known examples involve capillary phenomena (Rowlinson and Widom 2002; de Gennes *et al.* 2003), in which the energy of a liquid–fluid interface is given simply by a surface integral over a constant surface tension σ (see also Chapter 2, by John Bush, in this book). The resulting Hamiltonian is therefore

$$H = \int_{\Sigma} \mathrm{d}A\, \sigma, \tag{9.1}$$

where Σ is the surface domain and $\mathrm{d}A$ is an infinitesimal area element.

The Hamiltonian (9.1) not only is relevant to interfaces between pure phases (such as, water and water vapor) but also describes the behavior of soap films (Isenberg 1992). Soap films consist of a 5 nm–20 µm thick layer of water between two monolayers composed of amphiphilic soap ions (such as fatty acid anions; see Table 9.1). These ions can move laterally in the film, exactly as the lipids in a membrane do.

The fluidity of the surface implies that it cannot resist shear stresses. Hence, static shear is not a deformation which costs the surface any energy; in fact, it cannot even be defined. As we will restrict our considerations to fluid membranes, this property holds throughout this chapter.

9.2.1.2 *Canham–Helfrich Hamiltonian*

A bending deformation in the direction normal to the interface, however, *can* contribute to the total energy of a fluid surface. In a harmonic expansion, the energy due to any deformation of the interface is proportional to the square of the deformation (remember Hooke's law). In the particular case of bending, one has to consider quadratic functions of the curvature.

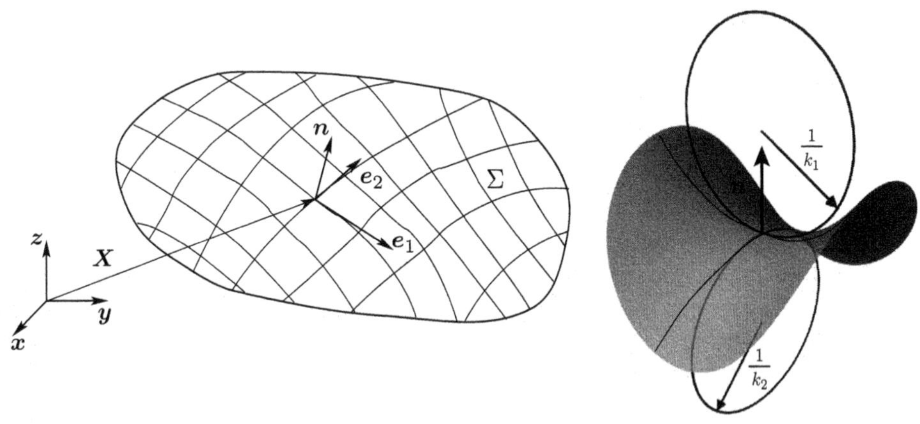

(a) Parameterization of the surface (b) Principal curvatures k_1 and k_2

Fig. 9.4 Continuum model of a membrane as a two-dimensional surface.

In order to describe the surface and its curvature properties, a differential geometric language can be used: a surface Σ is described locally by its position $\boldsymbol{X}(\xi^1, \xi^2) \in \mathbb{R}^3$, where the ξ^a $(a \in \{1,2\})$ are a suitable set of local coordinates on the surface (see Fig. 9.4(a)). The tangent vectors of Σ, $\boldsymbol{e}_a = \partial \boldsymbol{X}/\partial \xi^a = \partial_a \boldsymbol{X}$ $(a, b \in \{1,2\})$, form a local coordinate frame; together with the extrinsic unit normal vector $\boldsymbol{n} = \boldsymbol{e}_1 \times \boldsymbol{e}_2/|\boldsymbol{e}_1 \times \boldsymbol{e}_2|$, they define two geometrical tensors on the surface: the metric $g_{ab} = \boldsymbol{e}_a \cdot \boldsymbol{e}_b$ and the extrinsic curvature $K_{ab} = \boldsymbol{e}_a \cdot \partial_b \boldsymbol{n}$.

At every point, a surface has two principal curvatures k_1 and k_2, which are the eigenvalues of the local curvature tensor $K_a^{\ b}$ (see Fig. 9.4(b)).[6] The corresponding eigenvectors define two orthogonal principal directions. One therefore has to include two independent terms in the expression for the energy that depend on a quadratic combination of the two curvatures and are, furthermore, invariant scalars. One convenient choice is a combination of the Gaussian curvature $K_{\text{G}} = k_1 k_2$ and the squared trace $K^2 = (k_1 + k_2)^2$ of the curvature tensor $K_a^{\ b}$. Including surface tension, the complete Hamiltonian is then, according to Canham and Helfrich (Canham 1970; Helfrich 1973)

$$H = \int_\Sigma \mathrm{d}A \left[\sigma + \frac{\kappa}{2}(K - K_0)^2 + \bar{\kappa} K_{\text{G}}\right], \tag{9.2}$$

where κ and $\bar{\kappa}$ are proportionality constants called the *bending rigidity* and the *saddle-splay modulus*, respectively. The constant K_0 is the spontaneous curvature. Its value determines how much the surface prefers to be bent in its minimal-energy state. For up–down-symmetric membranes, K_0 vanishes.[7] Owing to the Gauss–Bonnet theorem,

[6] As usual, indices are lowered or raised with the metric or its inverse, respectively, and a repeated index a (one up, one down) implies a summation over $a = 1, 2$. More background on differential geometry can be found, for instance, in Do Carmo (1976); Spivak (1976), and Kreyszig (1991).

[7] A term linear in K can also occur if there is an area difference between the two layers of the membrane.

the term involving K_G can be written as the sum of a topological constant and a line integral over the boundary of the surface. It can thus often be neglected, as we will also see in the following.

A series of experimental and theoretical studies has shown that the energy associated with a membrane is mainly dominated by bending (see, for instance, Brochard and Lennon 1975; Faucon *et al.* 1989; Dai and Sheetz 1995; Lipowsky and Sackmann 1995; Seifert 1997; Cuvelier *et al.* 2005). The reason for this is that this is the softest mode: characteristically, weak bending is a deformation which costs significantly less energy than, for instance, stretching.

On length scales much larger than the bilayer thickness, the Hamiltonian (9.2) is suitable for describing a fluid membrane.[8] For typical phospholipid membranes, κ is of the order of a few tens of k_BT, where k_BT is the thermal energy (Seifert and Lipowsky 1995). Furthermore, membranes exhibit tensions σ in a broad range from 0 up to about 10 mN/m (Morris and Homann 2001). At even higher values, they rupture. The Gaussian bending rigidity $\bar{\kappa}$ is rather difficult to measure. Its value is usually negative and also smaller than that of κ in the same system (Gompper *et al.* 2004, A3.24). In Section 9.3, we will discuss in detail how these parameters can be measured experimentally.

9.2.2 Bulk energies

The membrane does not live in a vacuum but in an aqueous environment. Under these conditions, any *open* patch of membrane costs a large amount of energy because, at the edge, the hydrophobic core of the bilayer is in direct contact with water. Therefore, free membrane patches usually do not exist in the cell. They close to form vesicles (see Fig. 9.1(c)).

If the volume V is fixed at a constant value V_0, a term $-P(V - V_0)$ has to be added to the energy functional H for the free surface:

$$H_P = H - P(V - V_0), \tag{9.3}$$

where the pressure difference P is the Lagrange multiplier for the enclosed volume. A similar term may also be used to fix a pressure difference between the two sides of the surface: $H_P = H - PV$.

Both ensembles (constant volume and constant pressure) yield the same ground states for the shape of the surface. Questions of stability depend on which of the two variables is fixed: a spherical soap bubble, for example, is stable if its volume is fixed but would be unstable if the pressure was fixed.

9.2.3 Membrane energy in the Monge parameterization

The Hamiltonian (9.2) describes the energy of an open membrane in a parameterization-free way, *i. e.*, without defining the position vector $\boldsymbol{X}$ and its corresponding coordinates ξ^1 and ξ^2 explicitly. If one is interested in quantitative

[8]It should be mentioned that the Hamiltonian (9.2) is also valid for other interfaces, such as those containing block copolymers.

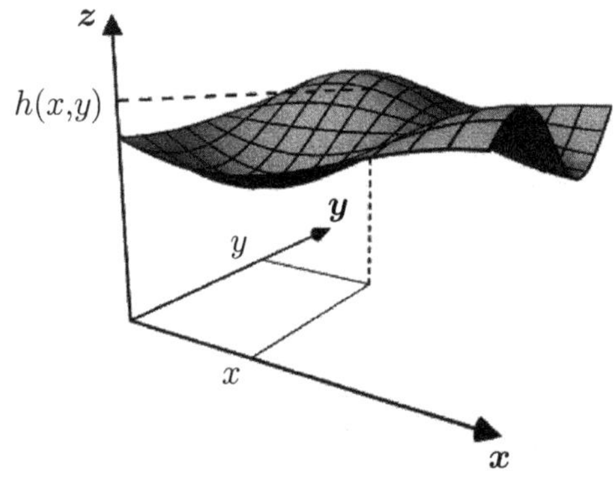

Fig. 9.5 Monge parameterization in Cartesian coordinates.

results, however, one needs to pick a surface parameterization. The choice followed in essentially all existing calculations in the literature is the "Monge parameterization" and—for analytical tractability—its linearized version. It is suitable for surfaces free of "overhangs". In such a case the surface can be described in terms of its height $h(x, y)$ above some reference plane, which we take to be the (x, y) plane (see Fig. 9.5). Notice that x and y thus become the surface coordinates. The tangent vectors on the surface can then be expressed as $\boldsymbol{e}_x = (1, 0, h_x)^{\mathrm{T}}$ and $\boldsymbol{e}_y = (0, 1, h_y)^{\mathrm{T}}$, where the $h_i = \partial_i h$ $(i, j \in \{x, y\})$ are the partial derivatives with respect to the surface coordinates. Let the symbol $\boldsymbol{\nabla}$ denote the two-dimensional nabla operator in the flat reference plane. In Cartesian coordinates, it is given by $\boldsymbol{\nabla} = \boldsymbol{x}\,\partial_x + \boldsymbol{y}\,\partial_y$.

In the following, we will assume that the membrane is nearly flat, which implies that the gradient term $|\boldsymbol{\nabla} h|$ is small compared with 1. This linearized version of the Monge parameterization is also called the small-gradient approximation. To rewrite the Hamiltonian (9.2) term by term in this approximation, let us first consider the infinitesimal area element $\mathrm{d}A$. This is given by the area of the parallelogram spanned by the two surface tangent vectors times the infinitesimal area element of the (x, y) plane: $\mathrm{d}A = |\boldsymbol{e}_x \times \boldsymbol{e}_y|\,\mathrm{d}x\,\mathrm{d}y$.

Exercise. Show that the area element in the Monge parameterization is given by $\mathrm{d}A = \sqrt{1 + (\boldsymbol{\nabla} h)^2}\,\mathrm{d}x\,\mathrm{d}y$. Expand this expression in the small-gradient approximation (*i. e.*, $|\boldsymbol{\nabla} h| \ll 1$).

The terms involving the curvatures are a bit harder to get. After some calculations, which can, for instance, be found in Appendix A of Müller (2007), one obtains the following for the curvature K:

$$K = -\boldsymbol{\nabla} \cdot \left(\frac{\boldsymbol{\nabla} h}{\sqrt{1 + (\boldsymbol{\nabla} h)^2}} \right), \tag{9.4}$$

which, in the small-gradient approximation, is simply $K = -\Delta h$. The symbol $\Delta = \boldsymbol{\nabla}^2$ denotes the Laplace operator on the flat reference plane.

Neglecting the term involving the Gaussian curvature K_{G} for the moment, the Hamiltonian (9.2) for an up–down-symmetric membrane in the small-gradient approximation becomes

$$E = \frac{1}{2}\int \mathrm{d}x\,\mathrm{d}y\,\left[\sigma(\boldsymbol{\nabla}h)^2 + \kappa(\Delta h)^2\right]. \tag{9.5}$$

Note that this expression is the 2+1-dimensional equivalent of the energy of a worm-like chain filament under tension (see Chapter 5, by Fred MacKintosh, in this book).

9.2.4 The shape equation

In Section 9.1, we have seen that membranes in cells are not isolated, free objects. They experience a variety of external forces and torques. For example, proteins adhere to them and impose deformations, actin filaments of the cytoskeleton push and pull at them, and pressure differences between the inner and outer sides may even cause their rupture. Furthermore, lipid rafts have a higher rigidity than their surroundings, and certain mechanosensitive membrane channels may open and close in a way that is gated by the tension of the membrane.

All these examples show that the mechanical properties of membranes are vital for the functioning of biological cells. A first step to understanding these properties is to determine what shape the membrane adopts if it is subjected to some specific boundary conditions.

To determine this equilibrium shape, the functional (9.2) has to be minimized with respect to deformations of the surface. These deformations are described by a change of the position function $\boldsymbol{X} \to \boldsymbol{X} + \delta\boldsymbol{X}$. The response of the Hamiltonian, δH, has to vanish in equilibrium, which allows us to isolate the corresponding Euler–Lagrange (or shape) equation. This nonlinear partial differential equation, finally, has to be solved with the appropriate boundary conditions to find the shape of the membrane.

To understand the procedure, we will determine the shape equation in the small-gradient approximation. There, the change in the position function corresponds to a small variation of the height function $h(x,y) \to h(x,y) + \delta h(x,y)$. The energy (9.5) responds in the following way:

$$\begin{aligned}
\delta E &= \frac{1}{2}\int \mathrm{d}x\,\mathrm{d}y\,\left[\kappa(\Delta h + \Delta\delta h)^2 + \sigma(\boldsymbol{\nabla}h + \boldsymbol{\nabla}\delta h)^2\right] - \frac{1}{2}\int \mathrm{d}x\,\mathrm{d}y\,\left[\kappa(\Delta h)^2 + \sigma(\boldsymbol{\nabla}h)^2\right] \\
&= \int \mathrm{d}x\,\mathrm{d}y\,\left[\kappa\Delta h\,\Delta\delta h + \sigma\boldsymbol{\nabla}h\cdot\boldsymbol{\nabla}\delta h\right] \\
&= \int \mathrm{d}x\,\mathrm{d}y\,\left[\kappa\Delta\Delta h - \sigma\Delta h\right]\delta h + \int \mathrm{d}x\,\mathrm{d}y\,\boldsymbol{\nabla}\cdot\left[(\kappa\,\Delta h)\boldsymbol{\nabla}\delta h + (\sigma\,\boldsymbol{\nabla}h - \kappa\,\boldsymbol{\nabla}\Delta h)\delta h\right] \\
&\overset{!}{=} 0.
\end{aligned} \tag{9.6}$$

(where the last line means that the left-hand side is set equal to zero).

In the final step, we have used integration by parts several times to isolate an area integral involving the variation δh and an area integral over a divergence. The latter integral can be recast as a boundary integral with the help of the divergence theorem. In equilibrium, both integrals have to vanish separately. This allows us to determine the differential equation which h has to satisfy. Since the variation δh is arbitrary, the shape equation in the linearized theory is given by

$$\Delta(\Delta - \lambda^{-2})\, h = 0, \tag{9.7}$$

where $\lambda := \sqrt{\kappa/\sigma}$ is a characteristic length which separates the short length scales over which bending energy dominates from the long scales over which tension does. For typical values of κ and σ, it is of the order of a few tens of nanometers.

We have neglected the energy term involving the Gaussian curvature so far, but including it would not change anything here: the variation (9.6) stays the same, since the term containing $K_{\rm G}$ is a constant as long as the topology of the surface and the curvature of the boundary do not change. If one wants to include a pressure difference P between the upper and lower sides of the membrane, one has to consider the variation of the Hamiltonian (9.3). It turns out that one only has to replace the right-hand side of eqn (9.7) by P to get the correct result.

In contrast to the shape equation of the fully covariant theory, eqn (9.7) is linear and can be treated much more easily mathematically. In the next section, we will consider one particular example where we will solve this equation in an axisymmetric setup.

Exercise. What is the general solution of the shape equation (9.7) for:

1. Translational symmetry, $h = h(x, y)$?
2. Rotational symmetry, $h = h(r)$, where r is the radial coordinate of the reference plane?

9.3 Measurements of elastic properties

Measurements with an atomic force microscope offer a direct way to probe the elastic properties of lipid bilayer membranes locally; provided the underlying stress–strain relation is known, material parameters such as the bending rigidity may be deduced. The saddle-splay modulus cannot be determined from bulk properties, but estimates from systems with topology changes exist.

In the previous section, we got a first glimpse of the mathematical toolbox which has been developed over the last thirty years for describing the mechanical properties of lipid bilayer membranes. To predict the behavior of a membrane in "real life", however, *quantitative* knowledge of its material parameters—most notably the bending modulus κ—is necessary.

9.3.1 The bending rigidity κ

Several methods for the experimental determination of κ have been proposed, such as:

- flicker spectroscopy: the spectrum of thermal undulations is monitored via light microscopy (Brochard and Lennon 1975; Faucon *et al.* 1989);
- micropipette aspiration: the relative area change of vesicles under micropipette aspiration is analyzed (Evans and Rawicz 1990; Rawicz *et al.* 2000);
- tether pulling: the force required to pull thin tubes, called "tethers", from a membrane is measured (Dai and Sheetz 1995; Hochmuth *et al.* 1996; Dai and Sheetz 1999; Maeda *et al.* 2002; Cuvelier *et al.* 2005).

These methods are essentially standard techniques nowadays and have been applied successfully in computer simulations as well (Deserno 2009). Without going too much into the details, let us just have a (simplified) look at the first one.

9.3.1.1 *Flicker spectroscopy*

Consider a membrane spanning a square planar frame of size $L \times L$ with periodic boundary conditions.[9] We expand its height $h(\boldsymbol{r}) = h(x, y)$ in a Fourier series:

$$h(\boldsymbol{r}) = \sum_{\boldsymbol{q}} h_{\boldsymbol{q}} e^{i\boldsymbol{q}\boldsymbol{r}}, \qquad \boldsymbol{q} = \frac{2\pi}{L}\begin{pmatrix} n_x \\ n_y \end{pmatrix}, \qquad n_x, n_y \in \mathbb{Z}. \tag{9.8}$$

Using $h_{-\boldsymbol{q}} = h^*_{\boldsymbol{q}}$, since $h(\boldsymbol{r})$ is real, and inserting eqn (9.8) in the small-gradient approximation (9.5) for the energy, one obtains

$$E = L^2 \sum_{\boldsymbol{q}} |h_{\boldsymbol{q}}|^2 \left(\frac{1}{2}\kappa q^4 + \frac{1}{2}\sigma q^2\right). \tag{9.9}$$

Since the modes are harmonic, we can simply use the equipartition theorem to get

$$L^2 \langle |h_{\boldsymbol{q}}|^2 \rangle \left(\frac{1}{2}\kappa q^4 + \frac{1}{2}\sigma q^2\right) = \frac{1}{2}k_{\mathrm{B}}T \Leftrightarrow \langle |h_{\boldsymbol{q}}|^2 \rangle = \frac{k_{\mathrm{B}}T}{L^2(\kappa q^4 + \sigma q^2)} \; . \tag{9.10}$$

This static structure factor $\langle |h_{\boldsymbol{q}}|^2 \rangle$ can be measured experimentally. Fitting it to eqn (9.10) yields the bending modulus and surface tension of the membrane.

9.3.1.2 *Atomic force microscopy*

With the possible exception of the tether experiments, all techniques mentioned so far are *global* in nature, i.e., they supply information averaged over millions of lipids, if not over entire vesicles or cells. However, this may be insufficient in a biological context (Heidemann and Wirtz 2004). In Section 9.1.1 we have, for instance, seen that membrane properties such as lipid composition and bilayer phase (and thus mechanical rigidity) may potentially vary on submicroscopic length scales. In the following, we

[9] Flicker spectroscopy is usually performed on closed membrane vesicles. This complicates the mathematical treatment but does not change the general idea.

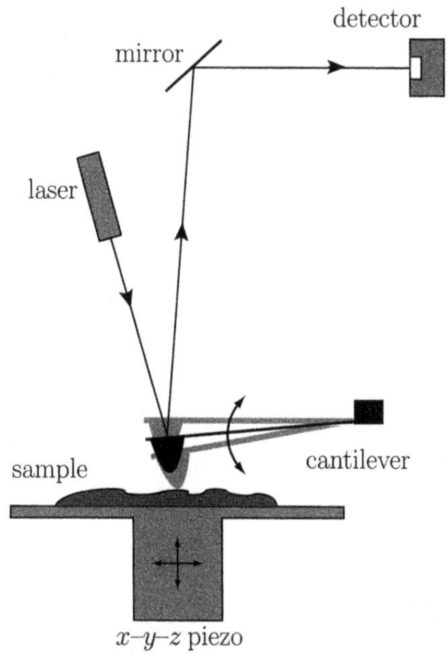

Fig. 9.6 Simplified scheme of an atomic force microscope (AFM) (Steltenkamp 2006).

will discuss an atomic force microscope (AFM) experiment which is able to probe the elastic properties of a membrane locally with the necessary spatial resolution.

In such an AFM measurement, the sample is scanned by a tiny tip which is mounted at the end of a cantilever spring (Fig. 9.6). The relative motion between the tip and the sample is controlled by a piezoelectric actuator. During the scan, the force between the tip and the sample changes and causes the spring to bend. To detect its deflection, a laser beam is focused on the back of the cantilever: a small movement of the tip changes the position of the reflected light on a split photodetector, hence giving information about the position of the tip.

One important mode of operation is the *contact mode.* This mode is comparable to the functional principle of a gramophone: the tip is in gentle touch with the sample, exerting a constant vertical force during the horizontal scan. Its position changes owing to interactions with the surface, yielding a topographic image of the sample.

The AFM can also be used for force spectroscopy: in this type of measurement, the tip is moved towards the sample in the normal direction. The position of the tip and the deflection of the cantilever are recorded and can be converted to force-versus-distance curves, or, briefly, force–distance curves. These curves contain information about the mechanical properties of the sample; one may thus try to extract local material parameters from them.

Indeed, the AFM has been used to probe the elastic properties of cells (such as their Young's modulus) (Radmacher 2002; Kosta 2004). Yet, obtaining truly *local* information still poses a formidable challenge. Apart from several complications associated with the inhomogeneous cell surface and intracellular structures beneath the lipid bilayer, one particularly notable difficulty emerges: the boundary conditions of the cell membrane away from the spot where the AFM tip indents are not known. This precludes a quantitative interpretation of the measured force, i.e., a clean way to translate this force into local material properties.

Appropriate model systems with defined geometry and membrane composition are thus required. One possible realization which complies with these requirements is the "nanodrum" setup.

9.3.1.3 The nanodrum setup

The nanodrum setup consists of a lipid bilayer membrane which is spread over an adhesive substrate featuring circular pores of well-defined radius (Steltenkamp *et al.* 2006; Steltenkamp 2006). The boundary conditions at the rim of the drum are known. Centrally indenting it with the AFM tip yields force–distance curves for a system with a precisely defined geometry.

The first measurements using the nanodrum setup were done on fluid DOTAP and gel-like DODAB membranes (see Figs. 9.2(b) and (c) in Section 9.1.1). As a substrate, planar gold-coated alumina was used, onto which a 3-mercaptopropionic acid (MPA) monolayer was chemisorbed. The negatively charged MPA monolayer ensured that the positively charged membrane could spread evenly on the substrate (see Fig. 9.7(a)). The radius of the drum was either 33.5 nm or 90 nm.[10]

To obtain force–distance curves, a DOTAP or DODAB membrane was suspended over a pore and subsequently probed with an AFM tip. Remarkably, the curves were linear even for significant indentations, a finding in agreement with the initial regime of membrane tether pulling (Powers *et al.* 2002; Derényi *et al.* 2002). In the following, we will see how such a behavior can be understood with the help of the theoretical tools presented above.

9.3.1.4 Theoretical model

In our modeling, we consider a flat, solid substrate with one circular pore of radius R_{pore}. The lipid bilayer membrane is adsorbed evenly onto the substrate and spans the pore. We assume that the AFM tip has a parabolic shape with a curvature radius R_{tip} at its apex. Furthermore, we restrict ourselves to the static axisymmetric situation, in which the tip pokes the free-standing membrane exactly in the middle of the pore (Fig. 9.7(b)). It is then sufficient to consider a cross section of the setup. The point at which the membrane detaches from the tip is referred to as the "contact point". For a certain downward force $F > 0$, the membrane is indented to a corresponding depth $h_0 > 0$, which is measured from the plane of the substrate to the depth of the apex of the tip. Note that it is also possible to pull the membrane up with a force $F < 0$ in the opposite direction if attractive interactions attach the membrane to the tip.

[10] More experimental details can be found in Steltenkamp *et al.* (2006) and Steltenkamp (2006).

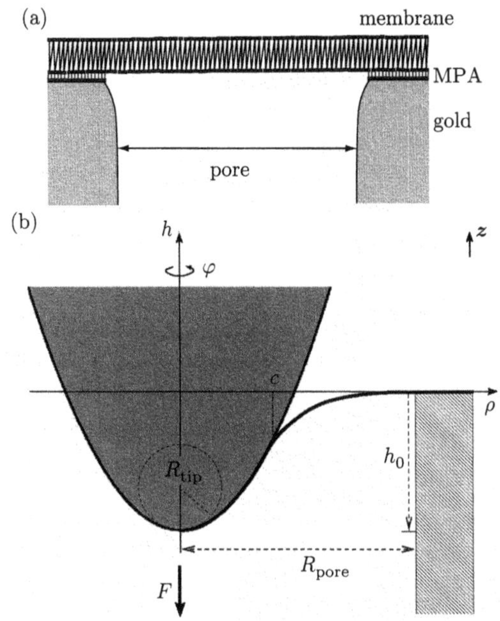

Fig. 9.7 (a) Lipid bilayer suspended over pores. In the experiment, it was subsequently probed with an AFM (Steltenkamp *et al.* 2006). (b) Illustration of the geometry. A parabolic tip with curvature radius R_{tip} indents a pore-spanning membrane with a force F to a certain depth h_0. The radius of the pore is R_{pore}. The membrane detaches from the tip at a radial distance $\rho = c$.

The total energy of the pore–tip system comprises two different contributions. First, the membrane is under a *lateral tension* σ. To pull excess area into the pore, work has to be done against the adhesion between the membrane and the flat substrate.[11] This work is given by σ times the excess area (David and Leibler 1991). Second, as we have seen before, a *curvature energy* is associated with the membrane. Summing the two contributions and assuming that the bilayer has no spontaneous curvature (which is true for the membranes used in the experiment), we arrive at the Canham–Helfrich Hamiltonian for an up–down-symmetric membrane (see eqn (9.2), with $K_0 = 0$). The Gaussian curvature term can be safely discarded again in the following, as it yields zero in our specific problem.

The route we want to follow here in order to find force–distance curves is to determine the equilibrium shape of the nonbound section of the membrane for a given indentation h_0.[12] To this end, the nonlinear shape equation has to be solved numerically. One may, however, additionally consider configurations where the membrane is

[11] The part of the system outside the pore just acts like a reservoir fixing the lateral tension. Its energetics do not have to be considered explicitly.

[12] The profile of the bound part is equivalent to the (known) shape of the tip, $h(\rho) = -h_0 + \rho^2/(2R_{\text{tip}})$, for $\rho \leq c$.

indented only a little and gradients are small. In this case, which we consider here, we can use the linearized shape equation (9.7).

The solution to this equation is a linear combination of the eigenfunctions of the Laplacian corresponding to the eigenvalues 0 and λ^{-2} (see the Exercise in Section 9.2.4). For rotational symmetry, it is given by $h(\rho) = h_1 + h_2 \ln(\rho/\lambda) + h_3 \, \mathrm{I}_0(\rho/\lambda) + h_4 \, \mathrm{K}_0(\rho/\lambda)$, where I_0 and K_0 are the modified Bessel functions of the first and the second kind, respectively (Abramowitz and Stegun 1970).

The constants $h_1, \ldots, h_4$ are determined from the appropriate boundary conditions,

$$h(R_{\text{pore}}) = 0, \quad h(c) = -h_0 + \frac{c^2}{2R_{\text{tip}}}, \tag{9.11a}$$

$$h'(R_{\text{pore}}) = 0, \quad h'(c) = \frac{c}{R_{\text{tip}}}, \tag{9.11b}$$

$$\text{and} \quad h''(c) = \frac{1}{R_{\text{tip}}}, \tag{9.11c}$$

where a dash denotes a derivative with respect to ρ. The equations (9.11a) follow simply from the requirement of continuity at the pore rim and the point where the membrane leaves the tip. Requiring a membrane that has no kinks and thus no diverging bending energy gives eqns (9.11b). Even though the differential equation is of fourth order, *five* conditions are required owing to its moving-boundary nature, *i. e.*, c is determined from an energy balance, which yields eqn (9.11c).

The solution of the boundary value problem (9.7), (9.11) can be used to calculate the force for a prescribed indentation in the following way: one can now insert the profile of the free membrane together with the shape of the bound part back into the functional (9.5) to obtain the energy of the equilibrium solution. This energy will then depend parametrically on the indentation h_0. Its derivative with respect to h_0 yields the force F.

The results one obtains with this theoretical model explain the linear behavior of the measured force–distance curves remarkably well (Norouzi *et al.* 2006). However, the linearity of the curves makes a unique extraction of the two main mechanical system properties, the lateral tension and the bending modulus, difficult. One possible way to overcome this impasse is by including the influence of adhesion between the tip and the membrane: numerical calculations show that force–distance curves then cease to be linear; hysteresis and nonzero detachment forces can show up. These rich characteristics offer one possibility to uniquely deduce membrane parameters such as the bending rigidity on the nanoscale (Norouzi *et al.* 2006).

However, changing the adhesion between the tip and the pore-spanning lipid membrane in a controllable way is experimentally challenging. An alternative way to learn more about the nanodrum system is by looking at membranes under high lateral tension. The bending of the membrane can then be neglected, and stretching will be the main contribution to the elastic behavior. In recent experiments, this case was considered and rupture tensions of the membrane could be determined (Mey *et al.* 2009). At the same time, the first experiments with biological cell membranes were made (Lorenz *et al.* 2009; Fine *et al.* 2009). Despite its recent introduction, the nanodrum setup has thus been proven useful for the determination of some of the elastic parameters of lipid bilayer membranes on the nanoscale.

9.3.2 The saddle-splay modulus $\bar{\kappa}$

The saddle-splay modulus, also called the Gaussian modulus in the literature, cannot be determined by the techniques of the previous section, since all of them rely on bulk properties. As mentioned before, the term containing $\bar{\kappa}$ in the energy functional (9.2) is equal to the sum of a topological constant and a boundary integral. One might therefore think that $\bar{\kappa}$ would be biologically irrelevant.[13] There are, however, cases where this term can play a role: in the Golgi apparatus, which is the organelle that sorts and distributes substances in the cell, topology changes occur all the time owing to the creation and absorption of small membrane vesicles.

Several approaches have been suggested to determine the value of $\bar{\kappa}$. A first estimate can be found by modeling the membrane as a thin homogeneous plate (Landau and Lifshitz 1986). One obtains $\bar{\kappa} = -(1-\nu)\kappa$, where ν is Poisson's ratio.[14] Its value lies between -1 and $\frac{1}{2}$. Consequently, $-2 \le \bar{\kappa}/\kappa \le -\frac{1}{2}$. This is not very exact, since membranes are inhomogeneous structures composed of two leaflets, but it yields an estimate of the range of possible values of $\bar{\kappa}$.

More reliable values can be obtained from systems with topology changes. Siegel and Kozlov (2004) found $\bar{\kappa} \approx -0.8\kappa$ by considering fusion intermediates at the phase transition between the fluid lamellar phase and the inverted cubic phase of a DOPE–Me membrane. Membrane vesicles that contain more than one lipid bilayer phase have also been used for measurements (Baumgart *et al.* 2005; Semrau *et al.* 2008). A change in the boundary between two phases will change the integrated Gaussian curvature of the system as well. This allows one to determine the difference $\Delta\bar{\kappa} := \bar{\kappa}_1 - \bar{\kappa}_2$ between the saddle-splay moduli $\bar{\kappa}_i$ $(i = 1, 2)$ of a two-phase system. In Semrau *et al.* (2008), vesicles consisting of cholesterol, sphingomyelin, and DOPC were considered. This system phase-separates into a liquid ordered and a liquid disordered phase (see Section 9.1.1 and the following section). Semran *et al.* found $\Delta\bar{\kappa} = 2.0 - 4.0 \cdot 10^{-19}\,J$, with $\kappa_{L_\mathrm{d}} = 20k_\mathrm{B}T$ and $\kappa_{L_\mathrm{o}} = 80k_\mathrm{B}T$. Apart from these few measurements, not much is known about the saddle-splay modulus. The development of further experiments remains a challenge for future research.

This section has presented some of the relevant techniques which allow one to determine the elastic parameters of a membrane. Once these parameters are known, problems of membrane mechanics can be addressed quantitatively. One particularly interesting problem in this context involves the budding and fission of model membrane systems, which will be discussed in the final section of this chapter.

[13] For example, closed vesicles have a fixed topology and do not have a boundary.

[14] Poisson's ratio is the ratio of the contraction of a material (perpendicular to the applied load) to the extension of the material (in the direction of the applied load).

9.4 Budding and fission of model membrane systems

Ternary mixtures of lipids including cholesterol show segregation of domains in artificial vesicles at room temperature. In the last decade, numerous experiments have tested the mechanical, thermal, and chemical stability of these inhomogeneous systems. Special interest has concerned situations close to in vivo cell processes such as absorption of PLA_2 proteins and osmotic shocks. We adapt the Canham–Helfrich model, extended to these systems.

9.4.1 Formation of lipid rafts

Internal organization is one of the most intriguing aspects of the cell. Living cells have to actively maintain gradients of all sorts. Compartmentalization and trafficking aid them in doing so, and both processes extensively use membranes. Not only are the various organelles in eukaryotic cells surrounded by membranes, but the basic intermediates in the intracellular transport pathways are also membrane structures such as tubes and vesicles (Alberts *et al.* 2002). The generation and properties of these structures have been extensively studied, and much is already known about their biology, biochemistry (Campelo *et al.* 2007; Taniguchi 1996) and biophysics (Seifert 1997; Roux *et al.* 2002). The emerging view is that the shape of the bilayer membrane *in vivo* is controlled not only by embedded and associated proteins but also to a large extent by the mechanical properties of the bilayer itself (Heinrich *et al.* 1999).

The existence of small membrane domains with a lipid composition that is markedly different from that of the rest of the membrane (sometimes referred to as "rafts", although considerable debate remains as to their precise interpretation) appears to be another key element of intracellular vesicular traffic (Edidin 2003). *In vivo*, it is difficult to observe lipid rafts, domains enriched in cholesterol (see also Section 9.1.1). Their size is estimated to be about a dozen nanometers, and it is highly probable that they are very unstable. Indeed, the plasma membrane of a cell is very inhomogeneous and very crowded, with attached proteins as numerous as the lipids. This explains why lipid rafts, supposed to be floating structures in a very crowded sea of lipids, are unstable. On the one hand, it is widely believed that they may be involved in several cell functions, such as signal transduction, lipid trafficking, cell migration, and protein displacements (Simons and Ikonen 1997). On the other hand, doubts concerning their existence have induced strong debates. In recent decades, these rafts have been observed experimentally in inhomogeneous lipid vesicles, having a size comparable to the vesicle radius, and their fabrication is now standard in giant unilamellar vesicles (GUVs) and also in lipidic tubes. Owing to their size, of the order of microns in such model systems, and to avoid useless debates, these cholesterol-enriched domains are called liquid ordered domains (or L_o domains), as opposed to liquid disordered (L_d) domains. From a microscopic viewpoint, a complete model of domain formation based on microscopic molecular interactions is still lacking, but a mathematical model based on a microscopic theory of coagulation and fragmentation has been developed to explain raft formation in lipid bilayer membranes (Richardson *et al.* 2007). From a mesoscopic point of view, L_o domains can be interpreted as resulting from a thermodynamic phase separation of the lipid species. Typically, this process triggers shape deformations in the multicomponent membrane, leaving striking and complex shapes (Jiang *et al.* 2000; Taniguchi 1996). More complete models incorporating coarse-grained molecular dynamics have been studied in order to explain the equilibrium phase separation in binary lipid mixtures at the liquid–gel phase coexistence point (Shi and Voth 2005).

Experimentally, phase separation has been confirmed in GUVs whose composition mimics those of cellular membranes (Dietrich *et al.* 2001; Veatch and Keller 2005). In spite of their "denser" structure, L_o domains conserve their fluidity, and the elastic description is similar for the two phases, each of them having its own set of physical constants, such as the bending modulus κ, the saddle-splay modulus $\bar{\kappa}$, and the surface tension σ. The Canham–Helfrich Hamiltonian (9.2) for each domain is then

$$H_i = \int_{\Sigma_i} \mathrm{d}A \left[\sigma_i + \frac{\kappa_i}{2}(K - K_{0i})^2 + \bar{\kappa}_i K_G\right]. \tag{9.12}$$

The index i refers to L_o or L_d domains, and the surface integral covers the entire surface of phase i, Σ_i, which may or may not be connected in the case of multidomain vesicles. The total energy for the vesicles becomes

$$H_{\text{tot}} = \sum_{i=1,2} (H_{L_o} + H_{L_d}) + \tau_C \int_C \mathrm{d}l - P \int \mathrm{d}V, \tag{9.13}$$

where C is the boundary between the phases. τ_C is the line tension between different domains containing different lipids; it results from the segregation of these lipids into different domains. The two layers of the membrane are generally assumed to be symmetric—both phases contain cholesterol molecules, which have a high flip-flop rate, and any stress due to area differences between the leaflets or to asymmetry of the layers is thus quickly relaxed. Very often, the spontaneous curvature K_{0i} is neglected. The interface between the two phases is described both by a jump in the values of the bending rigidities κ_i, $\bar{\kappa}_i$ and the surface tension σ_i and by a positive line tension τ_C at the interface C. As in the case of homogeneous vesicles, a variational approach allows one to determine the geometry of these domains and the global geometry of inhomogeneous vesicles and to study their stability to small perturbations.

Various experiments have shown that L_o domains can be extracted from an L_d domain. Such experiments have used various methods, such as tubular deformation (Allain *et al.* 2004), osmotic shocks (Baumgart *et al.* 2003), and absorption of external molecules such as proteins or detergents (Staneva *et al.* 2004, 2005). In all these processes, the line tension τ_C between domains plays a crucial role at the fission stage but can also drive the instability that produces the final shape transformation. An increase in line tension tends to reduce the interface length between domains, favoring fission and budding (Jülicher and Lipowsky 1996).

9.4.2 Budding and fission processes

Budding is a shape transformation process that occurs in both homogeneous and inhomogeneous lipid vesicles: a portion of the membrane is bent slightly to form a small bud, having the shape of a spherical cup. The resulting bud can either be separated from the larger vesicle in a multistep growth process, which normally ends with the expulsion of a *daughter vesicle*, or remain connected to it through a small membrane neck. Resembling a catenoidal segment, this neck has a vanishing mean curvature. This process can be triggered by various physical factors, such as the heating of the vesicle (Döbereiner *et al.* 1993), adsorption of proteins or detergent molecules at the interface between domains (Staneva *et al.* 2004, 2005), and changing the osmotic pressure (Döbereiner *et al.* 1993), which occurs at constant temperature. Domain formation and the subsequent budding process have been experimentally confirmed by means of fluorescence microscopy of GUVs (Baumgart *et al.* 2003, 2005; Bacia *et al.* 2005). The shapes adopted by lipid membranes can be classified into two regimes. In the *high-tension regime*, they look very similar to those of fluid interfaces, and are characterized by a constant mean curvature (Allain and Ben Amar 2006). In contrast, in the *low-tension regime* the shape is governed by bending elasticity. The morphological transitions in this case involve the formation of a small neck between the small protrusion and the *mother vesicle.* The presence of a membrane tension acts to suppress the budding process. The fission process is a later stage of budding and involves a striking topological change in the mother vesicle. To obtain a totally separated daughter vesicle, we need to destabilize the narrow neck produced during the budding transition.

Having determined some physical and biological properties of real fluid surfaces, let us now examine some examples of budding and fission (see Fig. 9.8). We focus on a budding process induced by proteins called PLA_2, which are known to play a major role in the traffic inside the cell.

9.4.3 Instability induced by adsorption of proteins in inhomogeneous vesicles

A spectacular budding process (Staneva *et al.* 2004) can be obtained by injection of PLA_2 (phospholipase A_2). Before injection, the GUV membrane is in a stable, nearly spherical state. However, high-quality images of vesicles reveal two spherical caps, one for each phase: the L_o and the L_d phase (Baumgart *et al.* 2003). These two caps have radii of the same order of magnitude (about 5μm, depending on the

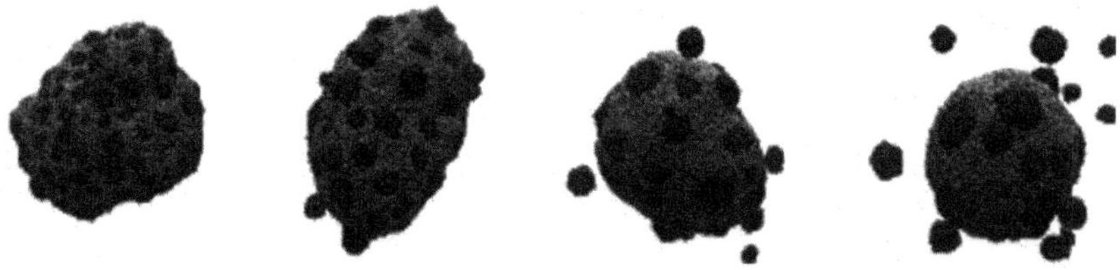

Fig. 9.8 Numerical evidence of budding and fission processes in a phase-separating vesicle. Reprinted figure 2 with permission from (Laradji and Sunil Kumar, 2004). Copyright 2004 by the American Physical Society.

experimental conditions) and are separated by a discontinuous interface. A few seconds after injection of PLA_2 with a micropipette in the vicinity of one L_o domain, a rather strong destabilization of the initial shape is observed: the L_o domain tries to rise. The discontinuity in slope at the interface between the two caps becomes more and more pronounced. The lifting can be strong enough to expel a raft from the vesicle. When two or three rafts are present initially, successive expulsions can be observed. The driving force for this deformation is the absorption of proteins, which locally deforms the membrane. There are no chemical reactions.

Let us examine how we can describe physically the absorption of these proteins by the outer leaflet of the membrane. Both phases (L_o and L_d) absorb the proteins as soon as they are introduced, but probably with different affinities. These proteins are not soluble in water, so we think that they remain localized on the membrane, and we neglect possible exchange with the surrounding bath. As a consequence, the number of these molecules remains constant. Moreover, we assume that the proteins cannot cross the interface. As suggested by Leibler (1986), the average curvature is coupled to the protein concentration, for two possible reasons. First, this could be due to the conical shape of the proteins, which locally causes a deformation of the membrane. Second, an osmotic pressure acting on the membrane results from the part of the protein in the water (Bickel *et al.* 2000). Whatever the microscopic effect, the proteins force the membrane to tilt nearby, and thus induce a local curvature. We define the energy due to proteins in the phase i as

$$H_{p_i} = \int_{\Sigma_i} \left[\Lambda_i K \phi + \lambda_i \phi + \left(\frac{\alpha_i}{2} (\phi - \phi_{\mathrm{eq}_i})^2 + \frac{\beta_i}{2} (\nabla \phi)^2 \right) \right] \mathrm{d}A, \tag{9.14}$$

where ϕ is the concentration of proteins on the surface. The coupling constant is Λ_i. In eqn (9.14), λ_i is a Lagrange multiplier which allows one to maintain the number of proteins constant in each phase. The model can easily be changed by considering λ_i as the chemical potential of the proteins. In this case, the proteins are free to move everywhere on the membrane, to cross the interfaces, or to go into the surrounding water. The last term in eqn (9.14) is a Landau expansion of the energy needed to absorb proteins on the surface near the equilibrium concentration ϕ_{eq}. The ϕ gradient indicates a cost in energy to be paid for a spatially inhomogeneous concentration. A linear perturbation analysis shows that there is a critical number of proteins absorbed in a domain, $R_i^2 \Phi_i$, which induces an instability in this domain (Allain and Ben Amar 2004):

$$R_i^4 \Phi_i^2 = \frac{2}{\alpha_i} \left(\sigma_i + \frac{1}{2} \phi_{\mathrm{eq}_i}^2 + \kappa_i q^2 - \frac{\Lambda_i^2 (q^2 - 2)}{\alpha_i + \beta_i (L_c / R_i)^2 q^2} \right), \tag{9.15}$$

q being a number close to unity. But, being linear, this approach concerns only the first stages of the instability, and we cannot draw conclusions about the final evolution of the vesicle: do we reach an equilibrium shape whose geometry is slightly modified by the inclusion of proteins, or does such a linear treatment explain the first stages of a catastrophic instability of the vesicle that ends with complete fission and the birth of a daughter vesicle, as shown by experiment? Indeed, the experiments suggest that both answers are possible, depending on the amount of proteins. Theoretically, one needs to perform a nonlinear dynamical calculation. In the next section, we suggest a model of fission in the case where the tension dominates over the elasticity, a limit valid experimentally and which allows one to understand the fission mechanism analytically.

9.4.4 Fission of an inhomogeneous vesicle

When capillarity dominates, one can find an easy solution for axisymmetric vesicles made of two domains or, eventually, three domains if axisymmetry is preserved (Allain and Ben Amar 2006). Assuming that the bending and saddle-splay rigidity are negligible (an approximation which can be corrected afterwards via a boundary layer analysis), the shape of a two-domain vesicle consists of two spherical caps defined by a set of four geometrical parameters: the radii of the two caps R_1, R_2 and the two angles at the boundary θ_1 and θ_2 (Fig. 9.9). The contact between the two caps gives the first continuity relation,

$$R_1 \sin\theta_1 = R_2 \sin\theta_2. \tag{9.16}$$

Mechanical equilibrium leads to a "Laplace" relation between the surface tension σ and the pressure P, $\sigma_i = R_i P/2$, and to a relation between the geometry and the line tension at the boundary between the domains,

$$R_1^2 \sin\theta_1 \cos\theta_1 = R_2^2 \sin\theta_2 \cos\theta_2 - \frac{2\tau_l}{P}. \tag{9.17}$$

The shape is controlled by the reduced line tension $\tau = \tau_l/P$ (homogeneous for a surface), which can be adjusted by changing the osmotic pressure. As an example, from Fig. 1(b) of Baumgart *et al.* (2003), we have calculated $\tau = 20.5\,\mu\text{m}^2$ from estimated values of R_1, R_2, θ_1, and θ_2. Notice that in Baumgart *et al.* (2003), the label 1 corresponds to the L_d domain and the label 2 to the L_o domain.

Solving eqns (9.16) and (9.17) for the above value of τ gives two possible solutions: $R_1 = 5.30\,\mu\text{m}$, $R_2 = 10.5\,\mu\text{m}$, $\theta_1 = 1.34$, and $\theta_2 = 0.514$, which are the measured values (Fig. 9.10(a)), but also $R_1 = 3.97\,\mu\text{m}$, $R_2 = 10.3\,\mu\text{m}$, $\theta_1 = 1.96$, and $\theta_2 = 0.364$ for the second solution (Fig. 9.10(b)). In order to explain why the first solution is preferred in the experiment, we calculated the energy, which was restricted here to two contributions: $F_{\text{TOT}} = -PV + \tau l$, where l is the perimeter of the interface C. Using a typical length scale $L_r = 10\,\mu\text{m}$, it is possible to construct a dimensionless energy $\tilde{F}_{\text{TOT}} = F_{\text{TOT}}/(\pi P L_r^3)$. Note that the value of L_r does not affect the physics of the problem, it is used only to have dimensionless lengths close to 1. So, we obtain

$$\tilde{F}_{\text{TOT}} = -\frac{V}{\pi L_r^3} + \frac{2\tau R_1 \sin\theta_1}{L_r^3}, \tag{9.18}$$

which gives -1.380 and -1.377, respectively, and shows that the experimentally observed solution is stable while the other one is unstable, as expected.

A systematic study of the pair of solutions for arbitrary values of τ is straightforward, and the results are presented in Fig. 9.11. Figure. 9.11(a) is a classical bifurcation diagram, where a pair of solutions with opposite stability appears. In this problem, τ is the control parameter and the energy $\tilde{F}_{\text{TOT}}$ is the order parameter. These two solutions differ geometrically, the unstable solution having a smaller neck compared with the stable one (obvious if $\tau = 0$). As τ increases, the two solutions become geometrically closer, up to a finite value of $\tau_c = R_1R_2/2$. Above the critical value ($\tau \geq \tau_c$), there is no connected solution of the Euler–Lagrange equations but the solution consisting of two separated spheres remains.

This bifurcation diagram describing a change in the topology of budding spheres is similar to that found for the catenoid problem, where a soap film is fixed on two parallel rings, separated by a small distance d compared with the radius of the rings. This is the same type of capillary instability, where there exist two similar solutions whose stabilities are governed by the energy. As the control parameter, here the effective line tension, is increased, the two solutions merge and a change of topology is expected at this point. We do not know if this change is irreversible, since fission

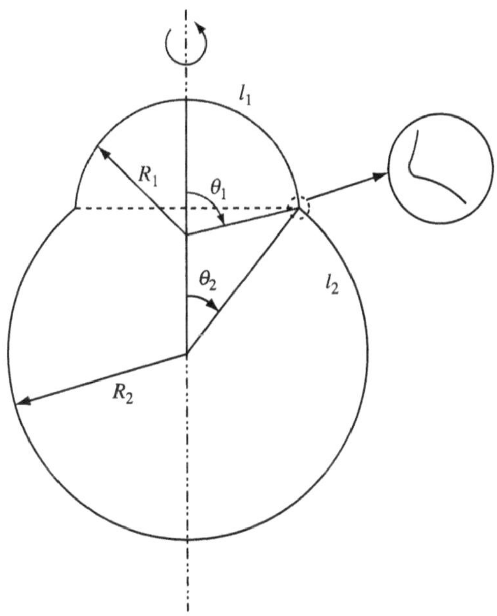

Fig. 9.9 Schematic representation of an axisymmetric vesicle, including the four parameters R_1, R_2, θ_1, and θ_2. The circle details the fold near the interface, where the elastic properties can no longer be neglected.

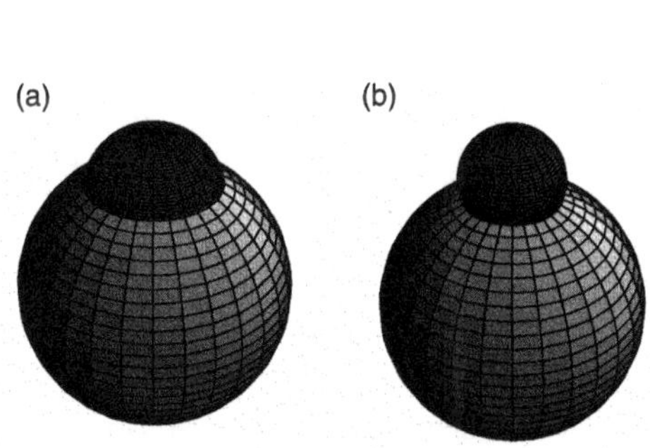

Fig. 9.10 The two solutions of eqns (9.16) and (9.17) for $\tau = 20.5\,\mu\text{m}^2$.

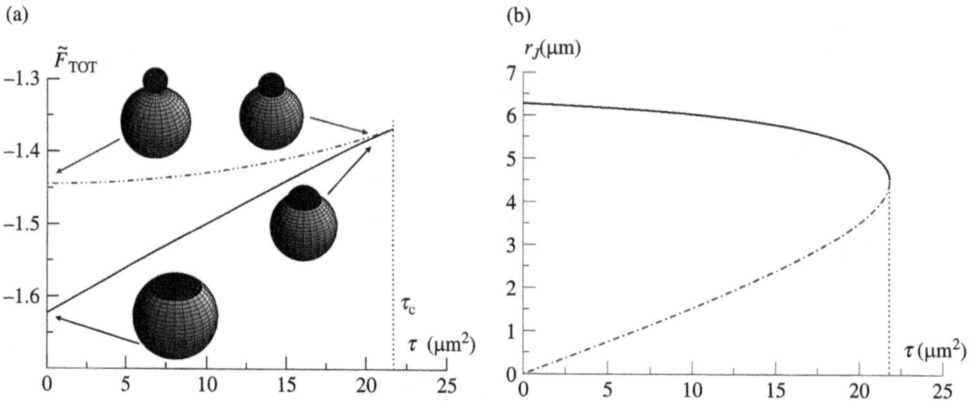

Fig. 9.11 Dimensionless energies $\tilde{F}_{\mathrm{TOT}}$ (a) and interface radius r_J (b) of the solutions of the Euler–Lagrange equations (eqns (9.16) and (9.17)) versus the control parameter $\tau = \sigma/P$. The calculat[illegible] was done with areas $A_1 = 136\,\mu\text{m}^2$ and $A_2 = 1296\,\mu\text{m}^2$. In both figures, the solid line cor[illegible]nds to the stable solution, experimentally observable; the dashed line corresponds to the [illegible]ble solution; and the dotted line corresponds to $\tilde{\tau} = \tilde{\tau}_c$, the critical value of the control [illegible]meter. For $\tilde{\tau} \geq \tilde{\tau}_c$, there is no longer a solution. Four pictures of vesicles have been added f[illegible] better understanding of the associated shapes.

requires a microscopic reorganization such as hemifission (Kozlovsky and Kozlov 2003). Experimentally, the daughter vesicles can remain connected by a small filament of lipids, but if the two vesicles move away from each other, the process is of course not reversible. The critical value τ_c is determined by eqn (9.17). Equation (9.17) gives the equilibrium of the forces in the radial direction (the r axis). The term in $\tau = \tau_l/P$ has the effect of pinching the membrane. The two others terms (in R_1^2 and R_2^2) are related to the pressure force on the membrane. The critical value τ_c is the value of the maximal force on the membrane. For a higher line tension (or smaller pressure), it is no longer possible to compensate the line tension, which splits the system into two independent vesicles. Such a treatment shows that the pressure is a crucial parameter, and osmotic shocks can induce the breakdown of inhomogeneous vesicles. It also shows that there is a region of high curvature (a fold) at the boundary between the domains, where elasticity becomes important. A boundary layer analysis shows that the capillary picture that we have is not modified by elasticity, which simply softens the cusp at the boundary at small scales, a result in agreement with our physical intuition and with experiments (Baumgart *et al.* 2003). Protein absorption does not change the zeroth-order shape equations of the stretched vesicle, which are dominated by tension. It modifies the shape of the fold near the interface, giving a new contribution to the effective line tension. If the absorption takes place in the external leaflet, it contributes to an increase in the line tension. This increase puts the system closer to the bifurcation point controlled by the parameter $\tilde{\tau} = (\tau_C + \tau_{\mathrm{cusp}})/P$ and induces a budding of the smaller phase: as $\tilde{\tau}$ increases, the neck radius decreases and the small domain seems to lift up. If the concentration is high enough that $\tilde{\tau} > \tau_c$, the budding is automatically followed by a fission process, creating two separated vesicles, one for each phase. If the concentration is not high enough, the lift-up will stop before the change of topology. In the meantime, it is possible that the vesicle may lose some of its molecules and retract to its initial configuration. This reversibility is impossible when fission is complete, for two reasons: first, the system relaxes the Gaussian elastic energy and two daughter vesicles are energetically favored, and second, owing to thermal fluctuations, the vesicles move away from each other and a coalescence process is unlikely. The fact that fission occurs proves that the timescale for fission is much smaller than that of the possible rearrangement of molecules between the leaflets.

9.4.5 Multiphase membrane tube

Lipid vesicles are not the only model system of interest for biology. Many lipid tubes are involved inside the cell, and also between cells. These tubes do not exist in equilibrium, and require either forces at both ends or at least a connection to a reservoir. This reservoir can be the membrane of the cell, the membrane of the organelle, or a vesicle for experiments *in vitro*. For tubular structures, mechanical effects play a major role. Biomimetic experiments (Roux *et al.* 2002) have shown that kinesin motors attached to microtubules can exert pulling forces on the membrane and prompt the formation of tubular structures that closely resemble similar structures identified in living cells. The physics of membrane tube formation from homogeneous vesicles has been studied both theoretically (Powers *et al.* 2002; Derényi *et al.* 2002) and experimentally (Heinrich *et al.* 1999). Experiments have been done to study the interplay between lipid domains and the behavior of tubes by pulling tubes from heterogeneous model membranes. Figure 9.12 illustrates one of the surprising conclusions of these experiments—a sequence of snapshots taken at regular intervals shows an initially homogeneous tube that first undergoes phase separation (triggered experimentally by photoinduced oxidation of cholesterol), and, after the passage of a certain amount of time, ruptures at precisely the phase boundary and disconnects. The two lipid phases are easily distinguished, once separation has occurred, by the

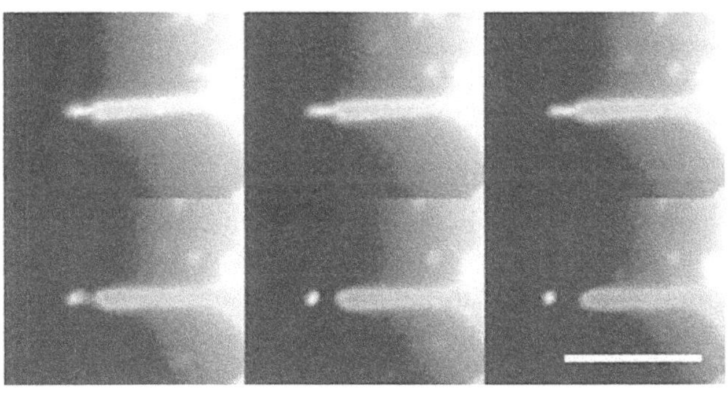

Fig. 9.12 Breakage of a heterogeneous membrane tube. Fission events occur at the sites of formation of small nonfluorescent domains resulting from phase separation. Scale bar: 10 μm.

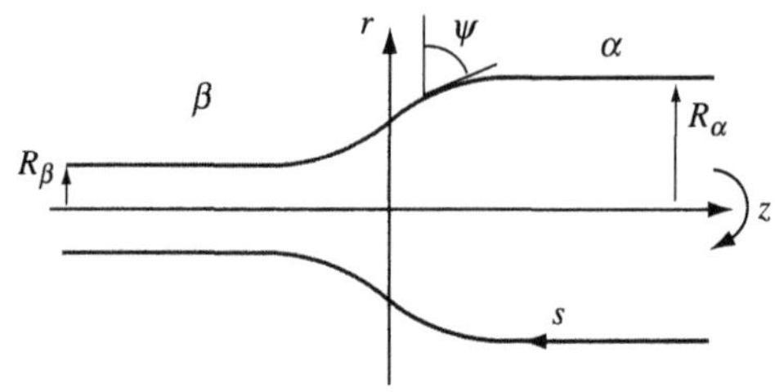

Fig. 9.13 Schematic representation of the junction.

use of a fluorescent marker that preferentially sits in the liquid disordered domains. Furthermore, the same experiments show that fission events such as these happen *only* in phase-separated tubes—tubes in which the lipids are mixed are essentially stable indefinitely. The dramatic loss of stability following phase separation can be understood from a mechanical point of view. To this end, theoretical models based on homogeneous tubes (Bukman *et al.* 1996; Božič *et al.* 2001) have been constructed that allow one to study a junction between two distinct phases, each of which has a tubular shape far away from the junction. The experiments suggest that phase separation occurs on a much faster timescale than fission, and that nucleation of the two phases leads to the formation of cylindrical domains between a more rigid and a less rigid phase. The tube radius and the junction length are generally small compared with the length of each phase domain. In order to minimize the interfacial energy between the adjacent domains, the interface rapidly becomes a circle perpendicular to the tube direction. The coarsening stage of the phase separation process proceeds very slowly, to eventually form two homogeneous phases in equilibrium, but in practice this slow relaxation is always preempted by tube fission. The experiments show that (at least to within optical resolution) tube fission occurs at the interface between the two domains. Assuming here that the tube and the junction are axisymmetric with respect to the direction along which the tube is pulled (the z-axis), we focus on one junction at the interface between two semi-infinite tubes ($\mathrm{z}=0$), each consisting exclusively of one of the phases.

The free energy of the elastic membrane must include the effect of the pulling force, whereas the pressure can be neglected:

$$H = \sum_{i=\alpha,\beta} \int_{\Sigma_i} \left[\sigma_i + \frac{\kappa_i}{2} K^2 + \bar{\kappa}_i K_G\right] \mathrm{d}A + \oint_C \tau \,\mathrm{d}l - \int f \,\mathrm{d}z. \tag{9.19}$$

The resulting Euler–Lagrange equation gives the radius of the tube at both plus and minus infinity, $R_i = (\kappa_i/2\sigma_i)^{1/2}$. In addition, mechanical equilibrium implies that the forces are equal and that $f = 2\pi(2\sigma_i\kappa_i)^{1/2}$, which imposes a relationship between the surface tensions and the bending rigidities $\sigma_\beta/\sigma_\alpha = \kappa_\alpha/\kappa_\beta$ $(= R_\alpha/R_\beta)$. The mismatch between physical constants such as the bending rigidities appears only in the boundary conditions, and strongly affects the interface shape. At the interface ($z = 0$), four boundary conditions must be satisfied (Allain *et al.* 2004; Campelo *et al.* 2007). Moreover, for a two-phase tube, the free energy $\mathcal{F}_{\mathrm{G}}$ associated with the Gaussian curvature gives exclusively a boundary contribution, as does the free energy associated with the line tension $\mathcal{F}_\tau$. $\mathcal{F}_\mathrm{G}$ is proportional to the difference in Gaussian rigidities; specifically, $\mathcal{F}_\mathrm{G} = 2\pi\,\Delta\bar{\kappa}\cos\psi(0)$ (ψ being the angle between the tangent and the z-axis); and $\mathcal{F}_\tau = 2\pi\tau r(0)$. Both processes promote the formation of a neck at the expense of the bending energy. The line tension creates a neck exactly at the interface, as it couples to the radius $r(0)$. The neck created by the discontinuity between the Gaussian rigidities is not located exactly at the interface but is slightly displaced into the phase with the lower Gaussian rigidity.

Figure 9.14 illustrates the different pinching effects that the line tension and Gaussian-rigidity differences individually have on the two-phase tube when the two bending rigidities are equal (and therefore $R_\alpha = R_\beta$). Fission can be induced either by a finite line tension or by a difference in Gaussian rigidity. A line tension has a stronger pinching effect on the tube, but a mismatch in Gaussian rigidity affects the shape of the boundary a lot. The experimental precision is not at a level where these experiments can be compared directly with calculated tube shapes (Campelo *et al.* 2007); however, the model predicts successfully a strong dependence of the fission dynamics on the elastic properties of the phases.

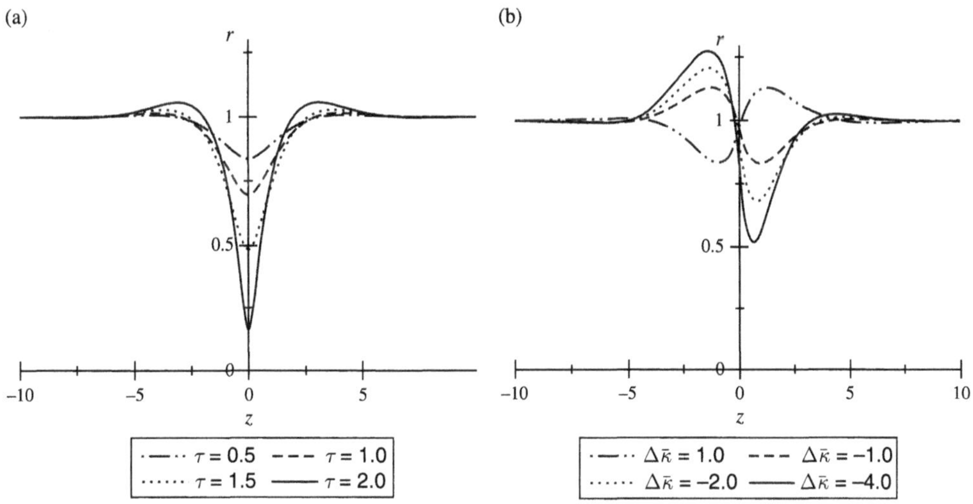

Fig. 9.14 (a) Numerical shapes of the junction for various line tensions in dimensionless units. The length scale is the radius of phase α, and the energy scale is the bending rigidity κ_α. The parameters are $R_\alpha = R_\beta = 1$, $\kappa_\alpha = \kappa_\beta = 1$, $\bar{\kappa}_\beta - \bar{\kappa}_\alpha = 0$, and $\sigma_\alpha = \sigma_\beta = 0.5$. The values of the line tension are $\tau = 0.5$, 1.0, 1.5, and 2.0. (b) Numerical shapes of the junction for various differences of Gaussian rigidity in dimensionless units. The length scale is the radius of the phase α, and the energy scale is the bending rigidity κ_α. The parameters are $R_\alpha = R_\beta = 1$, $\kappa_\alpha = \kappa_\beta = 1$, $\tau = 0$, and $\sigma_\alpha = \sigma_\beta = 0.5$. The values of the discontinuity in the Gaussian rigidity are $\Delta\bar{\kappa} = \bar{\kappa}_\beta - \bar{\kappa}_\alpha = 1.0$, -1.0, -2.0, and -4.0.

9.4.6 Periodic inhomogeneous tubular membranes

In this section, we ask the question of whether periodic, steady patterns exist for tubular membranes. Experimentally, for a composition which favors cholesterol-enriched phases, periodic tubes have been observed (Baumgart *et al.* 2003). Can we predict them in the framework of the Canham–Helfrich model, and explain their wavelength in terms of physical parameters such as the elastic coefficients and the surface and line tensions? As shown in the above section, nonperiodic inhomogeneous tubes have been experimentally observed (Roux *et al.* 2002), such as quasi-semi-infinite tubes, where a complete separation seems to be energetically preferred: each semi-infinite tube is made from a unique phase, separated from its neighbor by a unique junction. Let us assume that the chemical composition ratio is prescribed and, consequently, the ratio between the areas of the phases, A_{L_o}/A_{L_d} (see Fig. 9.15). For each phase, we integrate the free energy over the corresponding membrane area according to eqn (9.19). The length of each phase i measured along the z-axis is λ_i, and the total wavelength is thus $\lambda = \lambda_{L_d} + \lambda_{L_o}$. We choose as the unit of length the radius R_{L_o}, which is the value of the radius at its extremal value in the L_o phase (see Fig. 9.15). Our study will fix all the lengths of the pattern as a function of R_{L_o}, which is fixed by the total available mass of lipids. As previously, the interface between the two phases is described by a jump in the values of the bending and Gaussian rigidities and in the values of the surface tensions. Moreover, as is frequently done in standard experiments, we assume that at least one end of the tube is connected to a reservoir of lipids (a vesicle, for example) at infinity, so no force is required for an extension of the tube length. This imposes the condition that $\mathcal{H}_{L_o} = \mathcal{H}_{L_d} = 0$ and gives a first relation between the force f, the radius at the minimum $r_{\min} = R_{L_d}/R_{L_o}$, and the tensions $\sigma_{L_o} = \tilde{\sigma}_{L_o} R_{L_o}^2$ and $\sigma_{L_d} = \tilde{\sigma}_{L_d} R_{L_o}^2$. As f is a constant, we get

$$\begin{aligned}\frac{R_{L_o} f}{\pi} &= \frac{\kappa_{L_d}}{r_{\min}}\left[(1 - 2K_{0,L_d}\cdot r_{\min})^2 + 2\sigma_{L_d}\cdot\frac{r_{\min}^2}{\kappa_{L_d}}\right]\\ &= \kappa_{L_o}\left[(1 - 2K_{0,L_o})^2 + \frac{2\sigma_{L_o}}{\kappa_{L_o}}\right].\end{aligned} \tag{9.20}$$

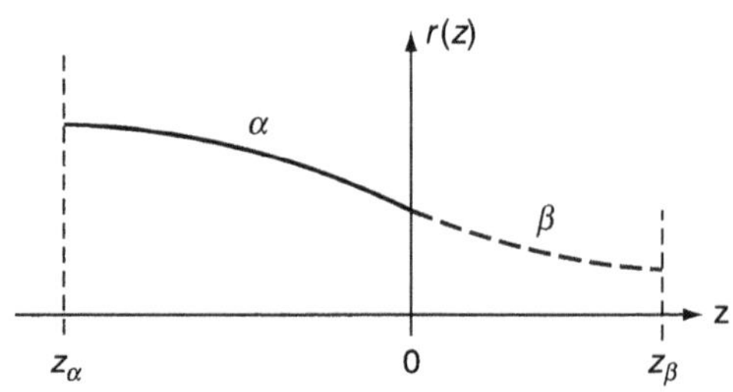

Fig. 9.15 Sketch of the geometry of a periodic biphasic membrane tube. The z-axis is the axis of symmetry of the tube. Half of the wavelength of both the L_d and the L_o phase is depicted.

The boundary conditions are due to either geometrical constraints such as the continuity of r and ψ or to mechanical equilibrium, our approach being purely static. It turns out that explicit solutions have been found, but they require a lipid membrane model incorporating both a spontaneous curvature and a pressure effect. A simplification arises for a periodic tube, since we do not have to consider nondivergent asymptotes at infinity and, in principle, we can truncate the solutions. As an example, the catenoid, which is a solution of the Euler–Lagrange equations, may be a good candidate to represent the concave phase of the tube. Exact solutions are known (Naito *et al.* 1995; Liu *et al.* 1999); they include as special limits catenoids and unduloids. Except for the catenoid, these solutions exist only in the case where the two layers of the membrane are asymmetric, the tension and the pressure being Lagrange multipliers that are functions of the spontaneous curvatures in the two phases. Nevertheless, it turns out to be impossible to satisfy all the boundary conditions for arbitrary physical elastic coefficients, since all the parameters of the solution are fixed by the spontaneous curvature of the bilayers. The only possibility remaining is the catenoid, $\sin\psi = r_{\min}/r$, which may represent the concave phase when the spontaneous curvature vanishes. It is possible to solve the Euler–Lagrange equations analytically for both phases by making a linear expansion for each phase in the vicinity of the cylindrical exact solution, $r(z) = R_i(1 + U_i(z))$, which gives a fourth-order linear differential equation. The solution has been given by Allain *et al.* (2004). The result of numerical integration of the nonlinear Euler–Lagrange equations, taking the correct boundary conditions into account, are shown in Fig. 9.16, which illustrate the effects induced in the geometry of the periodic tube by the mismatch of elastic constants between the two phases and the line tension.

A relevant example, to be compared with the experiments of Baumgart *et al.* (2003), is shown in Fig. 9.18. The ratio was fixed at $\kappa_{L_o}/\kappa_{L_d} \simeq 1.25$, and the line tension τ was approximately 9×10^{-13} N. The values of the Gaussian rigidities are hard to measure, since the Gaussian elastic energy manifests itself only in topological changes for homogeneous membranes and as boundary contributions for inhomogeneous membranes. We used the values $\bar{\kappa}_i = -0.83\,\kappa_i$. The chemical composition was set to keep the area ratio between the two phases equal to $A_{L_o}/A_{L_d} = 1.5$. We performed several different simulations for different values of the wavelength of the periodic tube. In all these calculations, the wavelength value was an input, which meant that we could choose the domain size arbitrarily. In other words, solving the Euler–Lagrange equations with the boundary conditions did not allow us to find the wavelength, a situation which is commonly found in studies of pattern formation. Since the wavelength can be chosen freely a priori, wavelength selection for a periodic tube may result from energy minimization. Fixing the surface ratio of the two phases to be equal to the chemical composition, we calculated the energy per unit length, incorporating the numerical solutions into eqn (9.19). The energy minimum corresponded to $\lambda = 4.4$ times R_{L_o}, the experimental value (Baumgart *et al.* 2003) being $\lambda \sim 5$ in the same units, so the two results are in agreement. Not surprisingly for a variational treatment, Fig. 9.17 shows that the competition between the elastic and capillary energies occurs for a wavelength of the order of the tube radius, short and very long wavelength compared with this length scale being eliminated. Recall

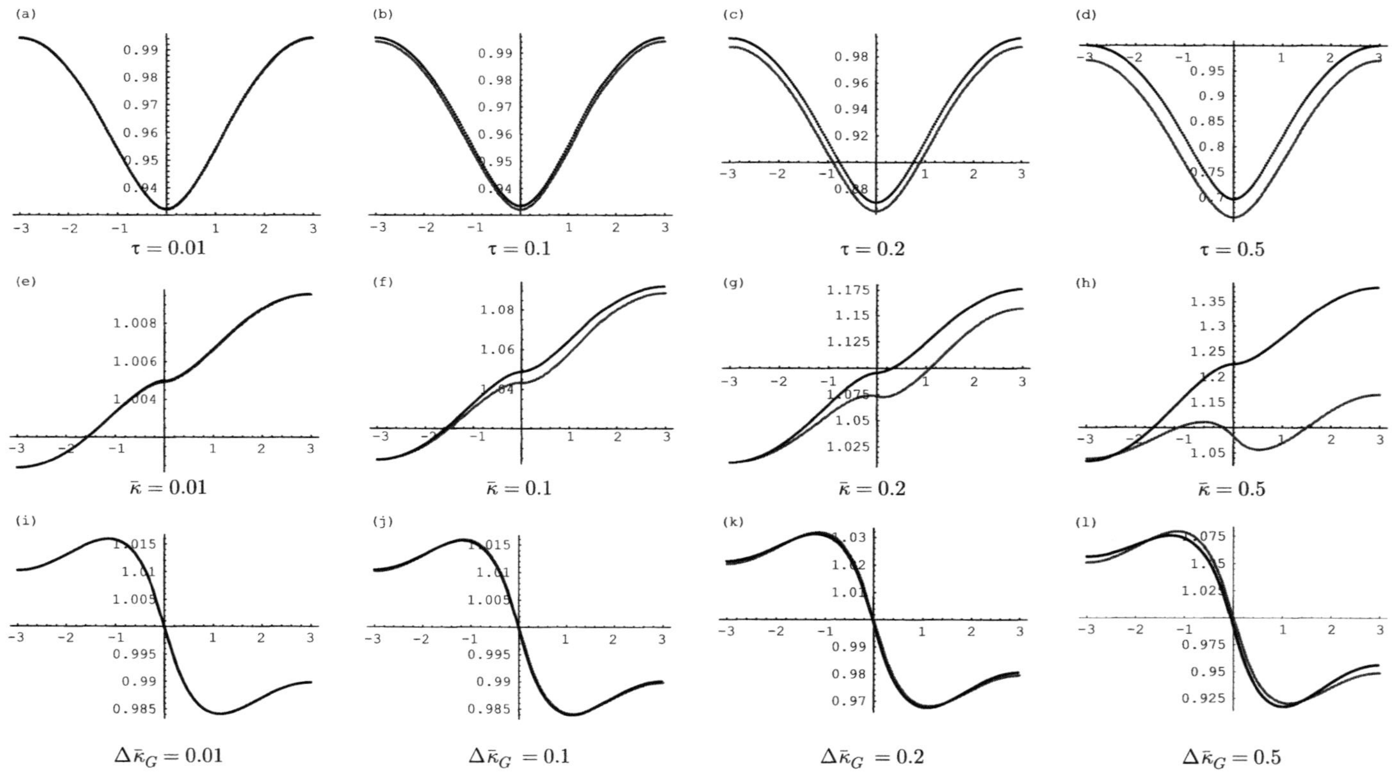

Fig. 9.16 Tube profiles $r(z)$ for a periodic tube where $z_{L_o} = 3$, $z_{L_d} = -3$, and $R_{L_d} = 1$ (see Fig. 9.15). The elastic coefficients of the L_o and L_d phases are the same, except where otherwise mentioned. In each figure, we present the solution of the linearized problem (dark curves) and of the complete nonlinear problem (light curves). We can see there that for small enough relative parameters, the two curves are almost the same in each case.

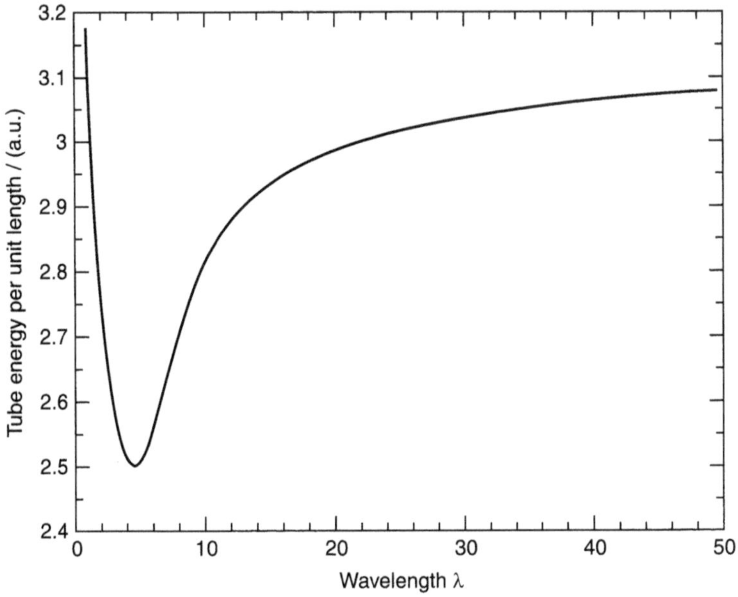

Fig. 9.17 Energy per unit length of a tube as a function of the imposed wavelength in units of R_{L_o}.

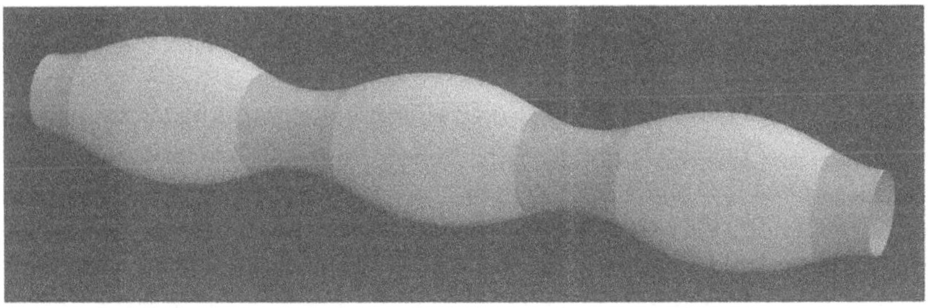

Fig. 9.18 Three-dimensional representation of the numerical tube, with the energetically selected wavelength. The light phase represents the L_o phase, and the dark phase the L_d phase.

that long wavelengths correspond to a semi-infinite tube, which was not selected for the parameters of Baumgart *et al.* (2003).

Periodic membrane tubes could be found numerically for arbitrary wavelengths once the physical parameters such as the chemical composition (see Fig. 9.19), the elastic rigidities, and the line tension were fixed. Periodic tubes are preferred to a complete phase separation as soon as the chemical composition does not favor too much of one phase, for a realistic set of physical constants. This means that the periodicity allows a decrease in the elastic energy, which compensates for the increase in capillary energy. The agreement between the only reported experiment and the

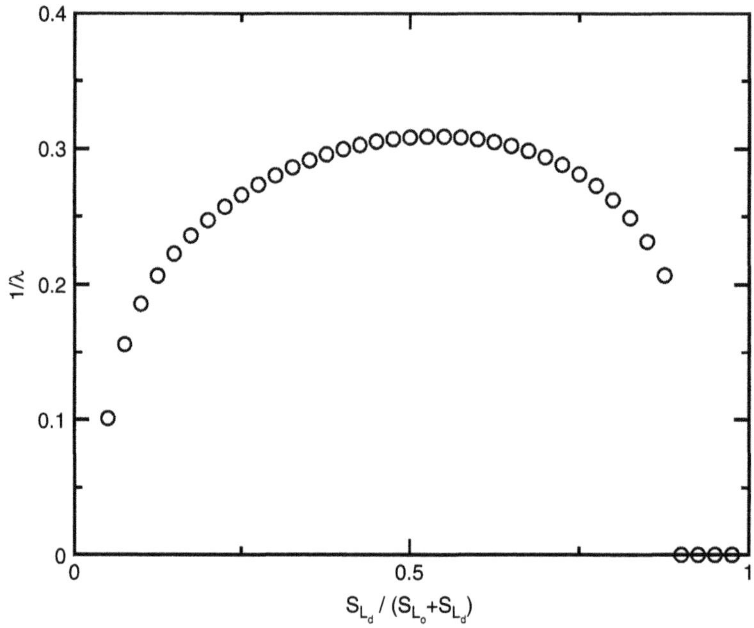

Fig. 9.19 Plot of the inverse of the selected wavelength as a function of the normalized chemical ratio. For extreme chemical ratios, semi-infinite tubes are preferred.

numerics confirms the validity of the model for tubular structures and the accuracy of the physical constants.

In conclusion, we have focused here on the strucural stability of inhomogeneous vesicles and tubes. The Canham–Helfrich model, once it has been adapted to the various phases present, is sufficient to explain both the existence of complex structures and their stability. In this chapter, we did not discuss the role of the spontaneous curvature, which is irrelevant most of the time in experiments using artificial GUVs. This spontaneous curvature has its origin in the chemical difference between the two leaflets of the membrane. It turns out to be important for cells but not for artificial vesicles. In a cell, the inner leaflet is in contact with the cytoskeleton and the cytoplasm, while the outer leaflet is in contact with the extracellular matrix, or other cells in the case of epithelium tissue. It turns out that this spontaneous curvature is responsible for a subtle modification of the boundary conditions between domains, as shown recently (Fournier and Ben Amar 2006; Trejo and Ben Amar 2011). It is known that L_o domains are thicker (approximately 6 nm thick) than L_d domains (approximately 5 nm thick). Such a relative difference, of order 10–20 percent, is huge and is ignored in the Canham–Helfrich model. This approach does not consider the membrane thickness, but assumes that the vesicle is simply a surface with physical properties. But such a mismatch between domains may have some importance for the structural and stability properties of multi-phase membranes. It turns out that only in the case of a breakdown of symmetry between the leaflets does this mismatch modify the physical properties of the lipid membrane. In fact, this mismatch is also responsible for an increase in line tension and for an apparent contact angle between domains at the interface between them. Nevertheless, after a distance of the order of the membrane thickness, one can forget this jump in thickness and the Canham–Helfrich model recovers its validity in each domain.

9.5 Conclusions

Research on the topic of this chapter has been done for about fifty years at least, but, during the last decade, a renewal of activity theoretically, numerically, and experimentally has been observed in the community. The scientific interest has moved from physicochemical processing of liposomes to theoretical biophysics with the aim of understanding better the elastic properties of these systems *in vitro* and also those of the plasma membrane *in vivo.* Lipid bilayer membranes play a major role in the biology of the cell; they are a selective barrier between the cell and the external world in the case of animal eukaryotes, and also play a major role inside the cell owing to the internal structure, made of interacting organelles. The physical flexibility of the bilayer is necessary for many crucial steps in the life of the cell, such as growth, motility, and mitosis. To describe this flexibility, a model has been proposed by Canham and Helfrich based mainly on simple physical assumptions and symmetries: the lipid membrane is replaced by a mathematical surface that also has physical properties described by an energy. Although this flexibility, which essentially means the ability to bend, is reminiscent of the bending of elastic sheets, there is no exact asymptotic derivation of the Euler–Lagrange equations for vesicles as there is for elastic sheets. A correct description of the bending of shells remains a theoretical challenge in the theory of elasticity. Nevertheless, the predictive power of the Canham–Helfrich model has been reinforced, being able to describe, for example, more complex experiments such as the budding of cholesterol-enriched domains. In parallel, sophisticated experiments have furnished the community with physical constants necessary for a complete test of the models.

We did not discuss here the transport of species across a membrane via the cooperative action of embedded proteins (see, for instance, Reynwar *et al.* (2007) and references therein) or transport inside or along the bilayer due to the fluidity of the membrane. Fluidity and elasticity are two sides of the same coin, although the physical effects and descriptions are completely different. Lipid membranes can react as a fluid or a solid depending on the deformation from the equilibrium state. This has nothing to do with the viscoelasticity known for bulk fluids, since the reaction of the membrane depends on the orientation of the displacement. The fluidity of lipid membranes is relevant in the plane of the membrane, whereas the elasticity involves the degree of freedom perpendicular to the membrane. The diffusive properties of lipid membranes have been exhaustively described in the classical books on cellular biology, mainly because of the diffusion of proteins in membranes. But experiments on GUVs involving L_o domains show obvious diffusion of these domains with a characteristic time of the order of a second, under thermal fluctuations. Membrane fluidity is thus an important property not only for lipids and proteins but also for domains. Diffusion of molecular species has been extensively studied experimentally, but the fluidity of domains has been studied much less. Nevertheless, a recent experiment has shown the possibility of a hydrodynamic viscous fingering instability in the border of a domain (Ben Amar *et al.* 2007), showing the viscous character of the lipid flow. With experimental and numerical techniques improving further, research on fluid membranes promises to lead to new developments in the near future.

Acknowledgments

This work would not have been possible without the help of our colleagues and collaborators over the past years. Specifically, we would like to thank Jean-Marc Allain, Miglena Angelova, Markus Deserno, Jean-Baptiste Fournier, Jemal Guven, Andreas Janshoff, Ingo Mey, Davood Norouzi, Ben Reynwar, Siegfried Steltenkamp, and Miguel Trejo, who have worked closely on these topics with us. Parts of this chapter are based on previous work of the authors and their collaborators.

References

Abramowitz, M. and Stegun, I. A. (eds.) (1970). *Handbook of Mathematical Functions* (9th edn.). Dover, New York.

Alberts, B., Johnson, A., Lewis, J., Raff, M., Roberts, K., and Walter, P. (2002). *Molecular Biology of the Cell* (4th edn.). Garland, New York.

Allain, J.-M. and Ben Amar, M. (2004). Biphasic vesicle: Instability induced by adsorption of proteins. *Physica A*, **337**(3–4), pp. 531–545.

Allain, J.-M. and Ben Amar, M. (2006). Budding and fission of a multiphase vesicle. *Eur. Phys. J. E* **20**(4), pp. 409–420.

Allain, J.-M., Storm, C., Roux, A., Ben Amar, M., and Joanny, J.-F. (2004). Fission of a multiphase membrane tube. *Phys. Rev. Lett.*, **93**(15), 158104.

Bacia, K., Schwille, P., and Kurzchalia, T. (2005). Sterol structure determines the separation of phases and the curvature of the liquid-ordered phase in model membranes intramembrane domains. *Proc. Natl. Acad. Sci. USA*, **102**(9), pp. 3272–3277.

Baumgart, T., Hess, S. T., and Webb, W. W. (2003). Imaging coexisting fluid domains in biomembrane models coupling curvature and line tension. *Nature*, **425**(6960), pp. 821–825.

Baumgart, T., Das, S., Webb, W. W., and Jenkins, J. T. (2005). Membrane elasticity in giant vesicles with fluid phase coexistence. *Biophys. J.*, **89**(2), pp. 1067–1080.

Ben Amar, M., Allain, J.-M., Puff, N., and Angelova, M. I. (2007). Stokes instability in inhomogeneous membranes: Application to lipoprotein suction of cholesterol-enriched domains. *Phys. Rev. Lett.*, **99**(4), 044503.

Bickel, T., Marques, C., and Jeppesen, C. (2000). Pressure patches for membranes: The induced pinch of a grafted polymer. *Phys. Rev. E*, **62**(1), pp. 1124–1127.

Božič, B., Heinrich, V., Svetina, S., and Žekš, B. (2001). Shapes of nearly cylindrical, axisymmetric bilayer membranes. *Eur. Phys. J. E*, **6**(1), pp. 91–98.

Brochard, F. and Lennon, J. F. (1975). Frequency spectrum of flicker phenomenon in erythrocytes. *J. Phys. (Paris)*, **36**(11), pp. 1035–1047.

Brown, D. A. and London, E. (1997). Structure of detergent-resistant membrane domains: Does phase separation occur in biological membranes? *Biophys. Res. Commun.*, **240**(1), pp. 1–7.

Brown, D. A. and London, E. (1998). Functions of lipid rafts in biological membranes. *Annu. Rev. Cell Dev. Biol.*, **14**, pp. 111–136.

Bukman, D. J., Yao, J. H., and Wortis, M. (1996). Stability of cylindrical vesicles under axial tension. *Phys. Rev. E*, **54**(5), pp. 5463–5468.

Campelo, F., Allain, J.-M., and Ben Amar, M. (2007). Periodic lipidic membrane tubes. *Eur. Phys. Lett.*, **77**(3), 38006.

Canham, P. B. (1970). The minimum energy of bending as a possible explanation of the biconcave shape of the human red blood cell. *J. Theor. Biol.*, **26**, pp. 61–81.

Cuvelier, D., Derényi, I., Bassereau, P., and Nassoy, P. (2005). Coalescence of membrane tethers: Experiments, theory, and applications. *biophys. J.*, **88**(4), pp. 2714–2726.

Dai, J. and Sheetz, M. P. (1995). Mechanical properties of neuronal growth cone membranes studied by tether formation with laser optical tweezers. *Biophys. J.*, **68**(3), pp. 988–996.

Dai, J. and Sheetz, M. P. (1999). Membrane tether formation from blebbing cells. *Biophys. J.*, **77**(6), pp. 3363–3370.

David, F. and Leibler, S. (1991). Vanishing tension of fluctuating membranes. *J. Phys. II (France)*, **1**(8), pp. 959–976.

de Gennes, P.-G., Brochard-Wyart, F., and Quéré, D. (2003). *Capillarity and Wetting Phenomena.* Springer, Berlin.

Derényi, I., Jülicher, F., and Prost, J. (2002). Formation and interaction of membrane tubes. *Phys. Rev. Lett.*, **88**(23), 238101.

Deserno, M. (2009). Mesoscopic membrane physics: Concepts, simulations, and selected applications. *Macromol. Rapid Commun.*, **30**(9–10), pp. 752–771.

Dietrich, C., Bagatolli, L. A., Volovyk, Z. N., Thompson, N. L., Levi, M., Jacobson, K., and Gratton, E. (2001). Lipid rafts reconstituted in model membranes. *Biophys. J.*, **80**(3), pp. 1417–1428.

Do Carmo, M. P. (1976). *Differential Geometry of Curves and Surfaces.* Prentice Hall, Upper Saddle River, NJ.

Döbereiner, H.-G., Käs, J., Noppl, D., Sprenger, I., and Sackmann, E. (1993). Budding and fission of vesicles. *Biophys. J.*, **65**(4), pp. 1396–1403.

Edidin, M. (2003). The state of lipid rafts: From model membranes to cells. *Annu. Rev. Biophys. Biomol. Struct.*, **32**, pp. 257–283.

Evans, E. and Rawicz, W. (1990). Entropy-driven tension and bending elasticity in condensed-fluid membranes. *Phys. Rev. Lett.*, **64**(17), pp. 2094–2097.

Faucon, J. F., Mitov, M. D., Meleard, P., Bivas, I., and Bothorel, P. (1989). Bending elasticity and thermal fluctuations of lipid-membranes–theoretical and experimental requirements. *J. Phys. (Paris)*, **50**(17), pp. 2389–2414.

Fine, T., Mey, I., Rommel, C., Wegener, J., Steinem, C., and Janshoff, A. (2009). Elasticity mapping of apical cell membranes. *Soft Matter*, **5**(17), pp. 3262–3265.

Fournier, J.-B. and Ben Amar, M. (2006). Effective creases and contact angles between membrane domains with high spontaneous curvature. *Eur. Phys. J. E*, **21**(1), pp. 11–17.

Gompper, G., Kaupp, U. B., Dhont, J. K. G., Richter, D., and Winkler, R. G. (eds.) (2004). *Physics Meets Biology.* Volume 19, *Matter and Materials.* Forschungszentrum Jülich GmbH, Jülich, Germany.

Gorter, M. D. and Grendel, F. (1925). On bimolecular layers of lipoids on the chromocytes of the blood. *J. Exp. Med.*, **41**(4), pp. 439–443.

Hancock, J. F. (2006). Lipid rafts: Contentious only from simplistic standpoints. *Nat. Rev. Mol. Cell Biol.*, **7**(6), pp. 456–462.

Heidemann, S. R. and Wirtz, D. (2004). Towards a regional approach to cell mechanics. *Trends Cell Biol.*, **14**(4), pp. 160–166.

Heinrich, V., Božič, B., Svetina, S., and Žekš, B. (1999). Vesicle deformation by an axial load: From elongated shapes to tethered vesicles. *Biophys. J.*, **76**(4), pp. 2056–2071.

Helfrich, W. (1973). Elastic properties of lipid bilayers: Theory and possible experiments. *Z. Naturforsch.*, **28c**, pp. 693–703.

Hochmuth, R. M., Shao, J.-Y., Dai, J., and Sheetz, M. P. (1996). Deformation and flow of membrane into tether extracted from neuronal growth cones. *Biophys. J.*, **70**(1), pp. 358–369.

Isenberg, C. (1992). *The Science of Soap Films and Soap Bubbles.* Dover, Mineola, NY.

Israelachvili, J. N. (1992). *Intermolecular and Surface Forces* (2nd edn.). Academic Press, London.

Jiang, Y., Lookman, T., and Saxena, A. (2000). Phase separation and shape deformation of two-phase membranes. *Phys. Rev. E*, **61**(1), pp. R57–R61.

Jülicher, F. and Lipowsky, R. (1996). Shape transformation of vesicles with intramembrane domains. *Phys. Rev. E*, **53**(3), pp. 2670–2683.

Kosta, K. D. (2004). Single-cell elastography: Probing for disease with the atomic force microscope. *Dis. Markers*, **19**(2–3), pp. 139–154.

Koynova, R. and Caffrey, M. (1998). Phases and phase transitions of the phosphatidylcholines. *BBA Biomembranes*, **1376**(1), pp. 91–145.

Kozlovsky, Y. and Kozlov, M. M. (2003). Membrane fission: Model for intermediate structures. *Biophys. J.*, **85**(1), pp. 85–96.

Kreyszig, E. (1991). *Differential Geometry.* Dover, Mineola, NY.

Landau, L. D. and Lifshitz, E. M. (1986). *Theory of Elasticity* (3rd edn.). Butterworth-Heinemann, Oxford.

Laradji, M. and Sunil Kumar, P. B. (2004). Dynamics of domain growth in self-assembled fluid vesicles. *Phys. Rev. Lett.*, **93**(19), 198105.

Leach, A. R. (2001). *Molecular Modelling: Principles and Applications* (2nd edn.). Pearson Education, Harlow, U.K.

Leibler, S. (1986). Curvature instability in membranes. *J. de Physique (France)*, **47**(3), pp. 507–516.

Lipowsky, R. and Sackmann, E. (eds.) (1995). *Structure and Dynamics of Membranes: From Cells to Vesicles.* Handbook of Biological Physics, Volume 1A. Elsevier Science, Amsterdam.

Liu, Q.-H., Haijun, Z., Liu, J.-X., and Zhong-Can, O.-Y. (1999). Spheres and prolate and oblate ellipsoids from an analytical solution of the spontaneous-curvature fluid membrane models. *Phys. Rev. E*, **60**(3), pp. 3227–3233.

Lodish, H., Berk, A., Matsudaira, P., Kaiser, C. A., Krieger, M., Scott, M. P., Zipursky, S. L., and Darnell, J. (2004). *Molecular Cell Biology* (5th edn.). W. H. Freeman, New York.

Lorenz, B., Mey, I., Steltenkamp, S., Fine, T., Rommel, C., Müller, M. M., Maiwald, A., Wegener, J., Steinem, C., and Janshoff, A. (2009). Elasticity mapping of pore-suspending native cell membranes. *Small*, **5**(7), pp. 832–838.

Maeda, N., Senden, T. J., and di Meglio, J.-M. (2002). Micromanipulation of phospholipid bilayers by atomic force microscopy. *BBA Biomembranes*, **1564**(1), pp. 165–172.

Mey, I., Stephan, M., Schmitt, E. K., Müller, M. M., Ben Amar, M., Steinem, C., and Janshoff, A. (2009). Local membrane mechanics of pore-spanning bilayers. *J. Am. Chem. Soc.*, **131**(20), pp. 7031–7039.

Morris, C. E. and Homann, U. (2001). Cell surface area regulation and membrane tension. *J. Membrane Biol.*, **179**, pp. 79–102.

Müller, M. M. (2007). *Theoretical Studies of Fluid Membrane Mechanics.* Dissertation, Johannes Gutenberg-Universität Mainz. `http://nbn-resolving.de/urn/resolver.pl?urn=urn:nbn:de:hebis:77-14750`.

Munro, S. (2003). Lipid rafts: Elusive or illusive? *Cell*, **115**(4), pp. 377–388.

Naito, H., Okuda, M., and Zhong-Can, O.-Y. (1995). New solutions to the Helfrich variation problem for the shapes of lipid bilayer vesicles: Beyond Delaunay's Surfaces. *Phys. Rev. Lett.*, **74**(21), pp. 4345–4348.

Nichols, B. (2005). Without a raft. *Nature*, **436**(7051), pp. 638–639.

Norouzi, D., Müller, M. M., and Deserno, M. (2006). How to determine local elastic properties of lipid bilayer membranes from atomic-force-microscope measurements: A theoretical analysis. *Phys. Rev. E*, **74**(6), 061914.

Powers, T. R., Huber, G., and Goldstein, R. E. (2002). Fluid-membrane tethers: Minimal surfaces and elastic boundary layers. *Phys. Rev. E*, **65**(4), 041901.

Radmacher, M. (2002). Measuring the elastic properties of living cells by the atomic force microscope. *Methods Cell Biol.*, **68**, pp. 67–90.

Rawicz, W., Olbrich, K. C., McIntosh, T., Needham, D., and Evans, E. (2000). Effect of chain length and unsaturation on elasticity of lipid bilayers. *Biophys. J.*, **79**(1), pp. 328–339.

Reynwar, B. J., Illya, G., Harmandaris, V. A., Müller, M. M., Kremer, K., and Deserno, M. (2007). Aggregation and vesiculation of membrane proteins by curvature-mediated interactions. *Nature*, **447**(7143), pp. 461–464.

Richardson, G., Cummings, L. J., Harris, H. J., and O'Shea, P. (2007). Toward a mathematical model of the assembly and disassembly of membrane microdomains: Comparison with experimental methods. *Biophys. J.*, **92**(12), pp. 4145–4156.

Roux, A., Cappello, G., Cartaud, J., Prost, J., Goud, B., and Bassereau, P. (2002). A minimal system allowing tubulation with molecular motors pulling on giant liposomes. *Proc. Natl. Acad. Sci. USA*, **99**(8), pp. 5394–5399.

Rowlinson, J. S. and Widom, B. (2002). *Molecular Theory of Capillarity.* Dover, Mineola, NY.

Seifert, U. (1997). Configurations of fluid membranes and vesicles. *Adv. Phys.*, **46**(1), pp. 13–137.

Seifert, U. and Lipowsky, R. (1995). Morphology of vesicles. In *Structure and Dynamics of Membranes: From Cells to Vesicles* (ed. R. Lipowsky and E. Sackmann), Handbook of Biological Physics, pp. 403–62, Volume 1A. Elsevier Science, Amsterdam.

Semrau, S., Idema, T., Holtzer, L., Schmidt, T., and Storm, C. (2008). Accurate determination of elastic parameters for multicomponent membranes. *Phys. Rev. Lett.*, **100**(8), 088101.

Shi, Q. and Voth, G. A. (2005). Multi-scale modeling of phase separation in mixed lipid bilayer. *Biophys. J.*, **89**(4), pp. 2385–2394.

Siegel, D. and Kozlov, M. (2004). The Gaussian curvature elastic modulus of N-monomethylated dioleoylphosphatidylethanolamine: Relevance to membrane fusion and lipid phase behavior. *Biophys. J.*, **87**(1), pp. 366–374.

Simons, K. and Ikonen, E. (1997). Functional rafts in cell membranes. *Nature*, **387**, pp. 569–572.

Spivak, M. (1976). *A Comprehensive Introduction to Differential Geometry* (2nd edn.). Publish or Perish, Boston, MA.

Staneva, G., Angelova, M. I., and Koumanov, K. (2004). Phospholipase A_2 promotes raft budding and fission from giant liposomes. *Chem. Phys. Lipids*, **129**(1), pp. 53–62.

Staneva, G., Seigneuret, M., Koumanov, K., Trugnan, G., and Angelova, M. I. (2005). Detergents induce raft-like domains budding and fission from giant liposomes. *Chem. Phys. Lipids*, **136**(1), pp. 55–66.

Steltenkamp, S. (2006). *Mechanische Eigenschaften von freitragenden artifiziellen und nativen Membranen.* Ph.D. thesis, Johannes Gutenberg-Universität Mainz.

Steltenkamp, S., Müller, M. M., Deserno, M., Hennesthal, C., Steinem, C., and Janshoff, A. (2006). Mechanical properties of pore-spanning lipid bilayers probed by atomic force microscopy. *Biophys. J.*, **91**(1), pp. 217–226.

Tanford, C. (1991). *The Hydrophobic Effect: Formation of Micelles and Biological Membranes* (2nd edn.). Krieger, Malabar, FL.

Taniguchi, T. (1996). Shape deformation and phase separation dynamics of two component vesicles. *Phys. Rev. Lett.*, **76**(23), pp. 4444–4447.

Thieringer, H. A., Jones, P. G., and Inouye, M. (1998). Cold shock and adaptation. *Bioessays*, **20**(1), pp. 49–57.

Trejo, M. and Ben Amar, M. (2011). Effective line tension and contact angles between membrane domains in biphasic vesicles. *Eur. Phys. J. E*, **34**(1), 9.

Veatch, S. L. and Keller, S. L. (2005). Miscibility phase diagrams of giant vesicles containing sphingomyelin. *Phys. Rev. Lett.*, **94**(14), 148101.

Part V

Morphogenesis

10
Modeling plant morphogenesis and growth

Richard S. SMITH

Institute of Plant Sciences, University of Bern, Switzerland

10.1 Modeling morphogenesis and growth

10.1.1 Morphogenesis and growth

Multicellular organisms develop from embryo to adult through a carefully controlled combination of morphogenesis and growth. Morphogenesis is usually thought of in terms of biochemical compounds, such as gene products and signaling molecules, that give cells the developmental cues required to differentiate and specialize into the various tissues required in the adult organism (Turing 1952). This idea of differentiation is critical in order to create complex form. If every cell was to do the same thing, then the complex shapes and specialized tissues that we see in the adult forms would not be possible. Since all cells in an organism typically contain the same genetic code, how do they decide to do different things?

Cells separated from two-cell mouse embryos can give rise to two identical adults, but if these same two cells are left together, they are able to coordinate their behavior, each becoming a founder cell for only a portion of the adult. Although this can only be done in animals at very early stages of development, plants are much more flexible, as entire plants can be regenerated from single cells taken from leaves or other tissues. As these single cells divide, how do their progeny collectively decide which ones will become shoot and which ones will become root? Furthermore, if larger pieces of tissue are taken, how is it that these groups of cells are able to organize into a single plant?

This question in the field of morphogenesis recurs repeatedly throughout development. For example, when a plant creates a new leaf, how does it decide which cells will become veins, and which will become intervening tissue? This is a question of *pattern formation* and, because of their innate flexibility, plants provide an excellent system in which to study this phenomenon. In fact, unlike animals, plants maintain undifferentiated stem-like cells throughout their lifetime at the shoot and root tips, in structures called meristems (Lyndon 1998). The undifferentiated cells at the tip of the shoot meristem are founder cells from which all of the aerial structures of the plant develop. A similar situation occurs in the root, with root apical meristems providing founder cells for the subsurface portion of the plant.

Another key difference between animals and plants is that plants are sessile and cells cannot move around. Thus the forms we see result from differential and polar growth. After the decision is made to differentiate, cells will become specialized for their new role in the organism, which often involves changes in the biochemical substances that control growth. Thus, in order to model many aspects of development, we require a biochemical model of the morphogenetic substances superimposed on a model of growth.

10.1.2 Dynamic systems with dynamic structure

Dynamic systems can be used to model many aspects of development. The chosen domain is characterized by a set of variables that represent the state of the model, with functions to describe their evolution over time. In very simple cases, it will be possible to find analytical solutions for the change in state, or trajectory, of the system over time. However, this is usually not possible, and numerical simulation on a computer is a more practical alternative. This usually requires the decomposition of the model domain into discrete components to facilitate computation. In some cases this discretization is somewhat arbitrary, such as when the finite element method is used to simulate the mechanical deformation of a section of a cell wall under the application of forces. The cell wall is considered as a continuous, although possibly anisotropic material, and, apart from numerical error, any discretization should yield the same result. In other cases, the domain may have a natural discretization which is not so easily captured with a continuum model. Models of pattern formation often involve cell-to-cell communication, making a discretization into cells a more suitable choice. Likewise, when modeling an entire plant, it might be convenient to discretize at the organ level, with each model component representing a leaf, flower, or internode.

Whatever the discretization chosen, developmental models often involve not only changes to the state of the system, but also changes to the underlying structure. Cells may divide into daughter cells, or plants may produce new organs. Models in which both the state and the structure of the model change over time are called *dynamic systems with dynamic structure* (Giavitto and Michel 2001). Although the theory of dynamic systems is well developed, relatively few methods have been proposed for simulating systems with dynamic structure.

10.1.3 L-systems, the one-dimensional case

At the level of entire plants, L-systems are particularly convenient for modeling development. Named after Aristid Lindenmayer, they were originally introduced as a formalism to describe the development of filamentous organisms (Lindenmayer 1968*a*,*b*). They have since been used to model a wide variety of phenomena that have a one-dimensional topology, such as lines of cells and branching structures; anything that can be represented by an acyclic graph. Prusinkiewicz and Lindenmayer (1990) presented an excellent introduction to the use of L-systems to model the development of plants.

L-systems are rewriting systems based on strings. The strings are made up of modules or letters from a given alphabet. Beginning with an initial string, called the axiom, rules are defined that map each module to one or more new modules. The symbols [ ] are used to indicate branches, with the modules contained within considered to be on the branch. Over the time evolution of the simulation, modules are replaced with more complicated groups of modules, causing the structure to grow (Fig.10.1). Parameters can be added to the modules to store data associated with a model component, such as the levels of biochemical substances. Context-sensitive rules can be used to simulate intermodule communication.

Note that the L-systems themselves do not specify the geometry, but only the topology, of the model. However, extensions to the theory, such as the Logo-style turtle interpretation of L-system strings (Prusinkiewicz 1986), provide a natural way to incorporate geometry directly into a model. Additional modules which are not model components are added to the string to control the movement of an imaginary turtle. Module parameters are added to control the size of model elements and to control the orientation between them. Because the underlying topology is that of an

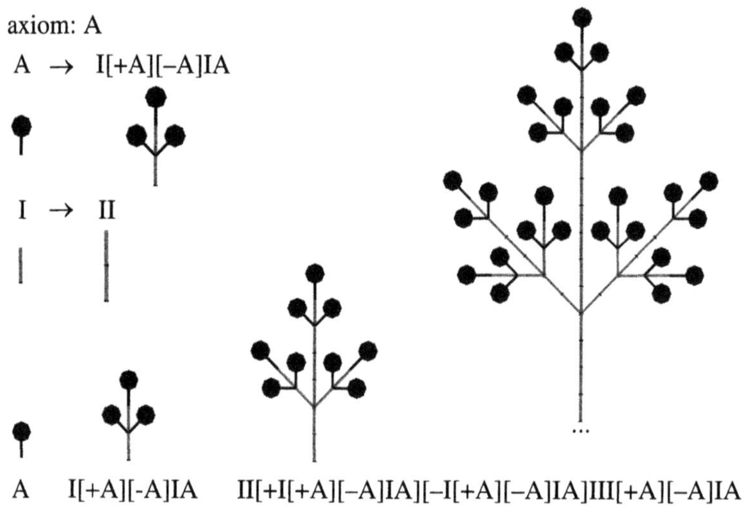

Fig. 10.1 This simple L-system demonstrates how development can be seen as a rewriting process. Modules are replaced by more modules at each step, causing the structure to grow. Plus and minus signs inside brackets indicate left and right branches. (Adapted from Prusinkiewicz *et al.* (1997)).

acyclic graph, it is always possible for modules to be inserted or deleted, or for the size or the orientation between them to change without destroying the overall coherence of the structure (although collisions may occur).

10.1.4 Extending L-systems beyond one dimension

Severals attempts have been made to extend the notion of L-systems to two or more dimensions. Map L-systems (Lindenmayer and Rozenberg 1979), cell systems (de Boer *et al.* 1992), and MGS (Giavitto and Michel 2001) can all be considered as rewriting systems that operate on graphs rather than strings. Although these formalisms have had some success, it has not proved to be easy to specify update rules on a more general graph structure. In order to uniquely specify context and ensure that neighborhood connectivity remains coherent, update rules can become extremely complex.

An alternative approach was taken by Smith *et al.* (2004), who introduced vertex–vertex systems (referred to as VV). VV was originally conceived as a modeling formalism to allow the local specification of mesh subdivision algorithms. The mesh was represented as a graph rotation system, a directed graph structure that maintains an ordering of each vertex's neighborhood. In geometric terms, this ordering allows the representation of orientable 2-manifolds (two-dimensional surfaces) embedded in three-dimensional space. Loosely speaking, the fact that the surface is orientable means that it has two sides, which can be distinguished at vertices by examining the ordering of their neighbors. This simple addition to a standard notion of a directed graph turns out to be surprisingly useful for this class of problems.

Unlike L-systems, however, VV is not implemented as a set of rewriting rules that update the state and structure of the model. Instead, it is implemented as a collection of language statements embedded in C++, although versions now exist based entirely on C++ class libraries. Although in some ways VV is not as elegant as L-systems or MGS, it is practical, and its procedural nature makes it convenient for handling complex models. In addition, the widespread availability of class and template libraries for C++ is also a great asset. The simulations presented in the rest of these lecture notes were all programmed using VV in the L-studio (Prusinkiewicz 2004) modeling environment.

10.2 Descriptive models of growing surfaces

10.2.1 Abstracting to two dimensions

Although the branching structure of plants is innately one-dimensional, the cellular makeup of most plant tissue has a three-dimensional topology. However, it is often possible to abstract the problem to a two-dimensional representation of the tissue. When exploring signaling events that occur primarily in a relatively thin layer of cells, it may be possible to abstract this layer to a single layer of cells. This approach has been used for the cambium layer in trees (Kramer 2002), the surface layer of cells in the shoot apex (Jönsson *et al.* 2006; Smith *et al.* 2006*a*; Barbier de Reuille *et al.* 2006), and the mesophyll layer in leaves (Mitchison 1980; Kramer 2004; Rolland-Lagan and Prusinkiewicz 2005; Barbier de Reuille *et al.* 2007; Stoma *et al.* 2008). Two-dimensional geometries may also be obtained by using a cross-sectional representation of the tissue. This has been proposed for both the shoot (Stoma *et al.* 2008; Bayer *et al.* 2009) and root (Grieneisen *et al.* 2007; Stoma *et al.* 2008) apices. Most cellular-level simulation models of plant morphogenesis and patterning use this simplification.

10.2.2 Modeling a growing shoot apex

The plant shoot apex consists of an external layer of cells, called the tunica, that surrounds the inner tissue, known as the corpus. The molecular processes leading to the positioning of new organs, which results in phyllotaxis patterning, are thought to occur in this outer layer (Reinhardt *et al.* 2003; Smith *et al.* 2006*a*). Thus it is sufficient to consider only the surface layer for simulation purposes.

This surface layer can be modeled as a surface of revolution, generated by rotating a planar curve around the longitudinal axis of the apex (Fig. 10.2). This planar curve is a B-spline (Foley *et al.* 1990), which can be defined interactively using a graphical editor available within L-studio, and can easily be changed in order to model apices with various profiles. A point on the apex surface is thus characterized by two coordinates (θ, a) where θ is the angle of rotation around the axis of the apex, measured with respect to a reference direction, and a is the distance from the apex tip, measured along the generating curve on the apex surface. In Cartesian coordinates, the (x, y, z) position of a surface point $S(\theta, a)$ is thus given by

$$S(\theta, a) = (x(a)\cos(\theta), y(a), x(a)\sin(\theta)). \tag{10.1}$$

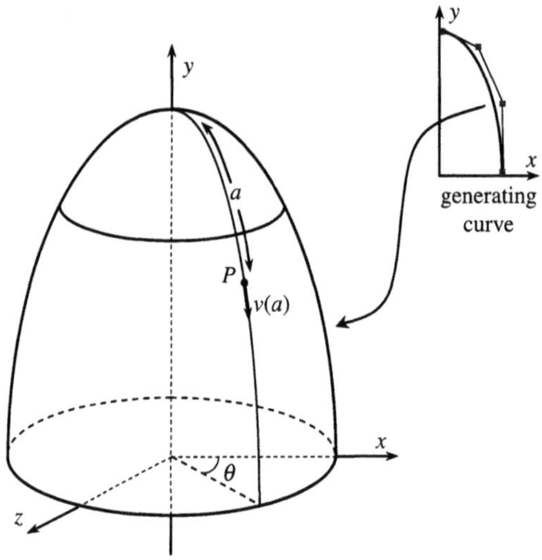

Fig. 10.2 A two-dimensional representation of the shoot apex. The apex shape is defined by a B-spline generating curve rotated around the longitudinal axis of the shoot apex. A sample point P on the surface with coordinates (θ, a) moves away from the apex tip with a velocity $v(a)$.

Growth is simulated by moving points away from the apex tip while preserving the overall shape of the reference surface. This motion is characterized by a function $RERG(a)$, which defines the *relative elemental rate of growth* (Hejnowicz *et al.* 1984) in the longitudinal direction (along the generating curve) at a distance a from the apex tip (Hejnowicz *et al.* 1984; Nakielski 2000). The velocity with which a point $S(\theta, a)$ moves away from the apex tip along the generating curve is then given by an integral:

$$v(a) = \int_0^a RERG(a)da. \tag{10.2}$$

Similarly to the generating curve, the growth function $RERG(a)$ can be defined graphically within L-studio, which makes it easy to specify various distributions of growth rates on the apex surface. It is often the case that growth is slower at the very tip of the apex, in the stem cell niche (Fig. 10.3). To speed computation, a lookup table for the values of the integral given in eqn 10.2 can be precomputed at startup.

10.2.3 Modeling a growing leaf

The growing-apex model in the previous section is suitable for modeling radially symmetric surfaces that grow at the tip, such as the shoot and root apices. Surfaces such as leaves do not show this symmetry, and have growth that is distributed throughout the surface. In this case it is possible to model the surface using a key-framing technique often used in computer graphics. The shapes of several stages of the growing leaf are specified by defining a sequence of Bézier surfaces that represent the leaf at different times (key frames) during the simulation (Fig. 10.4). These surfaces can

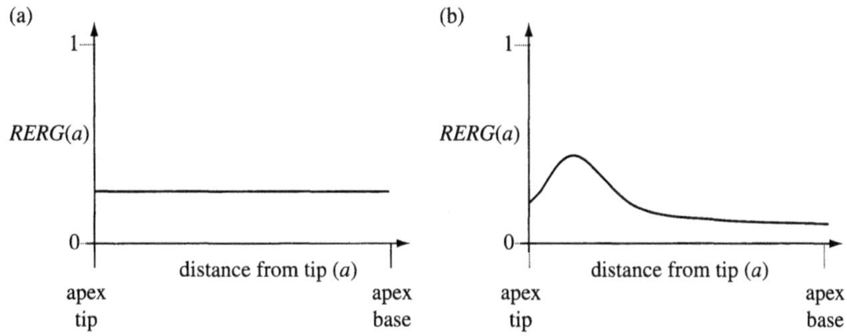

Fig. 10.3 Sample plots defining the relative elemental rate of growth $RERG$ as a function of the distance a from the apex tip. The functions are defined graphically, using an interactive B-spline editor. (a) Constant growth function. (b) Function with a decreased growth rate at the apex tip and flank.

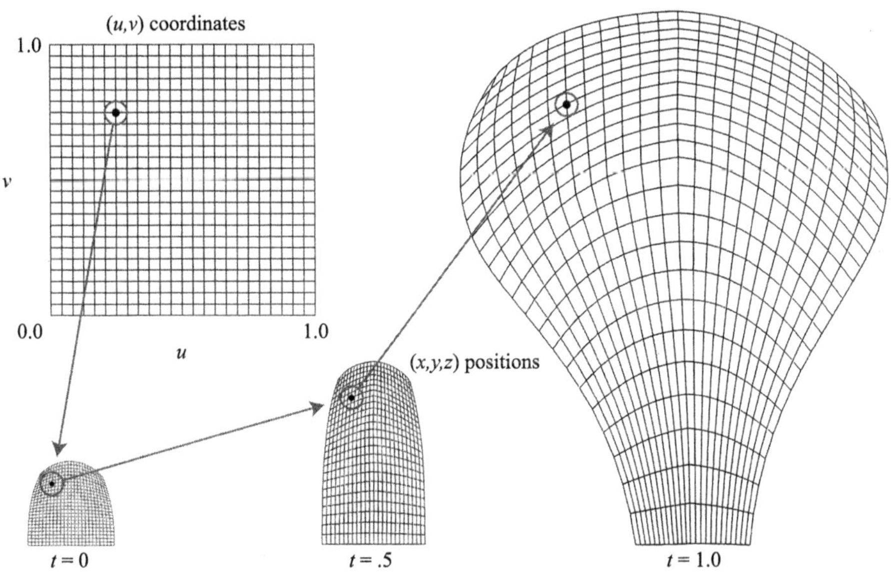

Fig. 10.4 Modeling a growing leaf by using Bézier surfaces. A sequence of Bézier surfaces represents the leaf at various times t during the simulation. Coordinates in the parameter space (u, v) shown at the top left are mapped into a three-dimensional space (x, y, z) as shown. The position of points at time steps in between the three key frames is found by linear interpolation.

be defined interactively using the surface editor provided with L-studio. Interpolation of the control points of the surfaces is used to produce the leaf at any desired point between the key frames. By advancing time in small steps, a smoothly growing leaf surface can be produced.

The Bézier surfaces used to model the leaf are two-dimensional parametric surfaces embedded in three-dimensional space. Points on the surface are represented by coordinates (u, v), which are mapped by Bézier surface evaluators (Neider *et al.* 1994) to positions (x, y, z). As the simulation progresses and the leaf changes shape, a (u, v) coordinate at the tip of the leaf will remain at the tip even though its actual position in space might change considerably. A surface point $S(u, v)$ is computed using the control points $\{P_{ij}\}$ as follows:

$$S(u,v) = \sum_{i=0}^{n} \sum_{j=0}^{m} \binom{n}{i} u^i (1-u)^{n-i} \binom{m}{j} v^j (1-v)^{m-j} P_{ij}. \tag{10.3}$$

The binomial coefficients in the above expression can be precomputed once at startup to speed the calculation.

10.2.4 Cells and cell division

Most static models of morphogenesis and patterning in plants have been implemented by using rectangular grids of cells. However, when one is considering growing and possibly irregularly shaped surfaces, rectangular cells are no longer practical. One possibility is to use Voronoi diagrams to model cells (Honda 1978; Jönsson *et al.* 2006), because of their irregular, cell-like appearance. They are not ideal for modeling growing plant tissues, however, since the rearrangement of Voronoi regions that inevitably occurs during cell division creates unrealistic motions of cell walls (Fig. 10.5). Such motions are not possible in plants owing to the rigidity of the interconnected cell walls that surround plant cells. More realistic divisions can be obtained by using map L-systems (Lindenmayer and Rozenberg 1979; Prusinkiewicz and Lindenmayer 1990) and cell systems (de Boer *et al.* 1992). However, these systems rely on physically based mass–spring simulations to produce realistically shaped cells, and are thus not

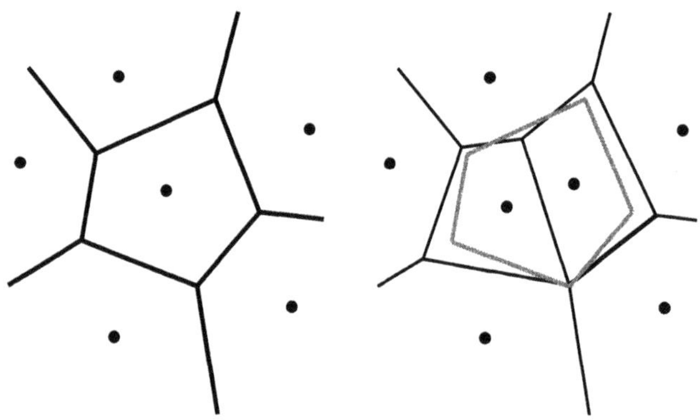

Fig. 10.5 Cell division in a tissue with cells represented by Voronoi regions. Unrealistic changes in the cells' geometry are introduced during cell division. Cell before division (left), and adjusted regions after division (right). Note the shortening and lengthening of some of the neighboring cell walls as wall junctions appear to "move".

directly applicable to the purely descriptive growth model described in Sections 10.2.2 and 10.2.3.

A model for cell division proposed by Nakielski (2000) for the shoot apex can be generalized to a much wider class of growing two-dimensional surfaces, including those described above (Smith *et al.* 2006*a*; Smith and Bayer 2009). Cells are modeled as polygons, with the positions of cell vertices changing over time as a result of surface growth. Cell division occurs when the cell size (polygon area) reaches a threshold value. Inspired by Errera's rules for cell divsion (Errera 1888), the shortest wall through the center of the cell is chosen for division. The position of the dividing wall may be adjusted to avoid four-way junctions, which are unusual in plant tissue. To produce more realistic cell shapes, the newly divided cell is "pinched" slightly by moving the

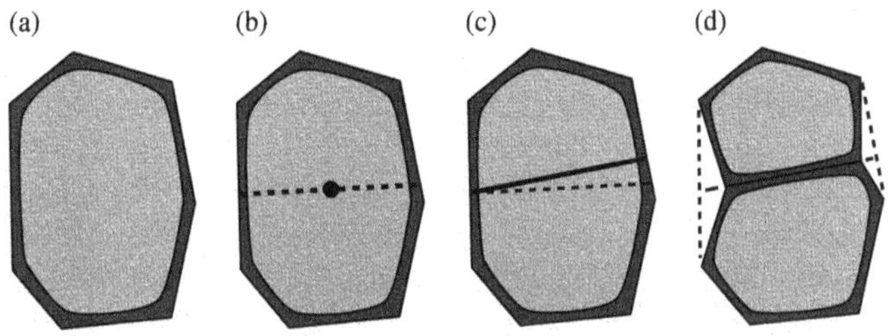

Fig. 10.6 Model of cell division. (a) A mother cell before division. (b) The chosen dividing wall is the shortest wall passing through the center of the mother cell. (c) The endpoints of the wall are displaced from any preexisting vertices of the mother cell in order to avoid four-way junctions. (d) The dividing wall is shortened. (Adapted from Smith *et al.* (2006*a*)).

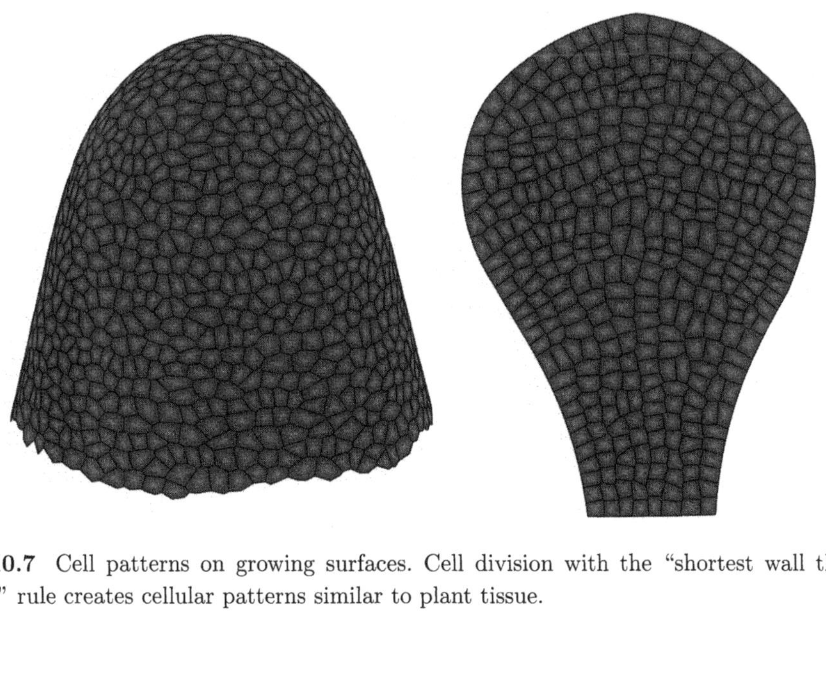

Fig. 10.7 Cell patterns on growing surfaces. Cell division with the "shortest wall through center" rule creates cellular patterns similar to plant tissue.

vertices of the dividing wall slightly towards each other (Fig. 10.6). These rules for cell division result in cellular patterns similar to those in observed in plant tissue, over a wide range of surface shapes and growth functions (Fig. 10.7).

Cell division requires the insertion of new vertices into the growing surface, as new wall junction vertices are created. Equations (10.1) and (10.3) are used to calculate the position in space of a surface point from an arc-length–angle pair in the case of the apex surface, or a pair of parametric coordinates in the case of the Bézier surface model. However, they do not lead easily to a closed formula to compute the reverse operation. In fact, the points of new vertices created during cell division will not in general lie on the surface, and instead the closest point on the surface must be found. This can be done by applying a Newton–Raphson root search technique to eqns (10.1) and (10.3) for each new point inserted into the surface.

10.3 A physically based model of the shoot apex

10.3.1 Physically based growth

In the growing-surface models discussed in the previous sections, the shape of the surface was provided as an input to the models. With physically based models, the shape can be an emergent property of the model. This opens up the possibility of modeling the processes that lead to these emergent shapes, and the interaction between morphogens, growth, and form. Since it is also widely believed that mechanics might play a direct role in morphogenesis, physically based models are becoming increasingly important. For a review of the effects of mechanical forces on plant morphogenesis, see Dumais (2007).

Several types of physically based models have been proposed for the shoot apex, and most are based on mass–spring systems. Although Hamant *et al.* (2008) have presented a finite element model for the shoot apex, this model does not include growth. Finite element models have advantages over mass–spring systems as they have a direct connection with continuum mechanics, allowing parameters such as Young's modulus and Poisson's ratio to be used directly in the model. It is not clear how to translate these parameters into equivalent spring constants, especially for elements with dimension greater than one. Nevertheless, mass–spring systems are able to model very complex behavior, often in real time, and are relatively easy to program (see Fig. 10.8).

In the simplest case, a surface is created by using vertices that represent only the cell centers connected by a network of springs (Jönsson *et al.* 2006). If required, the cell walls can be calculated by the finding the Voronoi regions of the cell centers, or the cells can simply be rendered as intersecting spheres. Alternatively, the cells can be defined by their walls as in the models described in Sections 10.2.2 and 10.2.3. In order to prevent the surface mesh from collapsing, a force normal to the surface in the outward direction is applied to each vertex. This force can be viewed as representing the combined effects of turgor pressure and other forces from the cells inside the shoot apex. Since the surface layer of the shoot apex is much stiffer than the inner tissue, the structure can be likened to the skin of a balloon that is holding back the pressure

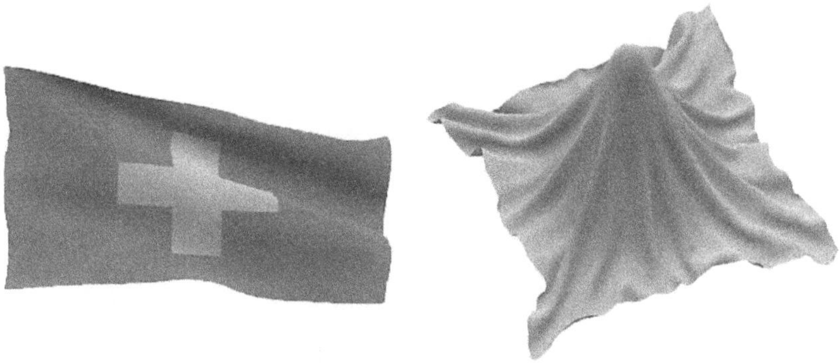

Fig. 10.8 Examples of two-dimensional surface simulations using mass–spring systems. Flag blowing in the wind (left) and a wrinkling cloth (right). Realistic-looking shapes can be created with very simple programs, often in real time.

within. In order to anchor the apex in space, the positions of the vertices at the bottom of the apex are fixed. Growth can be simulated by increasing the rest lengths of the springs, by subdividing and adding additional springs, or by a combination of these methods.

This type of model makes it possible to consider locally induced differences in growth rate, a property that is difficult to achieve in purely descriptive models. Biochemical substances can be used to control the local rate of increase in the rest lengths of the springs, simulating the actions of wall-relaxing enzymes thought to cause growth. Modeling the outgrowth of organs from the surface is straightforward, and it is possible to model shapes that are difficult to define descriptively (Fig. 10.9)

10.3.2 The mass–spring simulation

The mass–spring model of the shoot apex presented here replaces the descriptive surface growth model described in Sections 10.2.2 and 10.2.3 with a simple physics simulation. The cell walls are considered to be springs that are connected to point masses located at the junctions between walls. The point masses at all such junctions are considered to have an identical value in abstract units of 1. The force acting on the point mass of a vertex v located at a position p_v due to springs $\mathbf{F}_{s_v}$ is based on Hooke's law and is calculated as

$$\mathbf{F}_{s_v} = \sum_{u \in N_v} -k \left(1 - \frac{\|\mathbf{p}_u - \mathbf{p}_v\|}{l_{v \to u}}\right) \frac{\mathbf{p}_u - \mathbf{p}_v}{\|\mathbf{p}_u - \mathbf{p}_v\|}, \tag{10.4}$$

where N_v represents the neighbors of vertex v, $\mathbf{p}_u$ is the position of a neighboring vertex u, k is the spring constant per unit length, and $l_{v \to u}$ is the rest length the of spring joining vertex v to the neighboring vertex u. The norm symbol indicates the Euclidean norm, or distance, between the points. The expression within the brackets is positive when the distance between v and its neighbor u is less than the rest length, and negative if it is greater. Note that for a given difference from the rest length,

Fig. 10.9 Physically based model of a hypothetical flower apex with five bulging primordia. Darker cells are considered part of the growing tips of the apex and primordia. Locations of bulging primordia specified as model parameters.

the magnitude of the force is reduced as the rest length of the spring increases. This magnitude is then multiplied by the normalized vector $(\mathbf{p}_u - \mathbf{p}_v)/\,\|\mathbf{p}_u - \mathbf{p}_v\|$ and by $-k$ to give the forces. These forces are then summed over all the neighbors of the vertex.

In addition to the forces on a vertex due to springs, a uniform force representing the internal turgor pressure $\mathbf{F}_t$ is included in the model, which acts in the direction of the surface normal at each vertex. A real plant apex is able to control its growth direction, enabling a shoot to grow straight, to favor an upward growth direction, or to grow towards the light. These processes are not modeled directly but are accounted for by a directional force $\mathbf{F}_d$, which is applied to the vertices at the tip of the apex where directional growth occurs. This directional force is also applied at the tips of growing primordia, although the strength of the force may be different from that used for the main apex. Combining the various components, the total force from all sources $\mathbf{F}_{a_v}$ acting on a vertex v is

$$\mathbf{F}_{a_v} = \mathbf{F}_{s_v} + \mathbf{F}_{t_v} + \mathbf{F}_{d_v}. \tag{10.5}$$

Note that for cells in the apex which are not at the tip of a growing apex or primordium, the last term $\mathbf{F}_{d_v}$ is zero.

Once the forces on the vertices are calculated, the steady state of the system is determined (see Section 10.3.5), and growth is simulated by increasing the rest lengths of the springs. If a spring is longer than its rest length, then its rest length is increased in proportion to this difference. In the case of cell walls within the extent of an apex or primordium tip, this increase is larger, and the directional force $\mathbf{F}_d$ is also applied. This causes the apex and bulging primordia to grow quickly at the tips, while still increasing in size at a lesser rate at locations more distal from the tips.

10.3.3 Local determination of apex and primordium centers

In the previous section, the forces and increases in spring rest length were calculated differently for vertices near the tip of the apex. In general, it is not straightforward to determine the center of the apex tip, and which cells belong to it, in cellular models with emergent growth. The problem is even more difficult when one is trying to determine the centers of newly protruding leaf or flower primordia. In models on regular grids, a single vertex can be designated as the center of the tip of the apex or primordium if the subdivision scheme used keeps this vertex at the center as the simulation proceeds. In cellular models, however, a central cell may divide, causing the center of the tip of the apex or primordium to shift as one of the new daughter cells is chosen as the center. In addition, the irregularity of the wall lengths can cause cells to grow at different rates, shifting the location of a central cell of an apex or primordium tip.

To overcome this problem, a method is needed so that the position of an apex or primordium center can be determined on the basis of local information contained in the cells that make up the primordium itself. Ideally, this would involve modeling the molecular components that are responsible for determining the apex or primordium identity, extent, and boundary. In the model presented here, a somewhat simpler approach is taken, which nonetheless retains a local character.

For the main apex, the center is specified at the start of the simulation, whereas primordium centers are determined as the simulation proceeds. Along with the location of the center, an initial extent of the apex or primordium is specified, which may change as the apex or primordium ages. All of the vertices within the radius of this initial extent are considered as part of the growing tip. These vertices are subject to the directional force $\mathbf{F}_d$ discussed previously, and participate in the dynamic process of determining the center of the tip. At the end of each time step, after growth, the physics simulation, and any cell divisions that have occurred, all of the positions of the vertices belonging to the apex or primordium center from the previous step are averaged to determine a tentative center point for the next iteration. This average is weighted in favor of vertices closer to the previous center, and is in general not on the apex or primordium surface. A point on the surface is obtained by projecting this tentative center onto the surface layer of cells, in the direction of the force $\mathbf{F}_d$. All of the points that are within the extent of the apex or primordium are then updated with the distance to the new center, and will participate in the determination of the center at the next time step.

10.3.4 Calculation of distance on the apex surface

The maintenance of the apex and primordium centers, as well as the determination of which cells belong to a particular apex or primordium, requires the calculation of their distances to their respective centers. For the main apex itself, the Euclidean distance can be used, however, this can cause problems with protruding primordia. If the extent of a primordium at initiation reaches the other side of a tall, slender apex, vertices on the other side of the apex might be included. If the tip of one primordium comes too close to the flank of another, it can recruit vertices belonging to a completely different primordium.

This problem is solved by using the distance along the apex surface, instead of the distance through space, when determining how far a vertex is from an apex center. Although calculating the geodesic would give the most accurate measure of distance on the surface, the weighted graph distance (using the lengths of the edges), which can be calculated with Dijkstra's algorithm, has several advantages. Aside from being simple, fast, and easy to implement, graph distance might actually be a more accurate measure of distance as far as the diffusion and transport of molecular components is concerned. This is especially true if it is calculated using a separate graph containing only the centers of the cells.

10.3.5 The physics simulation

In mass–spring simulations, it is often desirable to visualize the dynamic behavior of a system as it progresses towards a steady state. This would be the case for a simulation of a bouncing ball, where the dynamic behavior of the physics simulation is of interest. In the case of the growing-apex model described here, each time step starts with the simulation at equilibrium. Growth, which changes the rest lengths of the springs, upsets this equilibrium, and the system must be solved for the equilibrium state under these new conditions. This means that in general, there will be many iterations of the physical simulation for each growth step of the model.

There are several methods by which the equilibrium state of the system can be determined. Given the total force $\mathbf{F}_{a_w}$ on a vertex w as defined earlier, the following equations describe the change in velocity $\mathbf{v}$ and position $\mathbf{p}$ over time:

$$\frac{d\mathbf{v}_w}{dt} = (\mathbf{F}_{a_w} - \varsigma \mathbf{v}_w) \quad dt, \tag{10.6}$$

$$\frac{d\mathbf{p}_w}{dt} = \mathbf{v}_w \, dt, \tag{10.7}$$

where ς is the damping constant. This assumes that the point masses at all of the vertices in the model are equal to 1. Using the forward Euler method, the values for the velocity and position at time $t + \Delta t$ can be calculated as

$$\mathbf{v}_{t+\Delta t} = \mathbf{v}_t + (\mathbf{F}_a - \varsigma \mathbf{v}_t) \quad \Delta t, \tag{10.8}$$

$$\mathbf{p}_{t+\Delta t} = \mathbf{p}_t + \mathbf{v}_t \quad \Delta t. \tag{10.9}$$

Although these formulas give acceptable results, if the velocity at time $t + \Delta t$ is substituted in eqn (10.9) for $\mathbf{v}_t$, then the step size can be increased by a factor of five or more in many cases. This method has been termed the *symplectic* forward

Euler method (Stern and Desbrun 2006). With this modification, the update formulas become

$$\mathbf{v}_{t+\Delta t} = \mathbf{v}_t + (\mathbf{F}_a - \varsigma \mathbf{v}_t) \quad \Delta t, \tag{10.10}$$

$$\mathbf{p}_{t+\Delta t} = \mathbf{p}_t + \mathbf{v}_{t+\Delta t} \quad \Delta t. \tag{10.11}$$

The physics simulation in each growth step is iterated until an equilibrium state is reached, which is determined by examining the maximum total force acting on any vertex. Since the total force will go to zero when the system is at a steady state, this can be used to apply a threshold to stop the iteration. In practice, it is convenient to also limit the number of iterations, as this improves performance and smooths out the simulation when primordia first appear and tend to "pop out" from the apex surface.

Since only the equilibrium state of the physics simulation is sought at each time step of growth, a relaxation method can be used that does not consider velocity. Vertices can simply be moved in the direction of the total force by some amount proportional to the magnitude of the force. The system is then iterated until the forces approach zero. Both methods produce similar results with similar execution times.

Simulations using a physically based growth model will, in general, require considerably more computation time than simulations using descriptive growth. This is because each growth step requires that the physics simulation be iterated until equilibrium is reached. Despite this drawback, physically based models of the shoot apex have advantages. Physically based models can be used to explore how biochemical signals interact with mechanics to produce the emergent shapes that we see as a plant develops. It is possible to explore the effects of mechanics on morphogenesis, as stress information is also available in the model.

10.4 Patterning in the shoot apex: Phyllotaxis

10.4.1 Phyllotaxis

Few patterns in nature are more conspicuous than the intersecting spirals seen in the head of a sunflower or on the side of a pine cone. In order to create these patterns, the plant must be able to position organs very precisely, sometimes to within a fraction of a degree. How is the plant able to accomplish this?

To examine this question we need to look at the shoot apex, at the tip of a growing plant shoot, for this is where new plant organs are formed. The central zone of the shoot apex consists of undifferentiated founder cells, surrounded by a relatively narrow band of cells called the peripheral zone. As the plant develops, the peripheral zone maintains an approximately constant distance from the tip of the apex. Only cells within the peripheral zone are competent to initiate organs. In some cases, it is convenient to use the notion of an *active ring* as an abstraction of the peripheral zone. This is defined to be a line on the apex surface that encircles the apex and is located at the center of the peripheral zone (Fig. 10.10). Since the radial position of plant organs changes little after initiation, it is in the shoot apex that the phyllotaxis patterns are formed. The radial position, and the timing of initiation of new leaves or florets, determines the macroscopic patterns we see.

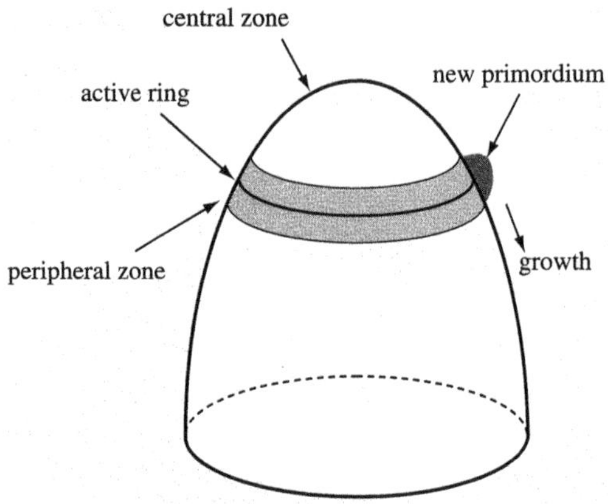

Fig. 10.10 Structure of the shoot apex. The central zone at the tip of the apex contains the stem cell niche, and is surrounded by a narrow band of cells called the peripheral zone, where new organ primordia initiate. In some cases, this band of cells is idealized as an active ring. Growth causes the new organs to move down the flank of the apex.

A simple observation of Hofmeister (1868) gives insight into what determines the site of a new primordium. He proposed that new plant organs initiate as far as possible from preexisting ones. Thus older primordia somehow inhibit the formation of new ones nearby. Several theories have been suggested for how this inhibition may occur. Existing primordia might release some sort of chemical inhibitor (Schoute 1913), or perhaps physical forces play a role (Green *et al.* 1996). Whatever the mechanism for this inhibition, phyllotaxis results from the positional information coming from existing plant organ primordia combined with growth of the shoot tip. Thus, in order to study phyllotaxis through simulation, a model of this inhibition and a dynamic model of a growing shoot apex are required.

10.4.2 An inhibition model of phyllotaxis

Can a simple spacing mechanism lead to the complex intersecting spiral patterns we see in plants? Although it is likely possible to treat this problem analytically, it can be easily addressed with computer simulation. Given the growing shoot apex structure as described in Section 10.2.2, the peripheral zone is modeled as a ring that is maintained at a fixed distance from the tip (Fig. 10.10).

Simulation of phyllotaxis proceeds in a sequence of time steps. At each step, points on the apical surface, including existing primordia, are moved away from the apex tip according to their velocities (eqn 10.2). The active ring is divided into equally spaced sampling points, and the inhibition from previous primordia is calculated. If one of these points drops below a given inhibition threshold, a new primordium is placed at the sampling point. In subsequent simulation steps, the new and the previous primordia move away from the active ring as a result of apex growth. This movement, combined with a possible decrease in the inhibiting influence of primordia with their age, reduces the inhibiting strength on the active ring. Over time, this inhibition drops sufficiently at some location to allow the formation of another primordium, and the process repeats.

It seems biologically plausible that the inhibition from previous primordia would depend inversely on their distance from sampling point on the active ring, and that inhibition would decay over time. This leads to the following function to calculate inhibition (Smith *et al.* 2006*b*). In a system containing n primordia, the combined inhibiting effect h on a sampling point S is calculated as the sum

$$h(S) = \sum_{i=1}^{n} \frac{1}{d(P_i, S)} e^{-bt_i}, \tag{10.12}$$

where $d(P_i, S)$ is the distance between primordium P_i and a sampling point S on the apex surface, t_i is the age of primordium i, and b controls the rate of the exponential decrease in inhibition over time. Initially, when the model is first started, all of the active-ring sampling points have zero inhibition, and one point is selected for the first primordium. Figure 10.11 shows how this simple spacing mechanism results in spiral phyllotaxis.

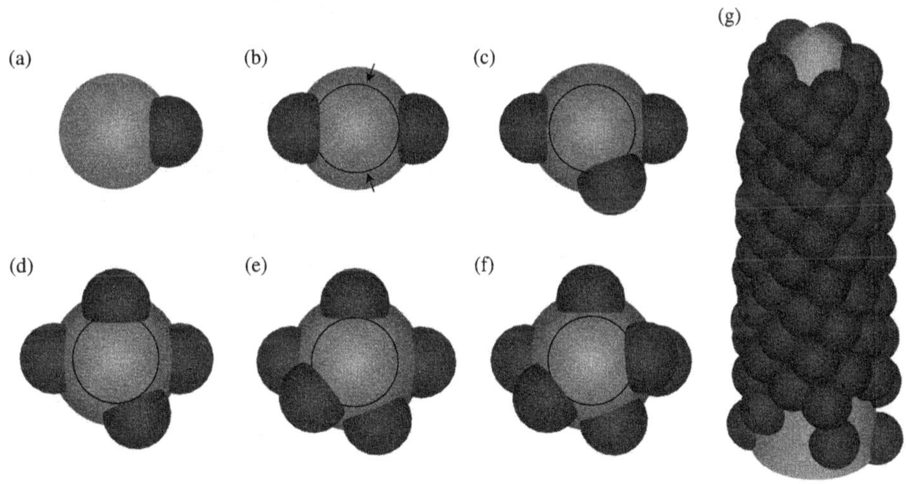

Fig. 10.11 Self-starting spiral Fibonacci phyllotaxis pattern created by using an active ring with 100 sampling points. (a) The initial primordium is arbitrarily placed by the model. (b) The second primordium appears in the area of least inhibition, at a divergence angle of 180° from the first. At this point there are two minima of inhibition on the active ring, both of which are closer to the older of the two existing primordia (arrows). (c) The third primordium appears at one of these minima, the choice determining the direction of the spiral. (d)–(f) Successive primordia appear, the position of each being determined by the inhibition from the previous few. This results in a spiral pattern. (g) Once the pattern is established, the inhibition threshold is raised, leading to a more densely packed pattern. Note the rows of spirals in alternating directions. (Adapted from Smith *et al.* (2006*b*)).

10.4.3 A spacing mechanism based on reaction–diffusion

Since phyllotaxis is a result of the superimposition of a simple spacing mechanism on a growing shoot apex, it is natural to ask how this spacing might occur in the plant in mechanistic terms. This leads back to the initial question of how some cells are selected for differentiation over others, even though they all operate with the same rules or genetic code. This question was addressed by Turing (1952), and his *reaction–diffusion* model has since become one of the most widely used models for pattern formation in biology (Meinhardt 1982; Murray 2002). Turing showed, using both analytical and computer simulation methods, that a pattern of peaks in concentration of two or more chemical substances, which he termed *morphogens*, could arise spontaneously in a ring of cells. Starting with homogeneous initial conditions, and using the same rules for changing morphogen concentration in each cell, slight perturbations in the initial conditions due to noise could upset an unstable equilibrium and lead to patterning.

To understand this process, it is instructive to look at a reaction–diffusion system proposed by Gierer and Meinhardt (1972). This model is based on two substances and is called an *activator–inhibitor* system, with equations that are more intuitive than those proposed by Turing. One substance, called the activator a, enhances its own production, as well as that of another substance h, termed the inhibitor. The inhibitor inhibits production of the activator. Such a system is easy to envision as a feedback loop in a genetic regulatory network (Fig. 10.12). On a ring of cells, the changes in concentration of the activator and inhibitor for each cell are given by

$$\frac{da_i}{dt} = \rho_{a_0} + \rho_a \frac{a_i^2}{1+h_i} - \mu_a a_i - D_a \sum_{j \in N_i} (a_i - a_j), \tag{10.13}$$

$$\frac{dh_i}{dt} = \rho_{h_0} + \rho_h a_i^2 - \mu_h a_i - D_h \sum_{j \in N_i} (h_i - h_j), \tag{10.14}$$

where a_i and h_i are the concentrations of activator and inhibitor in cell i, ρ_{a_0} and ρ_{h_0} control background production (small amounts), ρ_a and ρ_h control activator-enhanced production, μ_a and μ_h control decay, and D_a and D_h control the rates of diffusion from cell to cell, with the summations taken over all of the neighbors of cell i. Each cell stores single values for the concentrations of the activator and inhibitor, and the junction between cells is abstracted to a single interface. For stable peaks to form, the rate of activator diffusion must be much smaller than that of the inhibitor.

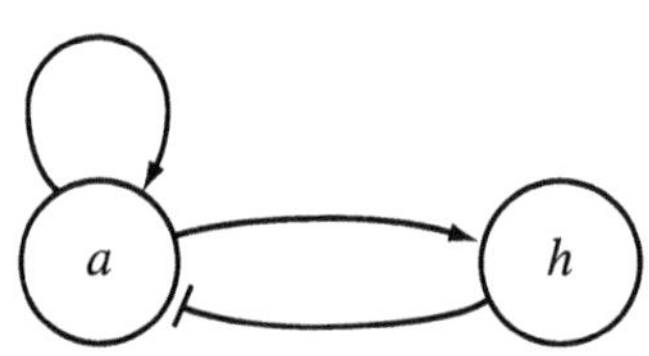

Fig. 10.12 Activator–inhibitor network. The activator a enhances its own production, as well as the production of an inhibitor h. The inhibitor suppresses the activator.

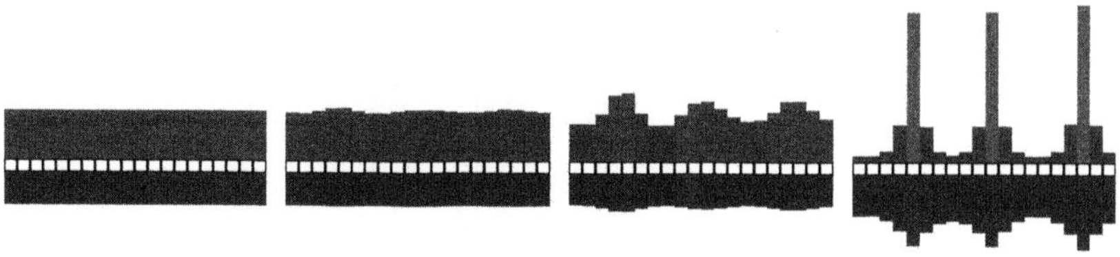

Fig. 10.13 Reaction–diffusion patterning using an activator–inhibitor system on a line of cells. The ends of the line are connected to each other to form a ring. The activator concentration is represented by the height of the bars on top, with the inhibitor below. The model starts with homogeneous initial conditions, with a small amount of noise added to the initial activator concentration. This noise destabilizes the system and the activator–inhibitor model produces relatively evenly spaced peaks of activator concentration. Note that both the activator and the inhibitor concentrations are high in the activated cells.

Small local maxima in activator concentration due to random variation lead to a local increase in production of both the activator and the inhibitor. The inhibitor diffuses away more quickly than the activator, reducing its effect on local activator self-enhancement, while suppressing activator self-enhancement nearby. In a system of identical cells, each operating with identical rules, this destabilization can lead to a spatial pattern of peaks in activator concentration (Fig. 10.13), which can trigger selective differentiation, leading to patterning. By changing the parameters in eqns (10.13) and (10.14), it is possible to vary the distance between the peaks. The interaction of multiple substances, combined with multiple cascading interactions, has been used to account for a wide variety of patterning processes observed in nature, including phyllotaxis (Meinhardt 1982). In the context of a growing plant apex, this reaction–diffusion mechanism can explain both the emergence of organs *de novo* and the thresholding mechanism of an inhibition model of phyllotaxis.

Figure 10.14 shows the result of applying activator–inhibitor model to the growing-leaf model described in Section 10.2.3. The diffusion term in eqns (10.13) and (10.14) need only be modified to take into account the differing areas of cells and lengths of cell walls. Initially, the leaf area is filled with fairly uniformly spaced peaks in the activator concentration. As the leaf grows, space opens up to create room for more peaks. In this case the spacing mechanism operates on a growing leaf, rather than a growing shoot tip. A similar mechanism is thought to regulate the spacing of trichomes in a leaf; however, the genetic regulatory network involves at least five substances and is substantially more complex (Barbier de Reuille *et al.* 2007). Nevertheless, the underlying principle is the same.

10.4.4 A spacing mechanism based on polar transport

A spacing mechanism based on reaction–diffusion seems like a plausible explanation for phyllotaxis patterning in the shoot apex. The idea that existing primordia emit a diffusing inhibitor to suppress organ formation nearby is almost 100 years old, but no such inhibitors have been found. Instead, the experimental evidence points to a

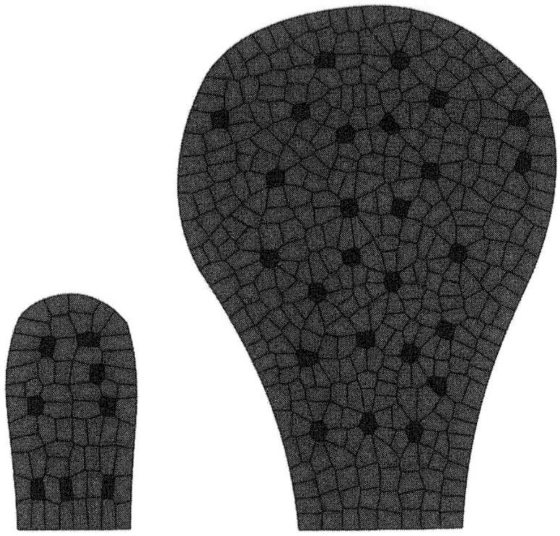

Fig. 10.14 Reaction–diffusion patterning in a cellular model of a growing leaf. The darker cells represent peaks in the activator concentration. As the leaf grows, space opens up in the leaf, allowing more peaks to form.

different type of patterning mechanism, based on the active transport of the plant hormone auxin (indoleacetic acid, IAA).

Plant tissue often displays a polarity with respect to auxin transport. When sections of tissue are excised and alternate ends are exposed to higher levels of auxin, then auxin will move faster in one direction than in the other. To explain this phenomenon, the *chemiosmotic model* of auxin transport was proposed (Rubery and Sheldrake 1974). Auxin is a weak acid, and in the neutral pH inside cells it is largely dissociated. In this ionic form, auxin is hydrophilic and unable to cross the plasma membrane. In order for auxin to leave a cell, it requires the activity of export carrier proteins located in the plasma membrane. In the lower pH of the extracellular space, auxin becomes protonated, and is able to reenter the cell, although plants also have specialized auxin import carriers to aid this process. Often, the export carriers display a polarity towards one side of the cell (see Fig. 10.15). When this polarity is coordinated at the tissue level, it results in a net directional flux of auxin through the tissue.

In the surface layer of cells in the shoot apex of *Arabidopsis*, the auxin export protein PIN1 is localized towords one side of the cell, pumping auxin towards the sites where new primordia will appear (Reinhardt *et al.* 2003). Mutant *Arabidopsis* plants missing the auxin exporter protein PIN1 do not initiate organs in their inflorescence meristems at all, hence the name PIN. However, Reinhardt *et al.* (2003) were able to restore organ formation by applying microdroplets of auxin directly to the peripheral zone. Based on these results, they proposed a model for organ initiation in which the local activation of cells is caused not by local self-enhanced production as is the case in reaction–diffusion models, but rather by the directed transport of auxin to organ initiation sites. Their model did not, however, explain what polarizes the exporter

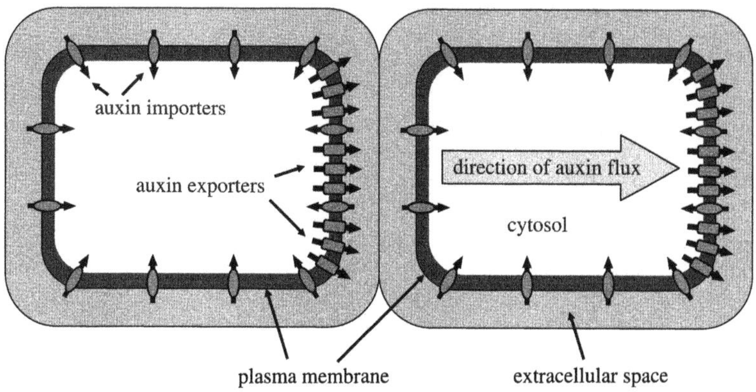

Fig. 10.15 Auxin transport in Arabidopsis cells. Export and import proteins sit in the plasma membrane and transfer auxin between the cytosol and extracellular space. Often, the export proteins are polarly localized, resulting in a net flux of auxin through the cells. (Adapted from Smith (2008)).

proteins. A hypothesis came from computer simulation studies (Jönsson *et al.* 2006; Smith *et al.* 2006*a*), leading to the discovery of a possible new patterning mechanism in plants.

In these simulation studies, it was proposed that cells could sense the concentration of auxin in neighboring cells, and polarize PINs preferentially towards cells with higher auxin concentration. This feedback of auxin on its own transport has an autocatalytic component: if one cell has a slightly higher auxin concentration, then this causes the PIN1 proteins in neighboring cells to orient preferentially towards the orginal cell with a higher concentration, causing a further increase in concentration. This increase then causes even more PINs to orient towards it, causing an even higher auxin concentration. In a tissue of cells, this can result in a *transport-feedback* spacing mechanism similar to Gierer and Meinhardt's activator-inhibitor system.

Again, a simple topology for looking at this patterning mechanism is a line of cells, connected at the ends to form a ring. There is no growth, and unlike the case with the activator–inhibitor system, it is not necessary to consider the production and decay of IAA and PIN1. Only the dynamics of the transport mechanism itself, the transport of IAA and the localization of PIN, are required for patterning.

Each cell stores an IAA concentration, and the orientation of a fixed amount of PIN1 export proteins, which can point towards the cell's left or right neighbors. The simulation begins with each cell containing an initial IAA concentration, which changes over time owing to diffusion and PIN1-mediated transport to and from neighboring cells. PIN1 proteins preferentially locate on the membranes facing neighboring cells with higher IAA concentrations according to the formula

$$[\mathrm{PIN}_{i\to j}] = [\mathrm{PIN}_i]\frac{b^{[\mathrm{IAA}_j]}}{\sum\limits_{k\in N_i} b^{[\mathrm{IAA}_k]}}. \tag{10.15}$$

where $[\mathrm{PIN}_{i\to j}]$ is the number of PIN1 proteins located in the membrane of cell i facing neighboring cell j, $[\mathrm{IAA}_j]$ is the concentration of IAA in the neighboring cell j, $[\mathrm{PIN}_i]$ is the total concentration of PIN1 proteins in cell i, and N_i are the neighbors of cell i. The orientation of PIN1 towards neighboring cells given by eqn (10.15) is an exponential function of IAA concentration, although linear or higher powers of the concentration will also produce patterning.

Active transport depends on the orientation of PIN1 proteins at the cell membranes. The effect of PIN1 on the efflux of auxin from cell i to a neighboring cell j is modeled using the formula

$$\mathrm{transport}_{i\to j} = T[\mathrm{PIN}_{i\to j}]\frac{[\mathrm{IAA}_i]^2}{1+\kappa_T[\mathrm{IAA}_j]^2}, \tag{10.16}$$

where T is the transport coefficient, and κ_T is the transport saturation coefficient. For a given number of PIN1 molecules near the wall separating cell i from cell j, the flux of auxin from i to j is thus assumed to increase with the concentration of auxin in cell i, and to saturate with increasing concentration of auxin in cell j. Note that this is different from the formula used by Jönsson *et al.* (2006), in which the saturation depends on the concentration in the source cell, not the destination cell. In eqn (10.16), transport is quadratic; however, linear or other powers, or even exponential functions, can be used.

The simulation proceeds in a sequence of time steps, with the PIN1 polarity calculated at the beginning of each time step. The entire content of PIN1 in a cell is allocated to the cell's membrane, with eqn (10.15) specifying the portioning-out of the available PIN1 to sections of the membrane facing each of the neighboring cells. The change in IAA concentration in each cell i is modeled by the following equation:

$$\begin{aligned}\frac{d[\mathrm{IAA}_i]}{dt} &= \mathrm{diffusion} + \mathrm{transport}\\ &= \sum_{j\in N_i} D([\mathrm{IAA}_j]-[\mathrm{IAA}_i])\\ &\quad+\sum_{j\in N_i}\left(T[\mathrm{PIN}_{j\to i}]\frac{[\mathrm{IAA}_j]}{1+\kappa_T[\mathrm{IAA}_i]} - T[\mathrm{PIN}_{i\to j}]\frac{[\mathrm{IAA}_i]}{1+\kappa_T[\mathrm{IAA}_j]}\right).\end{aligned} \tag{10.17}$$

Diffusion is assumed to take place directly between neighboring cells as in the reaction–diffusion model discussed previously. This is also the case for transport, and the auxin transported out of a cell is deposited directly into the neighboring cell, bypassing the extracellular space.

If the model is started in the absence of noise, all cells will begin with the same initial concentration, and therefore will have the same PIN1 orientation, transport, and diffusion. The concentrations of IAA in all of the cells will remain equal, and no pattern will form. This is similar to what happens in reaction–diffusion systems in the absence of the noise that is needed to "break symmetry". If even a small amount of noise is added to the model, for example by perturbing the initial cell concentrations, a relatively evenly spaced pattern of peaks in the IAA concentration

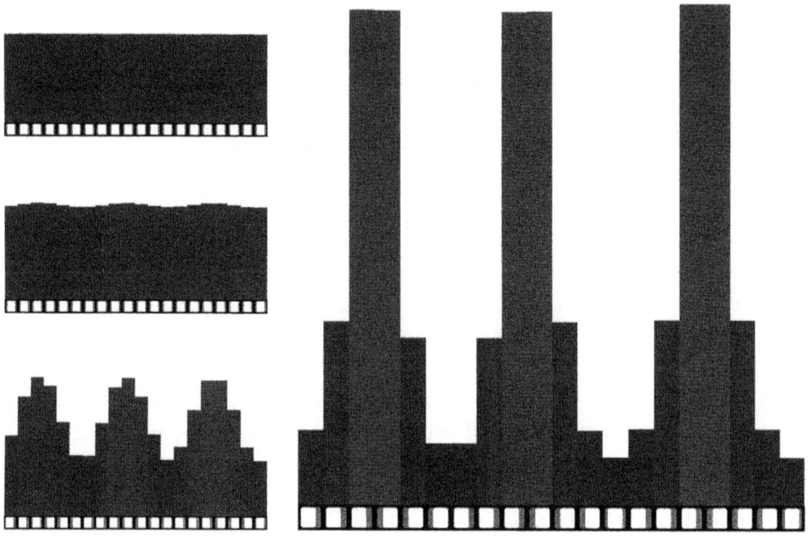

Fig. 10.16 Transport–feedback patterning on a line of 20 cells, with wrap-around boundary conditions (the leftmost and rightmost cells are considered neighbors). Taller (and brighter) bars indicate higher IAA concentrations. A small amount of noise present in the initial IAA distribution is required to break symmetry, and relatively evenly spaced peaks in IAA concentration are formed. Note the PIN polarization (lighter gray inside cells) towards the cells with higher auxin concentration.

will form (Fig. 10.16). Cells with small local maxima due to the initial conditions cause the PIN1 proteins in neighboring cells to be oriented preferentially towards them (very slightly), which causes the concentration of IAA at these maxima to increase. A positive feedback loop is then formed, which will recruit even more PIN1 proteins over time. The depletion of IAA in the cells surrounding a peak prevents the formation of other peaks nearby.

The amplitude of and spacing between these peaks can be controlled by manipulating the model parameters, with decreased transport leading to fewer peaks, and increased transport leading to more peaks. A reverse relationship is observed when the diffusion coefficient is mainpulated, as the spacing of the peaks is determined by a balance between diffusion and transport. If the diffusion coefficient is too high or the transport coefficient is too low, no peaks will form at all. For a given cell count and parameter set, the same number of peaks usually forms, regardless of the pattern of initial noise; however, the locations of the peaks vary.

10.4.5 A transport–feedback model of phyllotaxis

The transport–feedback spacing mechanism is able to make uniform peaks high in auxin concentration for a line of cells. When this model is implemented for a growing shoot apex, we would expect that as new space opens up in the center of the apex, it would be possible for new peaks to form. As in the inhibition model of phyllotaxis described in Section 10.4.2, the peaks would tend to form as far away from previous ones as possible, leading to phyllotaxis patterning.

In order to accommodate a growing structure, a source of auxin must be considered, as pattern formation can no longer be seen as simply redistributing auxin from some almost uniform initial state. Since it is unknown whether the auxin available in the shoot apex is supplied there by directed transport or by local production, the following simplifying assumption has been used. A uniform supply of auxin, modeled as local production, is provided to all cells of the apex which are outside of the central zone. These are defined to be all of the cells which are situated at a distance greater than a given distance from the tip. This distance, which also corresponds to the beginning of the peripheral zone, is provided as a model parameter and remains fixed throughout the simulation.

In addition to auxin production and decay, changes must be made to eqns (10.15)–(10.17) to reflect the different sizes of cells and cell walls. Since only the surface layer of cells is modeled, and it is assumed that all cells have the same thickness, quantities normally expressed as volumes can be expressed by areas, and areas can be expressed by lengths. This leads to the following formula for the model of auxin model on the cellular surface:

$$\begin{aligned}\frac{d[\mathrm{IAA}_i]}{dt} &= \text{production} - \text{decay} + \text{diffusion} + \text{transport}\\ &= \frac{\rho_{\mathrm{IAA}}}{1+\kappa_{\mathrm{IAA}}[\mathrm{IAA}_i]} - \mu_{\mathrm{IAA}}[\mathrm{IAA}_i] + \frac{D}{A_i}\sum_{j\in N_i} l_{i\to j}([\mathrm{IAA}_j]-[\mathrm{IAA}_i])\\ &\quad + \frac{T}{A_i}\sum_{j\in N_i}\left([\mathrm{PIN}_{j\to i}]\frac{[\mathrm{IAA}_j]^2}{1+\kappa_T[\mathrm{IAA}_i]^2} - [\mathrm{PIN}_{i\to j}]\frac{[\mathrm{IAA}_i]^2}{1+\kappa_T[\mathrm{IAA}_j]^2}\right), \qquad (10.18)\end{aligned}$$

where ρ_{IAA} controls the production of IAA, with saturation coefficient κ_{IAA}, and μ_{IAA} controls the decay of IAA, which is dependent on the IAA concentration. IAA transport is modeled as in eqn (10.16) for the line model. $l_{i\to j}$ is the length of the interface between cell i and cell j, and A_i is the area of cell i. No adjustment is made for the increasing cell area due to growth, or for the changes in cell area that occur as a result of the pinching of cells during cell division. The formula for PIN1 localization is similarly modified to account for the differing lengths of cell walls as follows:

$$[\mathrm{PIN}_{i\to j}] = [\mathrm{PIN}_i]\frac{l_{i\to j}b^{[\mathrm{IAA}_j]}}{\sum_{k\in N_i} l_{i\to k}b^{[\mathrm{IAA}_k]}}. \qquad (10.19)$$

Note that in eqns (10.18) and (10.19) the number of neighbors within the summation now varies from cell to cell. Note also that the length of cell walls does not explicitly enter the transport term in eqn (10.18), as it is already included when the PIN amount at the interface is calculated in eqn (10.19).

With eqns (10.18) and (10.19), the models described in Smith *et al.* (2006*a*) were able to produce reasonably spaced peaks of high auxin concentration on a growing shoot apex; however, the positioning was not reliable enough to produce stable spiral phyllotaxis patterns. This was also the case for the linear equations used by Jönsson *et al.* (2006). Smith *et al.* (2006*a*) were able to get reliable patterning by adding additional assumptions to their model that allowed cells in primordia to differentiate. These cells behaved differently from nonprimordium cells and used different rules governing auxin production and PIN polarization, with PINs preferentially polarizing towards the primordium center. Although it is quite reasonable to assume that cells in newly formed primordia can differentiate, it does complicate the model somewhat. It was found that a simpler solution, the use of an exponential function for transport similar to that used by Barbier de Reuille *et al.* (2006), could also produce stable patterning without any additional assumptions. In this case the transport term in eqn (10.18) becomes

$$\text{transport}_{i \to j} = T[\text{PIN}_{i \to j}] \frac{b^{[\text{IAA}_i]} - 1}{b^{[\text{IAA}_j]}}, \tag{10.20}$$

where the model parameter b is the base of the exponential transport function.

Figure 10.17 shows the result of this auxin model superimposed on the growing shoot apex described in Section 10.2.2 together with the dividing-cell model described in Section 10.2.4. As with the inhibition model, new peaks in auxin concentration form as space appears in the peripheral zone of the growing apex tip. The auxin transport–feedback spacing mechanism causes the new primordia to form as far as possible from each other, causing a spiral pattern to emerge.

In this model, the growing apex surface of Section 10.2.2 has been modified to allow the primordia to bulge from the surface slightly. Primordia are determined when

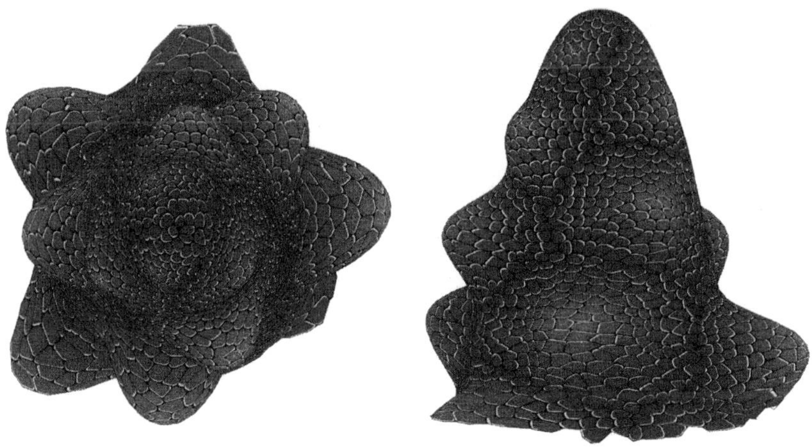

Fig. 10.17 Top (left) and side (right) views of results from a transport–feedback model of phyllotaxis on a growing cellular apex. The gray levels in cell interiors show auxin levels (brighter indicating more auxin), with the PIN1 polarization shown in white. Convergence points of auxin and PIN1 localization appear, and fill the space provided by the growing tip. The transport–feedback spacing mechanism causes new primordia to form as far as possible from previous ones, causing a spiral pattern to emerge.

two adjacent cells exceed a threshold concentration. The average of the cell centers is then taken as a primordium center, and projected back onto the apex surface. The radius and height increase with time, and are combined with a profile function that determines the shape of the bulging primordia. These are provided as model parameters.

10.4.6 Combining molecular and physically based models

The phyllotaxis simulation in the previous section uses a descriptive growth model, with the shapes of the apex and of the outgrowing primordia specified directly in geometric terms. This approach is simple and allows the modeler to focus on the patterning mechanism of phyllotaxis while abstracting from the processes involved in maintaining the shapes of the apex and outgrowing primordia. With physically based growth models, such as the one described in Section 10.3, the shapes of the apex and outgrowing primordia are an emergent property of the model. Although typically more expensive computationally, this approach allows the simulation of the mechanics of growth, and of the interaction of biomolecules with the physical properties of cells.

Fig. 10.18 Transport–feedback model of phyllotaxis on an apex model with physically based (mass–spring) growth. The gray levels in cell interiors show auxin levels (brighter indicating more auxin), with the PIN1 polarization shown in white. Convergence points of auxin and PIN1 localization initiate new growing tips in a spiral pattern. A slight tropism added to the primordium tip growth direction at each time step causes the tips to curve upwards.

Perhaps the simplest way the molecular-level phyllotaxis model and the physically based shoot apex model can be combined is to create new growing tips on the surface at the locations of primordium initiation. These tips are then treated in the same way as the growing apex tip, although the parameters for tip size, spring relaxation, and growth direction may be different. Figure 10.18 shows the result of such a simulation. Using the same auxin model as in the previous section, the model differs only in that auxin production is restricted to the peripheral zone, and PINs are given a preference towards the primordium center after primordium initiation. These modifications are required to prevent additional auxin peaks from forming on the flanks of primordia. In this model the primordia grow much larger, opening up space for additional auxin peaks to form.

With this model, it is quite straightforward to add tropisms to developing primordia or to have lateral apices which themselves produce primordia. In addition, it is possible to model primordia which are not radially symmetric, and develop into leaves, flowers, or other organs. All of these would be much more difficult to implement in the descriptive growth models presented earlier. In addition, a physically based model enables the possibility to study the mechanisms that determine primordium shape and extent, as well as any influence that mechanics may have on primordium positioning.

10.5 Leaf venation patterning

10.5.1 Sachs's canalization hypothesis

The intricate venation patterns seen in the leaves of vascular plants are in some cases almost as striking as spiral phyllotaxis. Equally striking is the fact that the same molecular components, auxin and its exporter PIN1, are the earliest known markers for vein initiation (Scarpella *et al.* 2006), and are key to the patterning process. However, the idea that leaf venation is a transport–feedback type of patterning process predates the discovery of the PIN1 transporter itself, although studies of the polar nature of auxin movement had predicted the existence of such transporters. This led to the chemiosmotic model of auxin transport (Rubery and Sheldrake 1974) (see Section 10.4.4).

Although a reaction–diffusion model of leaf venation was proposed by Meinhardt (1982), experimental evidence appears to favor Sachs's *canalization hypothesis* (Sachs 1981). Sachs proposed that the canalization of auxin into preferred routes of auxin flux occurs by auxin feeding back on its own transport. He drew an analogy with how water carves rivers in soft terrain. A cell's ability to transport auxin is assumed to increase with auxin flux, causing any initially dominant path to be reinforced. As in the case of river formation, as soon as the smallest canal begins to emerge owing to random variation, it will be accentuated and attract even more flow, causing the preferred path to strengthen. Simulation models of the canalization hypothesis on grids of cells have shown that this mechanism is indeed capable of selecting strands from a tissue of undifferentiated cells (Mitchison 1980; Rolland-Lagan and Prusinkiewicz 2005).

Mitchison proposed two variants of the canalization model in his simulations, facilitated diffusion and polar transport. In the facilitated diffusion model, a cell's ability to transport auxin in any direction increases with flux, regardless of the direction of the flux. Transport is passive, and auxin can move only down its concentration gradient. The increased transport ability of a file of cells due to flux can be seen as simply increasing the rate at which auxin diffuses through and/or between cells. Mitchison's second variant, the polar transport model, is more in line with the current understanding of auxin transporters, which are thought to be highly directional. In this case, auxin exporters or carriers, are allocated to sections of cell membrane facing neighboring cells, based on the net flux across the cell-to-cell interface. The following equations from Mitchison, reformulated by Rolland-Lagan and Prusinkiewicz (2005), are used to model carrier allocation at the cell membrane. The transport, or net flux, $\phi_{i \to j}$ from cell i to cell j is given by

$$\phi_{i \to j} = T\left([\mathrm{PIN}_{i \to j}][\mathrm{IAA}_i] - [\mathrm{PIN}_{j \to i}][\mathrm{IAA}_j]\right) + D\left([\mathrm{IAA}_i] - [\mathrm{IAA}_j]\right), \qquad (10.21)$$

where D is the diffusion coefficient, T is the transport coefficient, $[\mathrm{PIN}_{i \to j}]$ is the amount of carriers allocated to the section of the membrane in cell i facing cell j, and $[\mathrm{IAA}_i]$ is the concentration of auxin in cell i.

The carrier allocation, or the change in PIN1 at the section of the membrane in cell i facing cell j, is given by

$$\frac{d[\mathrm{PIN}_{i \to j}]}{dt} = \alpha\phi_{i \to j}^2 + \beta - \gamma[\mathrm{PIN}_{i \to j}], \qquad \text{if } \phi_{i \to j} > 0, \qquad (10.22)$$

$$\frac{d[\mathrm{PIN}_{i \to j}]}{dt} = \beta - \gamma[\mathrm{PIN}_{i \to j}], \qquad \text{if } \phi_{i \to j} \leq 0, \qquad (10.23)$$

where α controls the strength of the flux-driven carrier allocation, β controls a small amount of flux-independent background carrier allocation, and γ controls the decay of carriers from the cell membrane. If the net flux is zero or negative, only a background allocation and decay are considered.

The change in concentration of auxin with respect to time can now be written as

$$\frac{d[\mathrm{IAA}_i]}{dt} = \rho - \mu[\mathrm{IAA}_i] - \sum_{j \in N_i} \phi_{i \to j}, \qquad (10.24)$$

where ρ specifies the auxin production rate in cells, μ gives the auxin decay rate, and the summation is taken over all of the neighbors N_i of cell i.

A simulation of this model is shown in Fig. 10.19. In this simulation, ρ is zero everywhere except in the top row of source cells, and μ is the same in all cells except in the bottom row of sink cells, and whose auxin concentration is set to zero at every time step. Initally auxin flows uniformly from the line of source cells to the line of sink cells. However, some noise in the auxin levels is introduced into the model, which causes some paths to be very slightly favored. The autocatalytic reinforcement of these paths due to the nonlinear production of carriers based on flux causes a destabilization into discrete canals of flux. It is these canals which are then thought to go on to differentiate into veins in plant tissue.

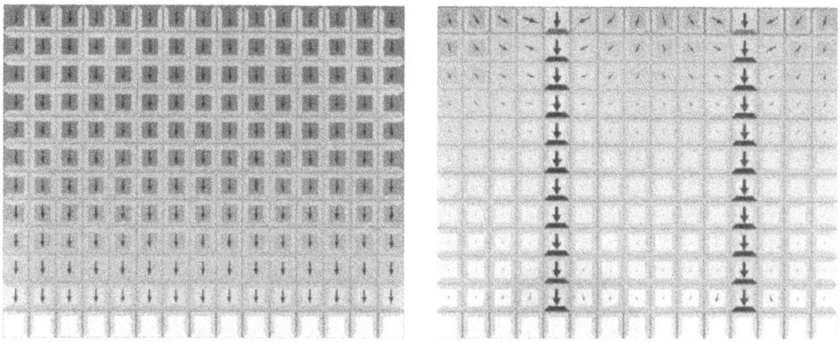

Fig. 10.19 The transport–feedback canalization mechanism for vein formation implemented on a grid of cells. The gray levels in cell interiors show auxin levels (darker indicating more auxin), with the PIN1 polarization shown in the trapezoids at the cell peripheries (darker indicates more PIN1). The black arrows represent the amount and the principal direction of the flux. Initially (left), a fairly uniform flux is present from the line of source cells (top row) to the line of sink cells (bottom row). A small amount of noise is added to the auxin concentrations, causing a destabilization and subsequent canalization into discrete auxin channels (right). Note the high flux but low concentration in the canals. Simulation after Rolland-Lagan and Prusinkiewicz (2005).

10.5.2 Combining models on a growing leaf

The auxin and PIN1 convergence point that triggers primordium initiation in the shoot apex is thought to simultaneously trigger the formation of the midvein in the newly developing leaf. Additional convergence points in the leaf margin are then thought to initiate the first secondary veins (Scarpella *et al.* 2006). This suggests a simple combination of the two transport–feedback models, for leaf venation and phyllotaxis. The simulation shown in Fig. 10.20 uses the "up-the-gradient" PIN polarization model, which makes peaks in the margin cells of the leaf, and the "with-the-flux" canalization model for strand formation in interior cells. The margin cells are separated from the interior cells; however, when their auxin concentration exceeds a threshold level, auxin is allowed to diffuse into the interior cells, causing veins to be initiated.

Again a small adaptation needs to be made to the canalization model equations to account for the different cell sizes. In addition, the canalization model in the previous section used a linear function of concentration for transport with no saturation, and the phyllotaxis simulations discussed previously used an exponential form with saturation. In this case it is convenient to use the same model for auxin transport for both processes, and a quadratic form is used. Thus the equations for the flux in both models become

$$\phi_{i\to j} = T[\mathrm{PIN}_{i\to j}]\frac{[\mathrm{IAA}_i]^2}{1+\kappa_T[\mathrm{IAA}_j]^2} + D([\mathrm{IAA}_i] - [\mathrm{IAA}_j]), \tag{10.25}$$

with symbols as defined previously. The carrier allocation in interior cells is given by eqns (10.22) and (10.23) (with-the-flux). The carrier allocation in margin cells is given

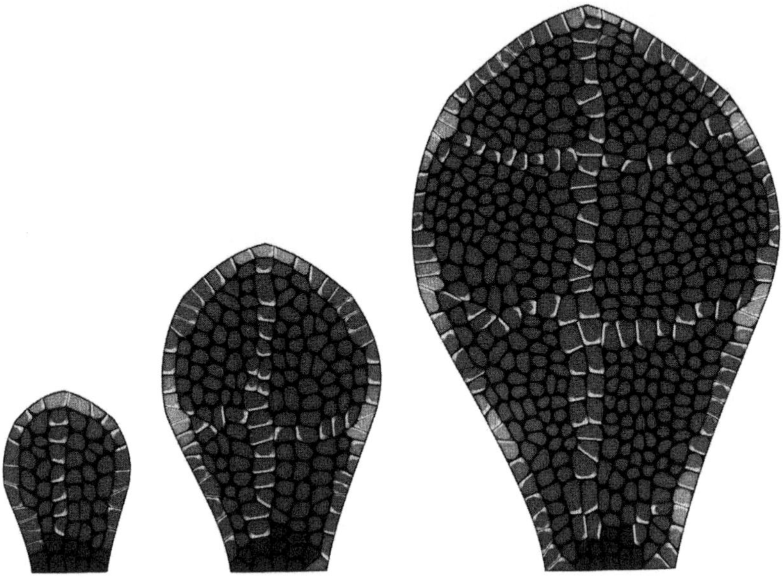

Fig. 10.20 Two transport–feedback models combined on a growing leaf surface.The gray levels in cell interiors show auxin levels (brighter indicating more auxin), with the PIN1 polarization shown in white. The margin cells use the phyllotaxis mechanism ("up-the-gradient") to orient the PINs, leading to the formation of peaks of auxin concentration in the margin. Upon reaching a threshold concentration, auxin begins to diffuse into the interior of the leaf, causing the peaks to act as auxin sources. The interior cells use the canalization mechanism ("with-the-flux") to orient PINs, causing strands to form that connect the sources to sink cells (dark cells) at the base of the leaf. Initially (left), there are only enough margin cells for one peak to form, resulting in a single midvein on the leaf. As the leaf grows and space opens up in the margin, more peaks form, initiating secondary veins.

by eqn (10.19) (up-the-gradient), except that we divide by the length of the cell walls since it will be accounted for in eqn (10.26). The auxin model for both cell types then becomes

$$\frac{d[\mathrm{IAA}_i]}{dt} = \frac{\rho_{\mathrm{IAA}}}{1+\kappa_{\mathrm{IAA}}[\mathrm{IAA}_i]} - \mu_{\mathrm{IAA}}[\mathrm{IAA}_i] - \frac{1}{A_i}\sum_{j\in N_i} l_{i\to j}\phi_{i\to j}, \tag{10.26}$$

where the auxin production constant ρ is set to zero in interior cells.

In the early stages of the simulation (Fig. 10.20) there is only room in the margin for one auxin peak. After the concentration of auxin at this peak has built up sufficiently, it diffuses into the interior, causing the midvein to initiate, connecting the auxin source to sink cells at the base of the leaf. These sink cells represent the effect of existing vasculature in the plant. As the leaf grows, space opens up for more peaks to form, initiating secondary veins that connect to the midvein.

10.5.3 A dual transport–feedback model for midvein formation

Although the model in the previous section is able to explain the formation of veins on a growing leaf, it has several drawbacks. Mitchison's canalization model predicts a high flux but a low concentration in developing canals, yet experimental work indicates that the concentration of auxin is likely to be high (Scarpella *et al.* 2006). In addition, the two methods for PIN1 polarization, and thus the two patterning mechanisms, operate independently in different types of cells in the previous model. However, experimental work suggests that both processes must operate in both cell types (Bayer *et al.* 2009). In the surface layer of cells, PINs first orient towards the convergence point that marks where the new organ primordium will emerge, but then, later the, cells at the center of this convergence point "switch" and orient towards the interior, initiating the midvein. The same is true for cells in the inner layers. Just below the site of primordium initiation, cells first orient their PINs apically towards the convergence point, but later orient downward as the convergence point extends into the inner layers to form the midvein. This suggests that both cell types are able to use both methods, with-the-flux and up-the-gradient, to orient PINs.

This leads to a model in which PINs can be allocated both by auxin concentration and by auxin flux (Bayer *et al.* 2009). The first equation of the model describes the total production and decay of PIN1 in each cell:

$$\frac{d[\mathrm{PIN}_i]}{dt} = \frac{\rho_{\mathrm{PIN}_0} + \rho_{\mathrm{PIN}}[\mathrm{IAA}_i]}{1 + \kappa_{\mathrm{PIN}}[\mathrm{PIN}_i]} - \mu_{\mathrm{PIN}}[\mathrm{PIN}_i], \tag{10.27}$$

where ρ_{PIN_0} represents PIN1 production independent of auxin levels, ρ_{PIN} specifies auxin-dependent PIN1 production, κ_{PIN} is a saturation coefficient, and μ_{PIN} is the decay coefficient. As the simulation proceeds, each cell develops a pool of PIN1 proteins available for allocation to cell membrane sections facing other cells. Since the surface layer is considered to be a different cell type from the inner tissue, the parameters for auxin-dependent and auxin-independent PIN production are different in the two cell types. The surface layer of cells always expresses PINs, whereas in interior cells, PINs are expressed only in cells that are high in auxin.

In the phyllotaxis model, all PINs are allocated instantaneously at the beginning of each time step, whereas in the canalization models, PINs accumulate at cell interfaces over time. In order to combine the two models, a variable $F_{i \to j}$ is added which records the flux history across the interface between cells i and j. This allows the instantaneous allocation of PINs for the canalization model. The equation to calculate this flux history is defined as

$$\frac{dF_{i\to j}}{dt} = B_{(i,j)} \cdot \frac{\rho_{F_0} + \rho_F \left(\max\left(0, \phi_{i\to j}\right)\right)^2}{1 + \kappa_F F_{i\to j}^2} - \mu_F F_{i\to j}, \tag{10.28}$$

where $\phi_{i\to j}$ is the auxin flux from cell i to j, ρ_{F_0} specifies a small default rate of increase, ρ_F specifies the auxin-flux-dependent increase in flux history, κ_F is a saturation coefficient, and μ_F specifies the rate of decay. Note that the measure of how much flux has crossed the interface recently is quadratically dependent on the auxin flux, and increases only if the flux is positive. This is similar to the rule for flux-based carrier allocation in Mitchison's simulation model of canalization. The term $B_{(i,j)}$ represents the separation between the surface layer and the interior cells. If the cells i and j are of the same type, $B_{(i,j)} = 1$, otherwise it is a lesser value which is specified as a model parameter. This gives the effect of reducing communication between the surface layer and interior cells in the meristem.

The next equation describes how the choice is made between flux-based and concen tration-based PIN1 allocation. A variable z_i is defined that is used to make this transition based on the auxin concentration of cell i:

$$z_i = \min\left(\frac{[\mathrm{IAA}_i]}{S_{\mathrm{PIN}}}, 1.0\right). \tag{10.29}$$

As the auxin concentration increases from 0 to a threshold concentration S_{PIN}, z_i increases linearly from 0 to 1. The variable $Z_{i\to j}$ is now defined to represent the allocation amount from the three possible sources: flux-based allocation, concentration-based allocation, and default or background allocation:

$$Z_{i\to j} = L_D + (1.0 - z_i)\, L_C[\mathrm{IAA}_j] + z_i L_F F_{i\to j}. \tag{10.30}$$

L_D specifies a small amount of background PIN1 allocation independent of auxin flux or concentration, L_C specifies an allocation based on the concentration in neighboring cells, and L_F specifies an allocation based on flux. The total amount of PIN1 allocated to each membrane section from the pool of PIN in each cell can now be specified as

$$[\mathrm{PIN}_{i\to j}] = [\mathrm{PIN}_i]\frac{B_{(i,j)} \cdot l_{i\to j} b^{Z_{i\to j}}}{\sum_{k\in N_i} B_{(i,k)} \cdot l_{i\to k} b^{Z_{i\to k}}}, \tag{10.31}$$

where $k \in N_i$ are the neighbors of cell i, $[\mathrm{PIN}_{i\to j}]$ is the amount of PIN1 on the membrane section of cell i facing cell j, $[\mathrm{PIN}_i]$ is the total amount of PIN1 in cell i, $l_{i\to j}$ is the length of the wall between cells i and j, and b is the strength of the exponential allocation, given as a model parameter. Again the term $B_{(i,j)}$ reduces, but does not eliminate, allocation across the boundary between the surface layer of cells and the inner layers. The complete auxin model can now be expressed as

$$\begin{aligned}\frac{d[\mathrm{IAA}_i]}{dt} &= \text{production} - \text{decay} + \text{diffusion} + \text{transport}\\ &= \frac{\rho_{\mathrm{IAA}}}{1+\kappa_{\mathrm{IAA}}[\mathrm{IAA}_i]} - \mu_{\mathrm{IAA}}[\mathrm{IAA}_i] + \frac{D}{A_i}\sum_{j\in N_i} B_{(i,j)} l_{i\to j}([\mathrm{IAA}_j] - [\mathrm{IAA}_i])\\ &\quad + \frac{T}{A_i}\sum_{j\in N_i} B_{(i,j)}\left([\mathrm{PIN}_{j\to i}]\frac{[\mathrm{IAA}_j]^2}{1+\kappa_T[\mathrm{IAA}_i]^2} - [\mathrm{PIN}_{i\to j}]\frac{[\mathrm{IAA}_i]^2}{1+\kappa_T[\mathrm{IAA}_j]^2}\right).\end{aligned} \tag{10.32}$$

Note that this is almost identical to eqn (10.18), except for the term $B_{(i,j)}$, used again to reduce communication between the different cell layers. Figure 10.21 shows a

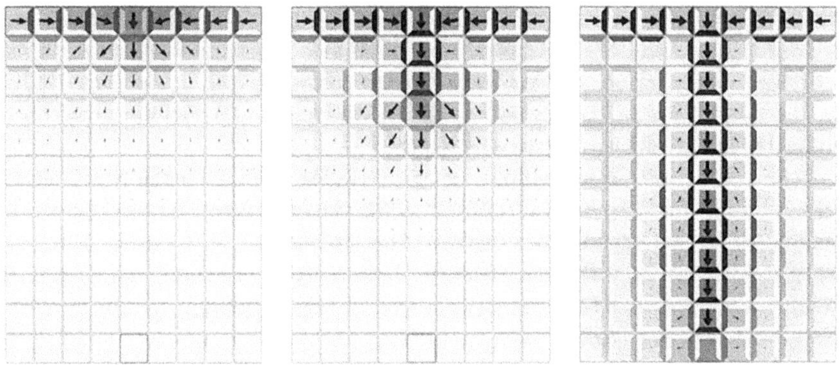

Fig. 10.21 Dual model of transport–feedback patterning on a grid of cells that represents a longitudinal section of a shoot apex. The gray levels in cell interiors show auxin levels (darker indicating more auxin), with the PIN1 polarization shown in the trapezoids at the cell peripheries (darker indicates more PIN1). The black arrows represent the amount and the principal direction of the flux. Auxin is supplied uniformly to the model in the top row of cells, which represent the surface layer of the meristem. Initially (left), up-the-gradient PIN1 polarization predominates in these cells, causing a convergence point of PIN polarity and high auxin levels to emerge. Note the apical polarization in the cells in the rows below. As the simulation proceeds, auxin levels increase, the polarization switches from up-the-gradient to with-the-flux (middle), and a strand emerges (right) connecting the convergence point to the single sink cell at the bottom center of the grid. Note the high level of auxin in the emergent strand.

simulation of eqns (10.27)–(10.32) on a grid of cells that is used as a two-dimensional abstraction of a longitudinal section of a shoot apex. The top row of cells is supplied with a uniform source of auxin and represents the surface layer. Initially, an auxin convergence point forms via the up-the-gradient mechanism for PIN1 polarization. As the simulation proceeds, auxin levels increase, causing a switch to flux-based polarization. This creates a strand of cells that connects the convergence point, which is acting as a source of auxin, to the sink cell below. This model recapitulates several experimental observations not seen in a pure canalization model. As the midvein initiates, the concentration of auxin in the strand is always elevated compared with surrounding tissues. This is consistent with the role of auxin as a signaling molecule that is likely to instruct differentiation in a concentration-based manner. Additionally, when the convergence point first appears, cells in the inner tissue polarize towards it, an experimentally observed phenomenon which is again difficult to reconcile with a pure canalization model. The same is true for the strong lateral polarization towards the vein center.

10.5.4 A dual transport–feedback model on a growing cellular template

A more realistic representation of the shoot apex can be made by digitizing a cross section of a real apex obtained from confocal images (Bayer *et al.* 2009). In the model that follows, this digitization was done using the Merrysim software package (Barbier de Reuille *et al.* 2005). Growth was added to this model by using the same Bézier surface interpolation as was used for the growing leaf in Section 10.2.3. In this way, the simulation of the formation of the midvein could include growth as the primordium bulged from the meristem surface.

When the model was implemented on an irregular cellular surface, it was found that in the model defined by eqns (10.27)–(10.32), the emerging strand did not always connect directly to the sink below. Instead, the tip seemed to wander owing to the irregular geometry of the cellular structure. This is not surprising, since high auxin levels in the vein imply a strong auxin gradient in the proximity of the growing tip. This sharp local gradient would easily overwhelm any much shallower global gradient of auxin in the tissue leading towards the sink. It is an open question then how auxin could both be high in the extending vein and simultaneously attract newly forming veins to each other via a "sink" effect (Heisler and Jönsson 2006; Bayer *et al.* 2009; Smith and Bayer 2009). In the model of Bayer *et al.* (2009), it was suggested that something else must guide veins to the sink, possibly via an innate tissue polarity, a property of plant tissue demonstrated by Sachs (1981). This polarity can be modeled in several ways; however, the simplest is by simply favoring allocation to cell walls facing the sink. A small change to eqn (10.30) to add a geometrical bias $V_{i\to j}$ in the flux allocation component is made as follows:

$$Z_{i\to j} = L_D + (1.0 - z_i)\, L_C[\mathrm{IAA}_j] + z_i L_F V_{i\to j} F_{i\to j}, \tag{10.33}$$

where $V_{i\to j}$ is set to 0 at the location of the cell most distal from the sink cells, and set to 1 at the location nearest the sink cells. Figure 10.22 shows the results of a simulation. As in the simulation on a grid of cells, the surface layer is provided with

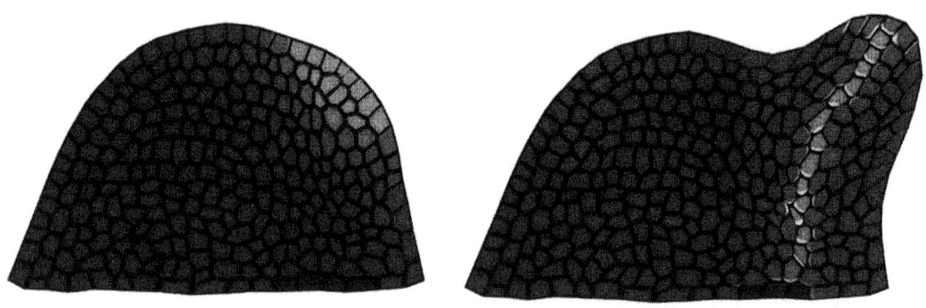

Fig. 10.22 Dual model of transport–feedback patterning on a growing cellular representation of a cross section of a shoot apex. The gray levels in cell interiors show auxin levels (lighter indicating more auxin), with the PIN1 polarization shown at the cell peripheries (lighter indicates more PIN1). Auxin is supplied uniformly to the model in a range of cells in the top layer (approximately 15 cells). Initially (left), up-the-gradient PIN1 polarization causes a convergence point to form in the surface layer, with apical polarization of the cells below. As the simulation proceeds, the primordium bulges from the surface as a strand emerges (right), connecting the convergence point at the center of the primordium to the sink cells (dark cells) below.

a uniform quantity of auxin in the cells near where the primordium will form. At low auxin levels, the surface cells form a convergence point for auxin and PIN polarity. This marks the center of where the primordium will form. As auxin levels increase, the polarization mechanism switches from up-the-gradient to with-the-flux, causing a strand of cells with high auxin and PIN expression levels to appear, connecting the convergence point to the sink cells below. Note that there is almost no background production of PINs in the inner layers of cells, and that PIN expression depends almost entirely on auxin-dependent-production of PINs. As with the model on the rectangular grid of cells, there is strong lateral polarization of PINs towards the center of the vein, as well as high auxin levels in the developing strand.

10.6 Conclusions

Perhaps the most successful formalism for modeling plant development has been L-systems, which are ideally suited for modeling growing branching structures at the tissue level. In recent years, as our knowledge of genes, proteins, and signaling molecules has expanded, there has been an increased interest in understanding plant development at the cellular level. This is understandable, as the cell is the fundamental building block of life and it is at this level that many key developmental decisions are made. Modeling development at this level, however, has some difficult challenges. Not only is it required to develop models of genetic networks and cell-to-cell signaling, but also these models must often be implemented on growing cellular structures in order to be relevant. Phyllotaxis patterning is a good example; without a growing structure with cells and cell division, any model proposed will always be perceived to be somewhat abstract. Except in some very special cases of filamentous organisms, most cellular structures are three-dimensional, although it is often possible to abstract to two dimensions. Most of the models presented in these notes use this simplification, and the formalism of vertex–vertex systems (VV) for their implementation. However, it is clear that methods need to be developed to conveniently represent growing, three-dimensional, cellular structures.

As computational power increases, it will be possible to explore a wider range of problems, such as physically based simulations of the growth of cellular tissues, which are currently pushing the limits of technology. It is these models, however, that enable the possibility to simulate a direct link between genes, growth, and form. Genes drive the chemistry that determines the physical properties of cells, which in turn determines how cells grow. In plants, it is the control of this growth that ultimately results in the emergence of form.

Acknowledgments

Support for this work was provided by the University of Bern, the Swiss SystemsX.ch project Plant Growth in a Changing Environment (PGCE), the Natural Sciences and Engineering Research Council of Canada (NSERC), the Alberta Informatics Circle of Research Excellence (iCORE), and the University of Calgary (Ruby doctoral scholarship) and is gratefully acknowledged. I would also like to acknowledge

Przemyslaw Prusinkiewicz, Cris Kuhlemeier, Emmanuelle Bayer, Soazig Guyomarcíh, Naomi Nakayama, Pierre Barbier de Reuille, Adam Runions, and Colin Smith for insightful discussions and for helping to formulate many of the ideas appearing in the co-authored works from which a substantial portion of these notes as derived.

References

Barbier de Reuille, Pierre, Bohn-Courseau, Isabelle, Godin, Christophe, and Traas, Jan (2005). A protocol to analyse cellular dynamics during plant development. *Plant Journal*, **44**(6), 1045–1053.

Barbier de Reuille, Pierre, Bohn-Courseau, Isabelle, Ljung, Karin, Morin, Halima, Carraro, Nicola, Godin, Christophe, and Traas, Jan (2006). Computer simulations reveal properties of the cell–cell signaling network at the shoot apex in *Arabidopsis*. *Proceedings of the National Academy of Sciences of the USA*, **103**(5), 1627–1632.

Barbier de Reuille, Pierre, Runions, Adam, Smith, Richard, Coen, Enrico, and Prusinkiewicz, Przemyslaw (2007). Trichome patterning on growing tissue. In *Proceedings of the 5th International Workshop on Functional-Structural Plant Models*, eds. Przemyslaw Prusinkiewicz, Jim Hanan, and Brendan Lane, pp. 365–370.

Bayer, Emmanuelle M., Smith, Richard S., Mandel, Therese, Nakayama, Naomi, Sauer, Michael, Prusinkiewicz, Przemyslaw, and Kuhlemeier, Cris (2009). Integration of transport-based models for phyllotaxis and midvein formation. *Genes and Development,*, **23**(3), 373–384.

de Boer, M., Fracchia, F., and Prusinkiewicz, P. (1992). A model for cellular development in morphogenetic fields. In *Lindenmayer Systems: Impacts on Theoretical Computer Science, Computer Graphics, and Developmental Biology*, eds. G. Rozenberg and A. Salomaa, pp. 351–370. Springer, Berlin.

Dumais, Jacques (2007). Can mechanics control pattern formation in plants? *Current Opinion in Plant Biology*, **10**(1), 58–62.

Errera, L. (1888). Über Zellfromen und Seifenblasen. *Botanisches Centralblatt*, **34**, 395–398.

Foley, J. D., van Dam, A., Feiner, S. K., and Hughes, J. F. (1990). *Computer Graphics: Principles and Practice* (2nd edn.). Addison-Wesley.

Giavitto, Jean-Louis and Michel, Olivier (2001). MGS: A ruled-based language for complex objects and collections. *Electronic Notes in Theoretical Computer Science*, **59**(4), 1–19.

Gierer, A. and Meinhardt, H. (1972). A theory of biological pattern formation. *Kybernetik*, **12**(1), 30–39.

Green, Paul B., Steele, C. S., and Rennich, S. C. (1996). Phyllotactic patterns: A biophysical mechanism for their origin. *Annals of Botany*, **77**(5), 515–528.

Grieneisen, Verônica A., Xu, Jian, Marée, Athanasius F. M., Hogeweg, Paulien, and Scheres, Ben (2007). Auxin transport is sufficient to generate a maximum and gradient guiding root growth. *Nature*, **449**(7165), 1008–1013.

Hamant, Olivier, Heisler, Marcus G, Jönsson, Henrik, Krupinski, Pawel, Uyttewaal, Magalie, Bokov, Plamen, Corson, Francis, Sahlin, Patrik, Boudaoud, Arezki, Meyerowitz, Elliot M., Couder, Yves, and Traas, Jan (2008). Developmental patterning by mechanical signals in arabidopsis. *Science*, **322**(5908), 1650–1655.

Heisler, Markus G. and Jönsson, Henrik (2006). Modeling auxin transport and plant development. *Journal of Plant Growth Regulation*, **25**, 302–312.

Hejnowicz, Z., Nakielski, J., and Hejnowicz, K. (1984). Modeling of spatial variations of growth within apical domes by means of the growth tensor. II. Growth specified on dome surface. *Acta Societatis Botanicorum Poloniae*, **53**, 301–316.

Hofmeister, Wilhelm (1868). *Handbuch der Physiologishen Botanik.* Engelmann, Leipzig.

Honda, Hisao (1978). Description of cellular patterns by Dirichlet domains: The two-dimensional case. *Journal of Theoretical Biology*, **72**, 523–543.

Jönsson, Henrik, Heisler, Marcus G., Shapiro, Bruce E., Meyerowitz, Elliot M., and Mjolsness, Eric (2006). An auxin-driven polarized transport model for phyllotaxis. *Proceedings of the National Academg of Sciences of the USA,* **103**(5), 1633–1638.

Kramer, Eric M. (2002). A mathematical model of pattern formation in the vascular cambium of trees. *Journal of Theoretical Biology*, **216**(2), 147–158.

Kramer, Eric M. (2004). PIN and AUX/LAX proteins: Their role in auxin accumulation. *Trends in Plant Science*, **9**(12), 578–582.

Lindenmayer, Aristid (1968*a*). Mathematical models for cellular interactions in development. I. Filaments with one-sided inputs. *Journal of Theoretical Biology*, **18**(3), 280–299.

Lindenmayer, Aristid (1968*b*). Mathematical models for cellular interactions in development. II. Simple and branching filaments with two-sided inputs. *Journal of Theoretical Biology*, **18**(3), 300–315.

Lindenmayer, A. and Rozenberg, G. (1979). Parallel generation of maps: Developmental systems for cell layers. In *Graph Grammars and Their Application to computer Science: First International Workshop*, eds. V. Claus, H. Ehrig, and G.Rozenberg, Lecture Notes in Computer Science 73, pp. 301–316. Springer, Berlin.

Lyndon, R. F. (1998). *The Shoot Apical Meristem.* Cambridge University Press.

Meinhardt, Hans (1982). *Models of Biological Pattern Formation.* Academic Press.

Mitchison, G. J. (1980). A model for vein formation in higher plants. *Philosophical Transactions of the Royal Society of London. B. Biological Sciences*, **207**, 79–109.

Murray, J. D. (2002). *Mathematical Biology I: An Introduction,* 3rd edition. Springer.

Nakielski, Jerzy (2000). Tensorial model for growth and cell division in the shoot apex. In *Pattern Formation in Biology, Vision and Dynamics,* eds. Alessandra Carbone, Misha Gromov, and Przemyslaw Prusinkiewicz, pp. 252–286. World Scientific.

Neider, Jackie, Davis, Tom, and Woo, Mason (1994). *OpenGL Programming Guide 9 (The Red Book).* Addison-Wesley.

Prusinkiewicz, Przemyslaw (1986). Graphical applications of l-systems. In *Proceedings of Graphics Interface 86 Vision Interface 86*, eds. Marceli Wein and Evelyn M. Kidd, pp. 247–253.

Prusinkiewicz, P. (2004). Art and science for life: Designing and growing virtual plants with l-systems. *Acta Hortica*, **630**, 15–28.

Prusinkiewicz, Przemyslaw and Lindenmayer, Aristid (1990). *Algorithmic Beauty of Plants.* Springer

Prusinkiewicz, Przemyslaw, Hammel, Mark, Hanan, Jim, and Mech, Radomir (1997). Visual models of plant development. In *Handbook of Formal Languages,* Vol.3, *Beyond Words,* eds. G. Rozenberg and. A. Salomaa, pp. 535–591. Springer.

Reinhardt, Didier, Pesce, Eva-Rachele, Stieger, Pia, Mandel, Therese, Baltensperger, Kurt, Bennett, Malcolm, Traas, Jan, Friml, Jir, and Kuhlemeier, Cris (2003). Regulation of phyllotaxis by polar auxin transport. *Nature*, **426**(6964), 255–260.

Rolland-Lagan, A. G. and Prusinkiewicz, P. (2005). Reviewing models of auxin canalization in the context of leaf vein pattern formation in *Arabidopsis. The Plant Journal*, **44**(5), 854–865.

Rubery, P. H. and Sheldrake, A. R. (1974). Carrier-mediated auxin transport. *Planta*, **118**(2), 101–121.

Sachs, T. (1981). The control of patterned differentiation of vascular tissues. *Advances in Botanical Research*, **9**, 151–262.

Scarpella, Enrico, Marcos, Danielle, Friml, Jir, and Berleth, Thomas (2006). Control of leaf vascular patterning by polar auxin transport. *Genes and Development*, **20**(8), 1015–1027.

Schoute, J. C. (1913). Beitrage zur Blattstellunglehre. I. Die Theorie. *Recueilde Travaux Botániques Néerlandais*, **10**, 153–339.

Smith, Richard S. (2008). The role of auxin transport in plant patterning mechanisms. *PLoS Biology*, **6**(12), e323.

Smith, Richard S. and Bayer, Emmanuelle M. (2009). Auxin transport-feedback models of patterning in plants. *Plant Cell Environment*, **32**(9), 1258–1271.

Smith, Colin, Prusinkiewicz, Przemyslaw, and Samavati, Faramarz (2004). Local specification of surface subdivision algorithms. In *Applications of Graph Transformations with Industrial Relevance: Second International Workshop, AGTIVE 2003* (ed. J. L. Pfaltz, M. Nagl, and B. Böhlen), pp. 313–327. Springer.

Smith, Richard S., Guyomarc'h, Soazig, Mandel, Therese, Reinhardt, Didier, Kuhlemeier, Cris, and Prusinkiewicz, Przemyslaw (2006*a*). A plausible model of phyllotaxis. *Proceedings of the National Academy of Sciences of the USA*, **103**(5), 1301–1306.

Smith, Richard S., Kuhlemeier, Chris, and Prusinkiewicz, Przemyslaw (2006*b*). Inhibition fields for phyllotactic pattern formation: A simulation study. *Canadian Journal of Botany*, **84**(11), 1635–1649.

Stern, Ari and Desbrun, Mathieu (2006). Discrete geometric mechanics for variational time integrators. In *Discrete Differential Geometry: An Applied Introduction*, full-day course at SIGGRAPH '06, (organizers Eitan Grinspun, Mathieu Desbrun, Konrad Polthier, and Peter Schroder)

Stoma, Szymon, Lucas, Mikael, Chopard, Jiérôme, Schaedel, Marianne, Traas, Jan, and Godin, Christophe (2008). Flux-based transport enhancement as a plausible unifying mechanism for auxin transport in meristem development. *PLoS Computational Biology*, **4**(10), e1000207.

Turing, A. M. (1952). The chemical basis of morphogenesis. *Philosophical Transactions of the Royal Society of London. B. Biological Sciences*, **237**, 37–52.

11
How cell mechanics shapes embryos

Michel Labouesse

Department of Cell and Developmental Biology,
IGBMC CNRS/INSERM/UdS, France

Summary

Embryonic morphogenesis refers to the stages during which embryos and their organs acquire their functional structure. It requires groups of cells with similar properties to organize themselves in space and adopt distinctive shapes. The physical mechanisms that control cell dynamics during that time are slowly beginning to emerge from the study of a few simple systems. Here, I will present some examples outlining the current understanding of cell mechanics from the physics point of view, emphasizing the balance between cortical tension and cell adhesion, the behavior of motors, and the response to tension. I will also outline how modeling is gaining prominence.

11.1 Introduction

Our cells are capable of a few individual behaviors, which are to divide, to grow, to differentiate, to die, to migrate, and to change their shape. This chapter will concentrate mainly on the latter behavior, with a focus on epithelial cells during embryonic development.

The development of all embryos, irrespective of their final appearance and size, follows a similar sequence of events. During an initial period of cell proliferation and growth, the basic body plan is laid down. Subsequently, cells within an organ primordium or in a specific area of the embryo start to express organ-specific or region-specific differentiation markers, and quickly thereafter start to change their organization. In so doing, they contribute to giving the organ or the embryo its final shape. Among all cells of the embryo, epithelial cells play a key role in driving tissue morphogenesis, because their intrinsic polarity introduces asymmetry [1, 2, 3].

Considerable progress has been made in the past ten years in our understanding of the cellular and genetic events that coordinate epithelial-cell shape changes. Three important features have contributed to this progress: the identification in vertebrate tissue culture cells of many components involved in cytoskeleton remodeling and intracellular trafficking [4, 5, 6, 7]; genetic analysis of the invertebrate model systems *Drosophila melanogaster* and *Caenorhabditis elegans* [3, 8], and progress in light microscopy. We have reached a stage where the main players capable of changing epithelial cell shapes are known. Future progress will necessarily incorporate the input of physics.

Epithelial cells are held in balance between two main forces, which ensure their adhesion or promote their reorganization [8]. Intercellular adhesion maintains epithelial cell integrity by providing a landmark for trafficking, and, at a more global level, maintains tissue integrity. Although fairly stable, adhesion complexes can be undone, in which case cells change their shape or, in more extreme cases, delaminate from the epithelium.

The main challenge in linking physics to cell biology and developmental biology is to define the forces that maintain or modify cell shapes. At a subcellular level, the challenge is to define the principles that orchestrate the self-organization and behavior of complex structures, in particular the molecular motors that drive morphogenesis. In other words, it is critical to recognize those phenomena that cannot be explained by the current understanding of cell biology. In addition, cells are not isolated but respond to external cues, including mechanical cues; it is thus important to define how they integrate those cues. In this chapter, as the influence of physics remains very recent in developmental biology, I provide a general overview of the field, trying to offer some general introduction to the developmental biology process behind it. I have deliberately chosen not to focus extensively on a single process, and to describe those events for which some modeling has been attempted.

11.2 Epithelial cells and the main forces exerted on them

Epithelial cells play a leading role in driving embryonic morphogenesis. Three fundamental properties of epithelial cells enable them to change their shape while maintaining intercellular adhesion and tissue integrity. First, epithelial cells are highly polarized, since their apical and basal plasma membranes contain distinct sets of lipids and proteins (Fig. 11.1) [1, 2]. Second, epithelial cells possess several cell–cell junctional complexes at the juncture between the apical and lateral basal membranes that allow them to maintain adhesion and polarity [2, 9]. The adherens junction (or *zonula adherens*), which includes E-cadherin/β-catenin/p120-catenin/α-catenin and nectin/afadin complexes, is critical for morphogenesis. It maintains tissue integrity and provides a platform for signaling and establishing polarity. Both E-cadherin and nectin/afadin complexes can tether actin microfilaments, acting also as scaffolds for

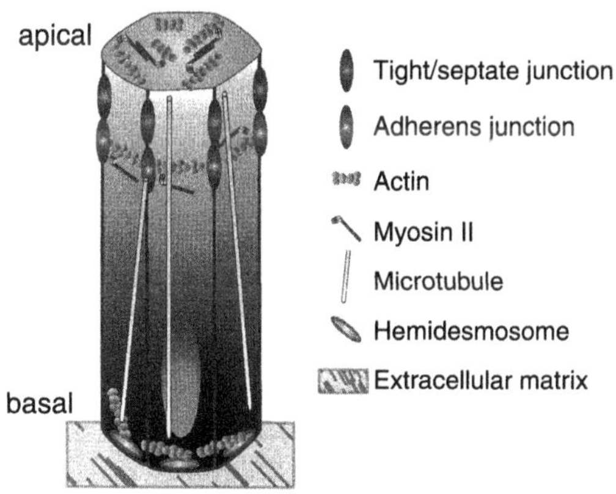

Fig. 11.1 Morphology of a typical columnar epithelial cell. The diagram shows the main components of a vertebrate polarized (gray shading) epithelial cell, with tight junctions, adherens junctions, hemidesmosomes, and cytoskeletal elements. Intermediate filaments are not represented for the sake of clarity, as they are not discussed in this chapter. The microtubules are generally oriented along the apico-basal axis. Focal adhesions can also be present to anchor the cell to the extracellular matrix. In *Drosophila*, septate junctions, which are found basal to adherens junctions, replace tight junctions; insects have no hemidesmosomes. In *C. elegans*, there is a functional equivalent of septate junctions, but it does not make an electron-dense structure; on the other hand, this organism has hemidesmosome-like junctions.

the assembly of numerous signaling proteins that control cytoskeleton remodeling. The second apical junctional complex, the tight junction (*zonula occludens* in vertebrates) or septate junction (in insects), acts as a fence to prevent the intermixing of apical and lateral membrane domains. Third, epithelial cells contain at their apical surface a network of actin-binding proteins, and at their basal surface other junctional complexes (hemidesmosomes and focal adhesions), which ensure adhesion to the extracellular matrix; they can also interact with actin microfilaments [3].

The mechanisms that specify and maintain epithelial polarity are the focus of intense research, but a priori do not directly impact on cell mechanics [10, 11, 12]. Rather, they do so indirectly, in that defects in cell polarity maintenance affect either the positioning of the E-cadherin complex or its stability. Note, however, that several polarity factors can feed back on adherens junction assembly.

More central to the present discussion of cell mechanics are forces exerted by the nonmuscle myosin II motor, which can create *tension* in areas where actin is present (mainly junctions and apical plasma membrane), and cell *adhesion forces*; these are intrinsic to cells. In addition, neighboring tissues can also apply forces, either a *pulling force* or a *pushing force*, on epithelial cells. Pulling forces can originate, for instance, from neighboring cells attached to the epithelial cells of interest when they extend in a given direction (e.g. from invaginating mesoderm in *Drosophila* pulling on extending germ band cells during gastrulation). Compressive forces can originate from hydrostatic pressure (e.g. in *C. elegans* embryos during embryonic elongation; see below) or if neighboring cells push on a tissue as part of a morphogenetic program (e.g. from *Drosophila* germ band cells pushing on stomodeum during germ band extension [13]). In static epithelial tissues, all forces are in balance, whereas in epithelia undergoing morphogenesis, the balance is displaced.

11.3 Methods used to investigate cell mechanics in embryos

Modern, fast imaging methods are playing an increasingly important role in dissecting the physical principles that control cell mechanics. In particular, semiautomated imaging methods coupled with image segmentation are providing quantitative data that are essential for building models and assessing strain levels and orientations [14]. Modeling that takes account of the balance of forces is also gaining prominence. In parallel, the use of genetic analysis (particularly in the case of the invertebrate species *D. melanogaster* and *C. elegans*) offers an unrivalled approach for introducing perturbations, identifying key players, and defining their properties while remaining in a physiologically relevant background.

Assessing the level of forces exerted on a given tissue or cell is important if one wishes to achieve a quantitative description of a given process and to be in a position to model the process. Direct force measurement poses practical challenges, as embryos are small, heterogeneous, and quite fragile, especially invertebrate embryos. The main approach used so far with *Drosophila* and *C. elegans* embryos has relied on laser nanosurgical ablation of entire cells. A junction or an actomyosin filament is considered as a spring under tension [15, 16, 17, 18, 19], and the speed with which junction remnants move apart after ablation provides an indirect measure of the tension exerted on the junction or filament. Injected nanoparticles can provide fiducial markers to be tracked in order to assess cytoplasmic viscosity [20]. Alternatively, if the nanoparticles are ferromagnetic, they can be used to apply a compressive force within the embryo when placed in a magnetic field [13].

More direct approaches are possible with larger embryos such as the frog *Xenopus* and the zebrafish *Danio rerio*, two classical models in developmental studies. One approach consists in using an atomic force microscope (AFM) setup to estimate cell–cell adhesion or cell cortex tension in single-cell force spectroscopy experiments, which can produce forces in the nanonewton range [21]. In one type of experimental setup, a cell is attached to the AFM cantilever and brought into contact with another cell; the force necessary to detach the two cells from each other provides an estimate of the adhesion force. In another type of experimental setup, a hard bead is attached to an AFM cantilever and brought into contact with a cell; the force necessary to deform that cell provides an estimate of the cell tension. The elastic modulus of a tissue explant can be approximately measured by placing it into a small compression chamber and applying a force in the 50 nN to 10 μN range; its resistance to compression and its stress relaxation properties can then be measured [22, 23].

11.4 Cell adhesion versus tension

Forces created by adhesion complexes (adherens junctions) and tension created by cortical actomyosin exert opposing forces on epithelial cells, which modulate the length of the contact area (Fig. 11.2(A)) [8]. Cells in a tissue show striking similarities to fluids and soap bubbles [24]. In a mature tissue, the two forces are in balance and cells remain static [8]. Any significant modification of this equilibrium will modify the cell shape or tissue/organ organization, and this accounts for most embryonic morphogenesis events.

In a series of famous experiments, Holftretter and colleagues showed that embryonic cells with different properties (e.g. future epidermis and future neurons), if mixed together, would sort out spontaneously and segregate into separate homogeneous subpopulations (Fig. 11.2(B)) [25]. By analogy with immiscible liquids, it was suggested that cells try to minimize their surface energy, which is related to surface tension, through the interplay of their respective cell–cell adhesion molecules [26]. This model for cell sorting was named the "differential adhesion hypothesis" (DAH) [26].

The strongest evidence supporting this hypothesis comes from studies in *Xenopus* and *Drosophila*. *Xenopus* cells with different levels of surface tension sort themselves into two subpopulations, with cells of lower surface tension enveloping cells of higher surface tension (Fig. 11.2(B)) [27]. Likewise, cells that express different levels of the same adhesion molecule will acquire different surface tension properties and sort according to the same rules, which can be described according to the Young–Laplace equation.

More recent analysis has challenged the importance of adhesion and given pre-eminence to tension in cell sorting, in particular during germ layer formation. All animals with an intestine, muscles, and a nervous system start by differentiating these three major derivatives (called endoderm, mesoderm, and ectoderm) at a very early stage of their development. When embryos gastrulate to generate the three germ layers,

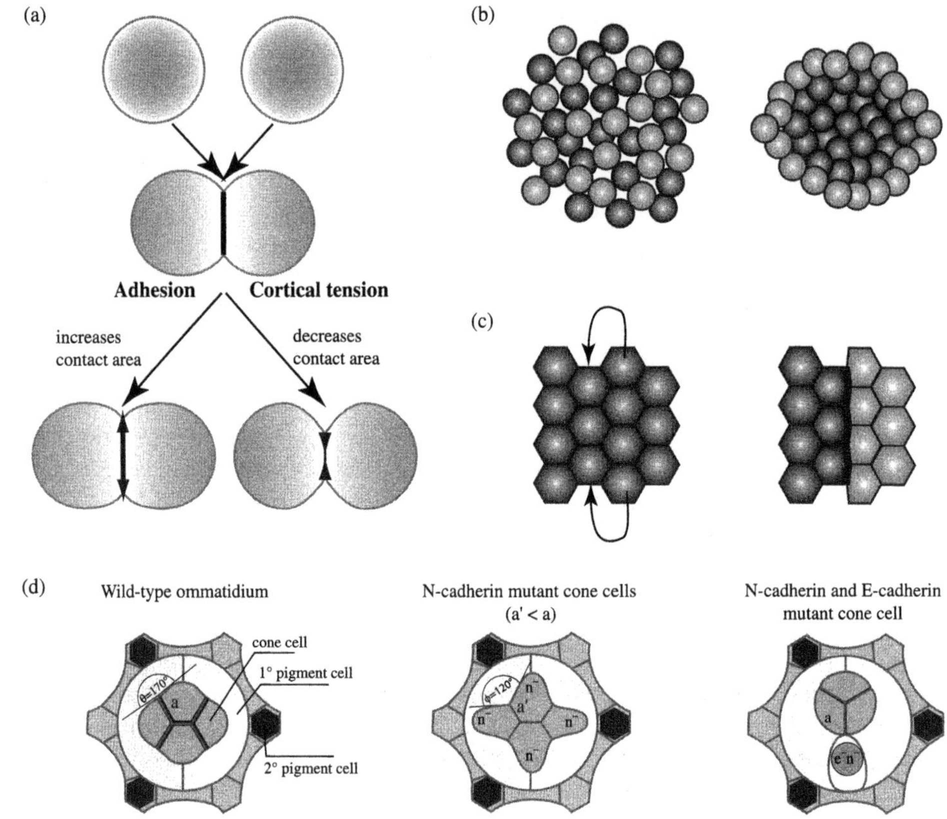

Fig. 11.2 Tension and adhesion in morphogenesis. (A) When two cells contact each other, cell adhesion tends to increase the area of contact, whereas cortical tension tends to decrease it. (B) When mixed together, two different cell types (dark and light gray) will spontaneously sort to form two distinct layers. Cells with the highest contractility (thick lines) end up at the center of the cluster. (C) A common process in embryogenesis is the formation of distinct domains/segments/compartments within an initially homogeneous field of cells (all dark gray, left) through specific signaling pathways (arrows). The formation of the boundary, which prevents any mixing between the two adjacent compartments (light and dark gray, right), involves the formation of an actomyosin belt (thick line) and higher myosin II activity. (D) A surface view of the wild-type *Drosophila* ommatidium, which includes cone cells, pigment cells, and additional support cells (left; photoreceptors, not shown, are found deeper). E-cadherin (light gray) mediates adhesion between cells; in addition, N-cadherin (dark gray) is also present at the interface between cone cells. When N-cadherin alone is missing (n^-, middle), the length of the cone–cone interface (a) becomes shorter (a') and the angle between the cells is shallower. When E-cadherin and N-cadherin are both missing from one or more cone cell ($\bar{e}\ \bar{n}$, right), the cell rounds up and becomes excluded from the central cluster. The other cone cells rearrange their shapes so as to minimize their surface energy, as would soap bubbles.

the ectoderm ends up surrounding the mesoderm, which itself surrounds the endoderm. Heisenberg and colleagues showed that tension plays an essential role in this sorting process [21]. They used atomic force spectroscopy to assay the adhesive strength of individual cells from each of the zebrafish embryonic germ layers, measuring the force necessary to separate two cells placed in contact. To obtain an approximation to the cortical tension, they also measured the resistance of each cell type to deformation by a rigid bead. Their results showed that the three germ layers have inverse adhesive and tensional properties:

$$\text{adhesion}_{\text{endo}} > \text{adhesion}_{\text{meso}} > \text{adhesion}_{\text{ecto}},$$

$$\text{tension}_{\text{endo}} < \text{tension}_{\text{meso}} < \text{tension}_{\text{ecto}}.$$

Cells from the three germ layers mixed together in culture dishes sort as predicted (except that ectoderm cells end up in the middle; in real embryos, there is an outer layer called the enveloping layer made of matrix protein that surrounds the embryo, and an internal yolk). However, they fail to sort when drugs are used to inhibit myosin II activity, which should affect tension but not adhesion [21].

A similar reappraisal of tension versus adhesion has occurred for the formation of the *Drosophila* ommatidium, the central entity forming the adult eye (there are roughly 800 ommatidia in the compound eye of the fly). One ommatidium contains four cone cells, found at the center, several types of pigment cells surrounding the cone cells; and photoreceptors (Fig. 11.2(D)). Proper adhesion between the primary and secondary/tertiary cone cells requires two heterophilic adhesion proteins, Hibris and Roughest, underlining the importance of cell adhesion [28]. Using genetic tricks, one can vary the number of cone cells and examine their spontaneous organization within an ommatidium [29]. Alternatively, one can examine how cone cells organize in the presence or absence of their two main adhesion molecules, N-cadherin and E-cadherin [29]. When the number of cone cells is varied, they form a cluster as would soap bubbles, in that they minimize the overall surface area of the group [29]. If both E-cadherin and N-cadherin are absent from a cone cell, that cell becomes excluded from the central group of cone cells (Fig. 11.2(D)). If N-cadherin alone is missing (it is normally found between cone cells only), the shape of the structure is altered, with a smaller angle $\phi = 120°$ at the interface with surrounding pigment cells (instead of $\theta = 170°$ as is found normally), and shorter cone–cone interfaces (a and a' in Fig. 11.2(D)). These observations support the idea that adhesion molecules play a critical role in minimizing cell surface energy [29].

This model has recently been revisited to incorporate the potential role of cortical tension. As indicated above, whereas adhesion tends to increase the contact length between two cells, cortical tension tends to decrease it. Carthew, Graner, and colleagues, when modeling cell shapes, could not properly predict cone cell shapes in various mutant combinations if they neglected cortical tension. In contrast, they could do so if they took cortical tension into account [30]. They introduced the following modified energy equation:

$$\epsilon = -\sum_{\text{interfaces}} J_{ij}P_{ij} + \sum_{\text{cells}} \left[\lambda_P(P_i - P_{0i})^2 + \lambda_A(A_i - A_{0i})^2\right]. \tag{11.1}$$

In eqn (11.1), J_{ij} is the adhesion energy between two cells i and j, P_{ij} is the length of an interface, λ_P is the perimeter modulus, P_{0i} is the target perimeter of cell i, A_i is the area of cell i, A_{0i} is the target area, and λ_A is the area modulus. The perimeter P_i is the sum of the lengths of the interfaces, $P_i = \sum_j P_{ij}$. The energetic cost δE associated with a change in surface area δS is related to the surface tension σ. In the present case, the change in the interfacial tension γ_{ij} of the membrane can be related to a change in the membrane length between cells i and j (see a and a' in Fig. 11.2(D)).

A recent genetic study further supports the notion that actomyosin contractility, and hence cortical tension, plays a key role in defining the length of cone cell interfaces. When the myosin II motor or one of its upstream regulators, the small GTPase Rho, is inactivated within a single cone or pigment cell, that cell increases its surface area [31].

11.5 Tension at compartment boundaries

The relative weights of tension and differential adhesion have also been a matter of considerable debate in efforts to understand how animals create boundaries between different compartments. To put this issue into context, the embryo must create not only different tissue types but also subdivisions or compartments within a given tissue, from which different subpopulations will emerge. Such compartments are separated by boundaries, which stop cells with different identities from mixing back. There are numerous examples of compartments, such as the vertebrate hindbrain rhombomeres, the *Drosophila* body segments, and the *Drosophila* wing compartments [32, 33]. Differential adhesion was until recently favored to account for boundary formation, in part because the adhesion molecules called Tartan and Capricious help maintain the dorsal/ventral boundary in *Drosophila* wing disks [34], but genetic analysis has failed to identify the relevant proteins that would mediate such a mechanism in other contexts. Two recent papers based on studies of *Drosophila* embryonic segments and wing anterior/posterior compartments have provided strong support for a role of cortical tension (Fig. 11.2(C)) [35, 36].

The latter two papers were based on a careful quantitative analysis of cell morphology across compartments to define whether cells had one of their faces aligned with the boundary. Laser ablation was used to assess tension away from and at the compartment boundary, and it was found that adherens junctions located at the boundary were exposed to significantly higher tensions (by a factor of 2 to 2.5). Consistently with a role of myosin II in maintaining the boundary, myosin II is enriched at the compartment boundary [35, 36, 37]. Local inhibition of myosin II using chromophore-assisted light inhibition (CALI) of a GFP-linked myosin regulatory light chain to inactivate myosin affects compartment boundary formation [36]. Genetic analysis was also used, where tension was reduced by mutations in the myosin II heavy chain (called Zipper in *Drosophila*) or by biochemical inhibition of myosin II regulatory light chain phosphorylation. In both cases, such treatments result in very irregular boundaries and in compartment intermixing, as do mutations affecting the upstream regulatory signaling pathways known to be important for defining compartment identities [32, 35, 36].

From the developmental standpoint, what remains to be defined is how myosin II becomes upregulated at the boundary, probably under the influence of signaling pathways defining compartment identities. From the biophysical standpoint, what needs to be clarified is the mechanism that ensures local myosin II recruitment. There might be a local positive feedback favoring further myosin II accumulation. For instance, tension might pull on the myosin heavy chain arm to stabilize the ADP-bound conformation that can bind actin. Indeed, it has been noted in other contexts that tension leads to recruitment of myosin II [38, 39, 40, 41].

11.6 Two elementary cell shape changes

11.6.1 Anisotropic cortical tension

The process of convergent extension is very common in embryogenesis for axis elongation in most species [42, 43]. In *Drosophila*, it occurs approximately midway in embryogenesis, when the ventral epithelial layer that forms the thorax and abdomen of the embryo, called the germ band, doubles in length [44]. Germ band extension involves mediolateral convergence of cells along the dorsoventral (D/V) axis, through an ordered intercalation of cells along the anterior–posterior (A/P) axis in the absence of cell division. Intercalation progresses because D/V-oriented junctions are remodeled in the plane of the cells, in such a way that cells alternately lose and gain neighbors (Fig. 11.3(A)) [45, 46]. Careful analysis has established that much of this process relies on "internal forces" produced by myosin II [45]. But how could myosin II produce anisotropic cell shape changes?

Two key features account for this process, a polarized distribution of myosin II, which is enriched along D/V-oriented junctions [45, 47], and the minimization of cell energy. In a beautiful and careful analysis, Lenne, Lecuit, and coworkers showed that myosin II is approximately 1.3 ± 0.25-fold more abundant at D/V junctions than at other junctions [17]. The mechanisms that bring about this polarized distribution remain to be defined, although it has been noted that the epithelial polarity factor Par3/Bazooka is conversely enriched along A/P junctions [47]. Using the same kind of equation as above (see eqn (11.1) and Fig. 11.3(A)), the elastic tension of the junction at the interface between two cells i and j can be written as

$$\gamma^{ij} = k_{\text{el}}(\Delta P_i + \Delta P_j); \tag{11.2}$$

and for two shrinking cells i and j, in which myosin is slightly enriched along the vertical D/V-oriented junctions, which generates a higher cortical tension β, as opposed to two cells i and k with a stable transverse A/P junctions, the tension can be written as

$$\gamma_{\text{v}}^{ij} = k_{\text{el}}\beta(\Delta P_i + \Delta P_j) \quad \text{and} \quad \gamma_{\text{t}}^{ik} = k_{\text{el}}\beta(\Delta P_i + \Delta P_k). \tag{11.3}$$

Using such equations for *in silico* modeling, Lenne, Lecuit, and coworkers showed that cycles of energy minimization and T1 transitions could predict the germ band extension observed *in vivo* [17]. By varying the parameter β, they found that the

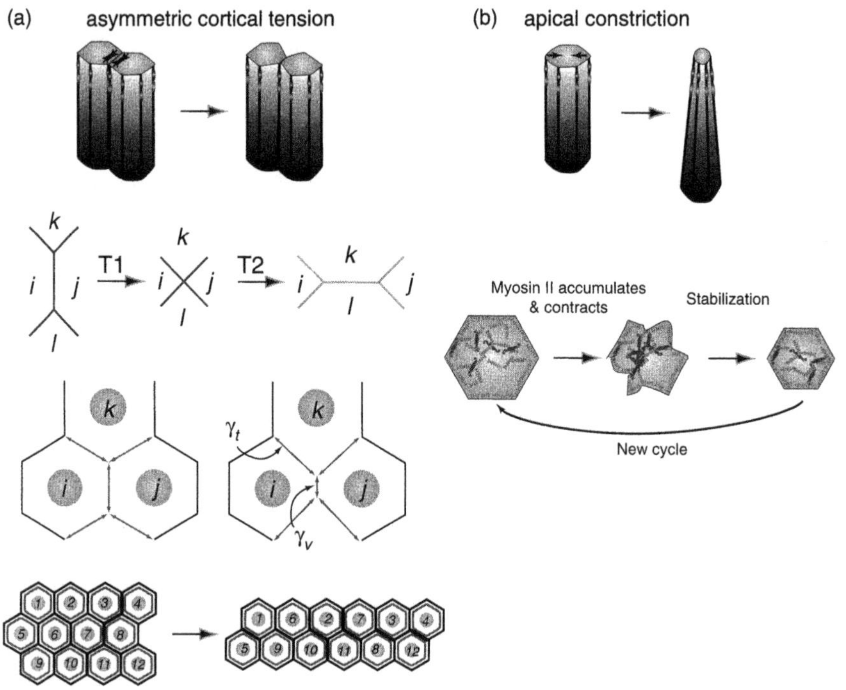

Fig. 11.3 Two elementary epithelial-cell shape changes. (A) Asymmetric cortical tension is a fundamental elementary cell shape change, whereby the junction between two adjacent cells shrinks, while the two cells remain adherent and other junctions remain largely unchanged (top). The junction that shrinks undergoes a series of transitions (second row): during T1, the junction shrinks to a single point; during T2, cells i and j, which were adjacent, become separated, while cells k and l, which were separated, establish an interface. Shrinking is initiated by the slight myosin II enrichment observed at the vertical junctions v compared with the transverse junctions t (third row). The ensuing increased tension at the shrinking interfaces γ_v compared with the nonshrinking interfaces can be expressed as outlined in the main text. This process is at the core of a common morphogenetic process called convergent extension, which allows a tissue layer to narrow in one direction and extend in the other (fourth row). Cells are shown viewed from the apical surface in the bottom three sets of diagrams; note that the almost perfect intercalation drawn here is shown just as a guide, and may not occur in this precise manner *in vivo*. (B) Apical constriction is another fundamental elementary cell shape change, whereby the apical cortex shrinks dramatically in an isometric fashion, while the cell elongates along its apicobasal axis. It involves cycles of myosin II accumulation at the apical cortex, which shrinks the apical surface by a small increment, followed by a stabilization phase, during which the apical surface does not change and the myosin II concentration decreases apically (second row; adapted from [52]). This process is at the core of morphogenetic processes such as *Drosophila* dorsal closure (Fig. 11.4(A)) and mesoderm invagination (Fig. 11.4(B)), and vertebrate neural tube formation.

value that best accounts for the *in vivo* extent and rate of extension is $\beta = 1.8$, which is in excellent agreement with *in vivo* observations recording actual cell shape changes. Interestingly, femtosecond-infrared-laser-cutting experiments to sever the actomyosin cortical network showed that the displacement speed of laser-operated junctions depends on the angle with the D/V axis. If ϕ is the angle between two junctions transverse to a D/V junction v, the variation law is close to

$$V_{\text{max, t}} \sim V_{\text{max, v}}\, 2\cos{(\phi/2)}. \tag{11.4}$$

These and additional data show that it is possible to associate carefully dissected quantitative data with predicted behavior derived from energy minimization and cell geometry.

11.6.2 Apical constriction

Apical constriction is another very common process, which can serve to create a hollow tube out of a flat epithelial layer, bring a cell layer within a preexisting cavity, and delaminate external cells internally [3, 48, 49]. In *Drosophila* embryos, this process has been thoroughly studied when the mesoderm invaginates internally during gastrulation and when the transient structure called the dorsal amnioserosa invaginates to help close the dorsal epidermis to which it is linked. In both cases, a genetic program initiates a constriction of the apical-most surface, which is coupled with cell elongation along the apicobasal axis [3, 49]. In both cases, myosin II activity is responsible for the constriction [50, 51]. The most surprising and unexpected observation, recently reported by two different laboratories, is that the process of constriction is pulsatile and progresses in a ratchet mode [52, 53]. First, myosin II accumulates apically and acts for a short period of time to locally and partially constrict the apex; second, another system, probably at the level of adherens junctions, maintains the cell constriction achieved after this first cycle; and third, the whole cycle starts all over again, with a period of 90–240 s. The mechanism behind this ratchet behavior remains unknown at this point. It might be related to the collective behavior of myosin II motors predicted 15 years ago by Jülicher and Prost and first observed *in vitro* by Riveline, Prost, and coworkers in 1998 [54, 55].

11.7 Integration of internal and external forces

Epithelial tissues rarely undergo morphogenesis in isolation, but instead remain connected to other cell layers or to adjacent epithelia. Using three examples, this final section surveys the little we know about how a given cell integrates external and internal forces.

11.7.1 The influence of *Drosophila* gastrulation on germ band extension

Germ band extension (see Fig. 11.3(A)) and gastrulation (Fig. 11.3(B) and Figs. 11.4(A) and (B)) occur concomitantly during *Drosophila* embryogenesis. The germ band extends along the anterior–posterior axis when dorsoventral junctions shorten and cells intercalate. Meanwhile, the mesoderm invaginates from the ventral side. Time

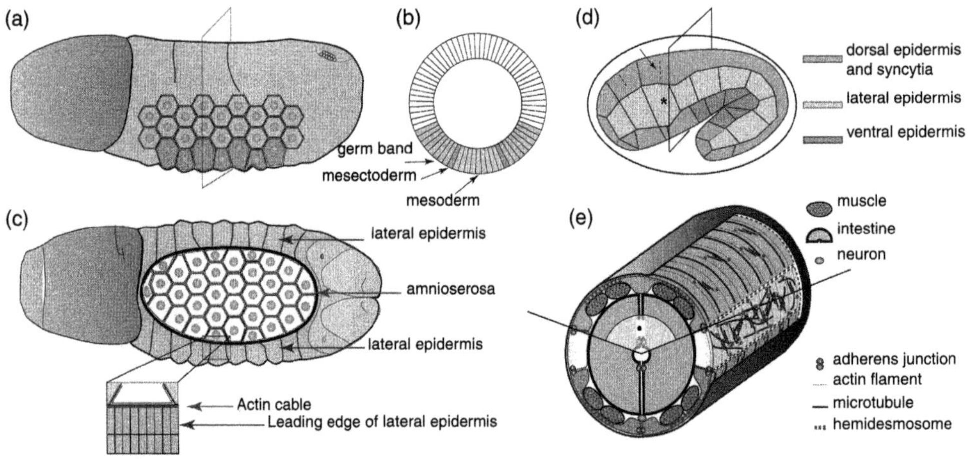

Fig. 11.4 An integrated view of three morphogenetic processes. Lateral view (A) and cross section (B) of a *Drosophila* embryo at the beginning of germ band extension and gastrulation (head, left; body, right). Germ band cells are represented as in Fig. 11.3(A) (three rows of hexagonal cells, plus stretched row at bottom); the mesoderm, which is not visible in the lateral view (at the bottom, i.e. ventral, in B), is connected to the germ band (lateral light gray cells) through one cell row of mesectodermal cells on each side of the embryo. Mesoderm cells constrict apically (with the apical surface on the outside border of the embryo) and subsequently undergo a movement of convergent extension. Their apical extension and convergent extension movements do not occur at the same rate throughout the length of the embryo, and are more prevalent posteriorly. (c) Dorsal view of the *Drosophila* embryo after germ band extension and retraction. A transient tissue, called the amnioserosa, covers its dorsal surface; the lateral epidermis extends dorsally to replace the amnioserosa during a process called dorsal closure. The inset at the bottom shows an enlarged view of the lateral epidermis/amnioserosa interface, which is characterized by the presence of an important actin cable maintaining the leading edge under strong tension (thick line). The apical surface of amnioserosa cells constricts (see Fig. 11.3(B)), which causes them to delaminate and eventually die, while pulling the leading edge dorsally. The two main forces driving dorsal closure are the tension at the lateral epidermal leading edge and the sum of the forces corresponding to the individual apical constriction movements of each amnioserosa cell. Lateral view (D) and transverse section (E) of a *C. elegans* embryo shortly after the beginning of elongation. The transverse section shows the three rows of epidermal cells (adapted from [64]). Myosin II is active mainly in the lateral seam cell row (central row) and is hardly active or completely inactive in the dorsal and ventral epidermal cells. The Lateral seam cells differ from the dorsal and ventral epidermal cells by three main criteria (E): the microtubule network in lateral seam cells is disorganized, lateral seam cells contain more myosin II, and the actin microfilaments form thick bundles in dorsal and ventral cells. The intestine, which occupies most of the inner part of the embryo, represents a hydrostatic incompressible skeleton.

lapse imaging and image segmentation have shown that germ band cells are subjected to an A/P-directed strain that induces those cells to change shape during the first 30 minutes of the 2-hour-long germ band extension process [14, 56]. Genetic analysis shows that this strain reflects a pulling force exerted by the mesoderm, which not only invaginates mainly along the D/V axis through apical constriction, but also appears to be under tension along the A/P axis after invagination [56]. How and at what level these tensile forces are integrated with the cell intercalation process described in Fig. 11.3(A) is unknown.

11.7.2 *Drosophila* dorsal closure

The dorsal surface of *Drosophila* embryos is covered with a transient tissue, called the amnioserosa, which disappears at the end of morphogenesis. It is replaced by cells of the lateral epidermis, which extend dorsally from each side, and then establish new junctions at the dorsal midline with their contralateral homologues (Fig. 11.4(C)). This process, known as dorsal closure, has been compared to wound healing; it is one of the best-known morphogenetic processes, for which many mutants have been identified [16, 57, 58]. It was also one the first for which laser ablation was used to probe force magnitudes, and the first for which systematic mathematical modeling was introduced [15, 16, 53]. Laser ablation of epidermal or amnioserosa cells has shown that three major forces are exerted on the lateral epidermis: (i) a pulling force P resulting from the pulsatile constricting amnioserosa cells; (ii) a tension T on the leading edge of the dorsal-most epidermal cells; and (iii) a resistive force R exerted by epidermal cells as they are being stretched [15, 16, 59]. Their magnitudes appear to obey the inequality

$$T \geq P \gg R\,. \tag{11.5}$$

Finally, an adhesive force Z keeps contralateral epidermal cells together as they zipper together [59]. The force Z coordinates the other forces (T, P, and R) driving closure and is essential for final sealing at the dorsal midline, but is not required for the bulk of the tissue movements during dorsal closure [59]. Interestingly, modeling indicates that the actomyosin cable present at the epidermal leading edge might locally lock the epidermal cells into a given shape, as the immediately neighboring amnioserosa cells are in the stabilization phase of their pulsatile cycle (see Fig. 11.3(B)) [53]. Furthermore, time lapse observation coupled with image segmentation and modeling suggests that there is a graded number of pulsatile cycles for amnioserosa cells, with central amnioserosa cells cycling for a longer time than those on the outer border in contact with the epidermis [53]. The mechanism behind this sequential arrest of the pulsatile behavior is unknown, though it might involve once again a mechanosensing device. In a parallel kinematic quantitative analysis, it was also concluded that there are different phases of amnioserosa contractions, and that there is a radial gradient of amnioserosa contraction, but a specific mechanism for its generation was not proposed [14, 60]. Dorsal closure thus highlights the existence of several tension-sensing mechanisms within epithelial cells that serve to stabilize cells in a given position or to synchronize their behavior.

11.7.3 *C. elegans* embryonic elongation

C. elegans embryos are very simple. The external epithelium surrounding the embryo includes five rows of epidermal cells aligned along the A/P axis, 11 lateral epidermal seam cells on each side, 11 epidermal cells on each ventral flank (which join at the ventral midline), and 22 epidermal cells on the dorsal side, many of which will fuse together to generate four syncytia (http://www.wormatlas.org/hermaphrodite/hypodermis/Hypframeset.html) (Fig. 11.4 (D) and (E)). During the final step of embryonic morphogenesis, which occurs in the absence of any cell division or cell migration, epidermal -cells elongate along the A/P axis [18, 61]. As a result of epidermal-cell shape changes, the embryo elongates fourfold along its A/P axis within 2 hours (at a rate of about 100 μm/hour), while its circumference decreases threefold. Each row of epidermal cells contains actomyosin filaments and microtubules, but their detailed organization in lateral seam cells differs from that in their dorsal and ventral neighbors [18, 62] (Fig. 11.4(E)). Genetic analysis suggests that myosin II is primarily required in the lateral seam cells during the first half of elongation [62], while it is kept mainly inactive in the dorsal and ventral epidermal cells [63], implying that the lateral seam cells play a central role in elongation.

The observations presented above are intellectually challenging because the dorsal and ventral cells also change their shapes, and would do so almost entirely in a passive way under the influence of the lateral epidermal cells. Ben Amar and coworkers, using finite element modeling, have proposed a solution to try to account for this apparent contradiction [64]. The embryo is considered as a homogeneous material, with both an active term defined by the myosin II motors and a passive viscoelastic response. The model takes account of the axisymmetric geometry of the embryo, and the following parameters:

- The active Cauchy stress, to take account of the chemical energy of ATP $\Delta\mu$ consumed by myosin II motors. Given the polarity p of actin, the theory of active polar gels [65] predicts that the active Cauchy stress σ_{act} can be written as a function of the motor activity, the circumferential angle θ (describing in polar coordinates the angle formed by the large dorsal epidermal surface), p (the unit vector in the circumferential direction; $p = e_\theta$), and ζ (a material coefficient of the cytoskeleton characterizing the actin–myosin activity, which is negative for contractile motors):

$$\sigma^i_{\text{act}}(\Delta\mu, p) = -\zeta\, \Delta\mu\, e_\theta \otimes e_\theta. \tag{11.6}$$

- The anisotropy of the microtubule network and its buckling properties, making use of viscoelastic theories at large deformations [66, 67, 68].
- Nonlinear elasticity, for the hydrostatic pressure exerted by internal cells considered as a soft skeleton [69, 70].

By taking these parameters into account and solving for the deformation of the radial angle between the two seam–dorsal junctions on each side of the embryo, the modeling predicts the following points:

- Elongation is possible with different levels of myosin II activity in nonseam cells (even no activity), but is more efficient with a significant myosin II activity in nonseam dorsal and ventral cells.
- The passive mechanical response of the microtubule network is essential for provoking a large finite stretch of the embryo with a contraction of circumferentially oriented myosin motors. In the absence of microtubules, theoretical modeling predicts a hysteresis behavior of the elongation curve, which is not observed *in vivo*.

The latter prediction is supported by studies using old pharmacological approaches to inhibit microtubule turnover, in which it had indeed been noted that microtubules were essential for elongation [18]. This earlier analysis should now be repeated with modern imaging methods, having the finite element model described above in mind, and where the myosin II activity is varied with genetic tricks.

Although I have focused mostly on *Drosophila* and *C. elegans* embryos, recent experiments suggest that in vertebrates too, local tension minima and differential stiffness can account for the formation of limb buds and axis elongation [71, 72].

11.8 Conclusions

The field of embryonic morphogenesis has rediscovered in the last few years the importance of physical laws, as put forth by D'Arcy Thompson a century ago [24]. The importance of tension is now acknowledged, in parallel with adhesion. We are still at a rather primitive stage, in which we can describe processes in quantitative terms but we are very far from having a physical understanding of them. In particular, we do not know how forces are sensed within the cell, which processes they influence, whether they induce conformational changes in proteins, and what is behind stiffness in living cells. The unexpected discovery of the ratchet apical-constriction phenomenon highlights to what extent the theoretical and experimental work performed for years by physicists to explain the collective and individual behaviors of proteins has not yet permeated the developmental biology field. Another very wide area that is beginning to surface is the signaling induced by mechanical forces [73, 74]. There is a need to know under what conditions such mechanotransduction signaling occurs and how it does so. In summary, the field of morphogenesis is ripe for a major contribution from physicists.

Acknowledgments

I am grateful to Daniel Riveline for critical reading of the manuscript. Work in my laboratory is supported by the CNRS, INSERM, and the Université de Strasbourg, as well as by funds from the European Union (STREP-FP6), the French Agence Nationale de la Recherche, and the Association pour la Recherche sur le Cancer.

References

[1] Mellman, I. and W. J. Nelson, Coordinated protein sorting, targeting and distribution in polarized cells. *Nat Rev Mol Cell Biol*, 2008. **9**(11): 833–45.

[2] Nelson, W. J., Adaptation of core mechanisms to generate cell polarity. *Nature*, 2003. **422**(6933): 766–74.

[3] Quintin, S., C. Gally, and M. Labouesse, Epithelial morphogenesis in embryos: asymmetries, motors and brakes. *Trends Genet*, 2008. **24**(5): 221–30.

[4] Revenu, C., *et al.*, The co-workers of actin filaments: From cell structures to signals. *Nat Rev Mol Cell Biol*, 2004. **5**(8): 635–46.

[5] Gardner, M. K., *et al.*, Microtubule assembly dynamics: New insights at the nanoscale. *Curr Opin Cell Biol*, 2008. **20**(1): 64–70.

[6] Howard, J. and A. A. Hyman, Dynamics and mechanics of the microtubule plus end. *Nature*, 2003. **422**(6933): 753–8.

[7] Kim, S. and P. A. Coulombe, Intermediate filament scaffolds fulfill mechanical, organizational, and signaling functions in the cytoplasm. *Genes Dev*, 2007. **21**(13): 1581–97.

[8] Lecuit, T. and P. F. Lenne, Cell surface mechanics and the control of cell shape, tissue patterns and morphogenesis. *Nat Rev Mol Cell Biol*, 2007. **8**(8): 633–44.

[9] Knust, E. and O. Bossinger, Composition and formation of intercellular junctions in epithelial cells. *Science*, 2002. **298**(5600): 1955–9.

[10] Nejsum, L. N. and W. J. Nelson, Epithelial cell surface polarity: the early steps. *Front Biosci*, 2009. **14**: 1088–98.

[11] Martin-Belmonte, F. and K. Mostov, Regulation of cell polarity during epithelial morphogenesis. *Curr Opin Cell Biol*, 2008. **20**(2): 227–34.

[12] Knust, E., Regulation of epithelial cell shape and polarity by cell–cell adhesion (Review). *Mol Membr Biol*, 2002. **19**(2): 113–20.

[13] Desprat, N., *et al.*, Tissue deformation modulates twist expression to determine anterior midgut differentiation in *Drosophila* embryos. *Dev Cell*, 2008. **15**(3): 470–7.

[14] Blanchard, G. B., *et al.*, Tissue tectonics: Morphogenetic strain rates, cell shape change and intercalation. *Nat Methods*, 2009. **6**(6): 458–64.

[15] Hutson, M. S., *et al.*, Forces for morphogenesis investigated with laser microsurgery and quantitative modelling. *Science*, 2003. **300**(5616): 145–9.

[16] Kiehart, D. P., *et al.*, Multiple forces contribute to cell sheet morphogenesis for dorsal closure in *Drosophila*. *J Cell Biol*, 2000. **149**(2): 471–90.

[17] Rauzi, M., *et al.*, Nature and anisotropy of cortical forces orienting Drosophila tissue morphogenesis. *Nat Cell Biol*, 2008. **10**(12): 1401–10.

[18] Priess, J. R. and D. I. Hirsh, *Caenorhabditis elegans* morphogenesis: The role of the cytoskeleton in elongation of the embryo. *Dev Biol*, 1986. **117**(1): 156–73.

[19] Williams-Masson, E. M., *et al.*, The cellular mechanism of epithelial rearrangement during morphogenesis of the *Caenorhabditis elegans* dorsal hypodermis. *Dev Biol*, 1998. **204**(1): 263–76.

[20] Daniels, B. R., B. C. Masi, and D. Wirtz, *Probing single-cell micromechanics in vivo*: The microrheology of *C. elegans* developing embryos. *Biophys J*, 2006. **90**(12): 4712–9.

[21] Krieg, M., *et al.*, Tensile forces govern germ-layer organization in zebrafish. *Nat Cell Biol*, 2008. **10**(4): 429–36.

[22] Davidson, L. and R. Keller, Measuring mechanical properties of embryos and embryonic tissues. *Methods Cell Biol*, 2007. **83**: 425–39.

[23] Forgacs, G., *et al.*, Viscoelastic properties of living embryonic tissues: A quantitative study. *Biophys J*, 1998. **74**(5): 2227–34.

[24] Thomson, D. W., *On Growth and Form.* 1961, New York: Cambridge University Press.

[25] Steinberg, M. S. and S. F. Gilbert, Townes and Holtfreter (1955): Directed movements and selective adhesion of embryonic amphibian cells. *J Exp Zool A Comp Exp Biol*, 2004. **301**(9): 701–6.

[26] Steinberg, M. S., Reconstruction of tissues by dissociated cells. Some morphogenetic tissue movements and the sorting out of embryonic cells may have a common explanation. *Science*, 1963. **141**: 401–8.

[27] Foty, R. A. and M. S. Steinberg, The differential adhesion hypothesis: A direct evaluation. *Dev Biol*, 2005. **278**(1): 255–63.

[28] Bao, S. and R. Cagan, Preferential adhesion mediated by Hibris and Roughest regulates morphogenesis and patterning in the *Drosophila* eye. *Dev Cell*, 2005. **8**(6): 925–35.

[29] Hayashi, T. and R. W. Carthew, Surface mechanics mediate pattern formation in the developing retina. *Nature*, 2004. **431**(7009): 647–52.

[30] Kafer, J., *et al.*, Cell adhesion and cortex contractility determine cell patterning in the *Drosophila* retina. *Proc Natl Acad Sci USA*, 2007. **104**(47): 18549–54.

[31] Warner, S. J. and G. D. Longmore, Distinct functions for Rho1 in maintaining adherens junctions and apical tension in remodelling epithelia. *J Cell Biol*, 2009. **185**(6): 1111–25.

[32] Vincent, J. P. and D. Irons, Developmental biology: Tension at the border. *Curr Biol*, 2009. **19**(22): R1028–30.

[33] Lawrence, P. A. and G. Struhl, Morphogens, compartments, and pattern: Lessons from *Drosophila?* *Cell*, 1996. **85**(7): 951–61.

[34] Milan, M., *et al.*, The LRR proteins capricious and Tartan mediate cell interactions during DV boundary formation in the *Drosophila* wing. *Cell*, 2001. **106**(6): 785–94.

[35] Landsberg, K. P., *et al.*, Increased cell bond tension governs cell sorting at the *Drosophila* anteroposterior compartment boundary. *Curr Biol*, 2009. **19**(22): 1950–5.

[36] Monier, B., *et al.*, An actomyosin-based barrier inhibits cell mixing at compartmental boundaries in *Drosophila* embryos. *Nat Cell Biol*, 2006. **12**(1): 60–5; supp. pp. 1–9.

[37] Major, R. J. and K. D. Irvine, Localization and requirement for Myosin II at the dorsal–ventral compartment boundary of the Drosophila wing. *Dev Dyn*, 2006. **235**(11): 3051–8.

[38] Fernandez-Gonzalez, R., *et al.*, Myosin II dynamics are regulated by tension in intercalating cells. *Dev Cell*, 2009. **17**(5): 736–43.

[39] Reichl, E. M., *et al.*, Interactions between myosin and actin crosslinkers control cytokinesis contractility dynamics and mechanics. *Curr Biol*, 2008. **18**(7): 471–80.

[40] Ren, Y., *et al.*, Mechanosensing through cooperative interactions between myosin II and the actin crosslinker cortexillin I. *Curr Biol*, 2009. **19**(17): 1421–8.

[41] Kee, Y. S. and D. N. Robinson, Motor proteins: Myosin mechanosensors. *Curr Biol*, 2008. **18**(18): R860–2.

[42] Keller, R., D. Shook, and P. Skoglund, The forces that shape embryos: Physical aspects of convergent extension by cell intercalation. *Phys Biol*, 2008. **5**(1): 15007.

[43] Keller, R., Shaping the vertebrate body plan by polarized embryonic cell movements. *Science*, 2002. **298**(5600): 1950–4.

[44] Irvine, K. D. and E. Wieschaus, Cell intercalation during *Drosophila* germband extension and its regulation by pair-rule segmentation genes. *Development*, 1994. **120**(4): 827–41.

[45] Bertet, C., L. Sulak, and T. Lecuit, Myosin-dependent junction remodelling controls planar cell intercalation and axis elongation. *Nature*, 2004. **429**(6992): 667–71.

[46] Zallen, J. A. and E. Wieschaus, Patterned gene expression directs bipolar planar polarity in Drosophila. *Dev Cell*, 2004. **6**(3): 343–55.

[47] Blankenship, J. T., *et al.*, Multicellular rosette formation links planar cell polarity to tissue morphogenesis. *Dev Cell*, 2006. **11**(4): 459–70.

[48] Keller, R., L. A. Davidson, and D. R. Shook, How we are shaped: The biomechanics of gastrulation. *Differentiation*, 2003. **71**(3): 171–205.

[49] Sawyer, J. M., *et al.*, Apical constriction: A cell shape change that can drive morphogenesis. *Dev Biol*, 2009.

[50] Dawes-Hoang, R. E., *et al.*, Folded gastrulation, cell shape change and the control of myosin localization. *Development*, 2005. **132**(18): 4165–78.

[51] Franke, J. D., R. A. Montague, and D. P. Kiehart, Nonmuscle myosin II generates forces that transmit tension and drive contraction in multiple tissues during dorsal closure. *Curr Biol*, 2005. **15**(24): 2208–21.

[52] Martin, A. C., M. Kaschube, and E. F. Wieschaus, Pulsed contractions of an actin–myosin network drive apical constriction. *Nature*, 2009. **457**(7228): 495–9.

[53] Solon, J., *et al.*, Pulsed forces timed by a ratchet-like mechanism drive directed tissue movement during dorsal closure. *Cell*, 2009. **137**(7): 1331–42.

[54] Riveline, D., *et al.*, Acting on actin: The electric motility assay. *Eur Biophys J*, 1998. **27**(4): 403–8.

[55] Jülicher, F. and J. Prost, Cooperative molecular motors. *Phys Rev Lett*, 1995. **75**(13): 2618–2621.

[56] Butler, L. C., *et al.*, Cell shape changes indicate a role for extrinsic tensile forces in Drosophila germ-band extension. *Nat Cell Biol*, 2009. **11**(7): 859–64.

[57] Harden, N., Signaling pathways directing the movement and fusion of epithelial sheets: Lessons from dorsal closure in *Drosophila*. *Differentiation*, 2002. **70**(4–5): 181–203.

[58] Jacinto, A. and P. Martin, Morphogenesis: Unravelling the cell biology of hole closure. *Curr Biol*, 2001. **11**(17): R705–7.

[59] Peralta, X. G., *et al.*, Upregulation of forces and morphogenic asymmetries in dorsal closure during *Drosophila* development. *Biophys J*, 2007. **92**(7): 2583–96.

[60] Gorfinkiel, N., *et al.*, Mechanical control of global cell behaviour during dorsal closure in *Drosophila. Development*, 2009. **136**(11): p. 1889–98.

[61] Chisholm, A. D. and J. Hardin, Epidermal morphogenesis. In *WormBook*, ed. the *C. elegans* Research Community, 2005: doi/10.1895/wormbook.1.7.1, http://www.wormbook.org.

[62] Gally, C., *et al.*, Myosin II regulation during *C. elegans* embryonic elongation: LET-502/ROCK, MRCK-1 and PAK-1, three kinases with different roles. *Development*, 2009. **136**(18): 3109–19.

[63] Diogon, M., *et al.*, The RhoGAP RGA-2 and LET-502/ROCK achieve a balance of actomyosin-dependent forces in *C. elegans* epidermis to control morphogenesis. *Development*, 2007. **134**(13): 2469–79.

[64] Ciarletta, P., M. Ben Amar, and M. Labouesse, Continuum model of epithelial morphogenesis during *Caenorhabditis elegans* embryonic elongation. *Philos Trans A Math Phys Eng Sci*, 2009. **367**(1902): 3379–400.

[65] Kruse, K., *et al.*, Generic theory of active polar gels: A paradigm for cytoskeletal dynamics. *Eur Phys J E Soft Matter*, 2005. **16**(1): 5–16.

[66] Brangwynne, C. P., F. C. MacKintosh, and D. A. Weitz, Force fluctuations and polymerization dynamics of intracellular microtubules. *Proc Natl Acad Sci USA*, 2007. **104**(41): 16128–33.

[67] Brangwynne, C. P., *et al.*, Microtubules can bear enhanced compressive loads in living cells because of lateral reinforcement. *J Cell Biol*, 2006. **173**(5): 733–41.

[68] Li, T., A mechanics model of microtubule buckling in living cells. *J Biomech*, 2008. **41**(8): 1722–9.

[69] Qiu, G. Y. and T. J. Pence, Remarks on the behaviour of simple directionally reinforced incompressible nonlinearly elastic solids. *J Elast*, 1997. **49**: 1–30.

[70] Qiu, G. Y. and T. J. Pence, Loss of ellipticity in plane deformations of a simple directionally reinforced incompressible nonlinearly elastic solid. *J Elast*, 1997. **49**: 31–63.

[71] Damon, B.J., *et al.*, Limb bud and flank mesoderm have distinct "physical phenotypes" that may contribute to limb budding. *Dev Biol*, 2008. **321**(2): 319–30.

[72] Zhou, J., H. Y. Kim, and L. A. Davidson, Actomyosin stiffens the vertebrate embryo during crucial stages of elongation and neural tube closure. *Development*, 2009. **136**(4): 677–88.

[73] Wozniak, M. A. and C. S. Chen, Mechanotransduction in development: A growing role for contractility. *Nat Rev Mol Cell Biol*, 2009. **10**(1): 34–43.

[74] Orr, A. W., *et al.*, Mechanisms of mechanotransduction. *Dev Cell*, 2006. **10**(1): 11–20.

The manufacturer's authorised representative in the EU for product safety is Oxford University Press España S.A. of el Parque Empresarial San Fernando de Henares, Avenida de Castilla, 2 – 28830 Madrid (www.oup.es/en or product.safety@oup.com). OUP España S.A. also acts as importer into Spain of products made by the manufacturer.

www.ingramcontent.com/pod-product-compliance
Ingram Content Group UK Ltd.
Pitfield, Milton Keynes, MK11 3LW, UK
UKHW052229270726
14060UKWH00004B/669

* 9 7 8 0 1 9 9 6 0 5 8 3 5 *